PERGAMON INTERNATIONAL LIBRARY
of Science, Technology, Engineering and Social Studies
*The 1000-volume original paperback library in aid of education,
industrial training and the enjoyment of leisure*
Publisher: Robert Maxwell, M.C.

THE PRINCIPLES OF
POLLINATION ECOLOGY

THIRD REVISED EDITION

THE PERGAMON TEXTBOOK
INSPECTION COPY SERVICE

An inspection copy of any book published in the Pergamon International Library
will gladly be sent to academic staff without obligation for their consideration for
course adoption or recommendation. Copies may be retained for a period of 60
days from receipt and returned if not suitable. When a particular title is adopted or
recommended for adoption for class use and the recommendation results in a sale
of 12 or more copies, the inspection copy may be retained with our compliments.
The Publishers will be pleased to receive suggestions for revised editions and new
titles to be published in this important International Library.

THE PRINCIPLES
OF
POLLINATION ECOLOGY

by

K. FÆGRI

*Professor of Systematic Botany and Plant Geography,
University of Bergen*

and

L. VAN DER PIJL

Emer. Professor of Botany, Catholic University of Nijmegen

THIRD REVISED EDITION

PERGAMON PRESS

OXFORD · NEW YORK · TORONTO · SYDNEY · PARIS · FRANKFURT

U.K.	Pergamon Press Ltd., Headington Hill Hall, Oxford OX3 0BW, England
U.S.A.	Pergamon Press Inc., Maxwell House, Fairview Park, Elmsford, New York 10523, U.S.A.
CANADA	Pergamon of Canada Ltd., 75 The East Mall, Toronto, Ontario, Canada
AUSTRALIA	Pergamon Press (Aust.) Pty. Ltd., P.O. Box 544; Potts Point. N.S.W. 2011, Australia.
FRANCE	Pergamon Press SARL, 24 rue des Ecoles, 75240 Paris, Cedex 05, France
FEDERAL REPUBLIC OF GERMANY	Pergamon Press, GmbH, 6242 Kronberg-Taunus, Pferdstrasse 1, Federal Republic of Germany

First edition 1966

Second revised edition 1971

Third revised edition 1979

British Library Cataloguing in Publication Data

Faegri, Knut
The principles of pollination ecology. — 3rd revised ed.
1. Fertilization of plants
I. Title II. Pijl, Leendert van der
581.1'6'62   QK926   78-40132

ISBN 0-08-021338-3 Hardcover
ISBN 0-08-023160-8 Flexicover

Printed in Great Britain by
William Clowes & Sons Limited
London, Beccles and Colchester

CONTENTS

Preface to the Third Edition   ix

Chapter 1.   A SHORT HISTORY OF THE STUDY OF POLLINATION
ECOLOGY   1
     1.1. The Sexuality of Plants   1
     1.2. Cross-pollination. The Foundation of Pollination Ecology   1
     1.3. The Pre-Darwinian Period   2
     1.4. Darwin and the Post-Darwinian Period   2
     1.5. The Modern Period   3

Chapter 2.   TECHNIQUES IN POLLINATION ECOLOGY   4

Chapter 3.   POLLINATION AS SPORE DISPERSAL   7

Chapter 4.   SPORE AND POLLEN DISPERSAL IN LOWER PLANTS AND
GYMNOSPERMS   9
     4.1. Spore Dispersal in Thallophytes   9
     4.2. Spore Dispersal in Mosses   9
     4.3. Spore Dispersal in Pteridophytes. Conclusion   10
     4.4. Pollination in Gymnosperms   10

Chapter 5.   POLLINATION IN ANGIOSPERMS   13
     5.1. Pollination and Adaptation   13
     5.2. Structure and Function of the Angiosperm Flower   16
     5.3. Cross- and Self-pollination   23

Chapter 6.   ABIOTIC POLLINATION   34
     6.1. Wind Pollination, Anemophily   34
     6.2. Water Pollination, Hydrophily   40

Chapter 7.   BIOTIC POLLINATION. PRINCIPLES   42

Chapter 8.   BIOTIC POLLINATION. PRIMARY ATTRACTANTS   55
     8.1. Primary Attractants. I: Pollen   58
     8.2. Primary Attractants. II: Nectar   63
     8.3. Primary Attractants. III: Oil   69
     8.4. Primary Attractants. IV: Other Substances   70

8.5. Primary Attractants. V: Protection and Brood-place   71

8.6. Primary Attractants. VI: Sexual Attraction   74

Chapter 9.   BIOTIC POLLINATION. SECONDARY ATTRACTANTS   77

9.1. Secondary Attractants. I: Odour   77

9.2. Secondary Attractants. II: Visual Attraction   81

9.3. Secondary Attractants. III: Temperature Attraction   86

9.4. Secondary Attractants. IV: Motion as an Attractant   86

Chapter 10.   STRUCTURAL BLOSSOM CLASSES   88

Chapter 11.   ANIMALS AS POLLINATORS   96

11.1. Insects (Invertebrates) as Pollinators   98

     11.1.1.   Beetles and Beetle-pollinated Blossoms. Cantharophily   99

     11.1.2.   Flies and Fly-pollinated Blossoms. Myophily   102

     11.1.3.   Hymenopters in General. Wasps   107

     11.1.4.   Ants   109

     11.1.5.   Bees and Bee-blossoms. Melittophily   110

     11.1.6.   Butterflies and Moths. Psychophily and Phalaenophily   115

     11.1.7.   Other Invertebrates   119

11.2. Vertebrates as Pollinators   121

     11.2.1.   Pollination by Birds. Ornithophily   123

     11.2.2.   Pollination by Bats. Chiropterophily   129

Chapter 12.   "RETROGRADE" DEVELOPMENTS   134

12.1. "Revertence" to Abiotic Pollination   134

12.2. Autogamy   135

12.3. Apomixis and Vegetative Propagation   140

Chapter 13.   THE DEVELOPMENT OF FLOWERS IN RELATION TO MODE OF POLLINATION   142

13.1. The Flag Blossom in Leguminosae   142

13.2. The Gullet Blossom in Tubiflorae   146

Chapter 14.   POLLINATION ECOLOGY AND SPECIATION   151

Chapter 15.   POLLINATION ECOLOGY AND THE BIOCOENOSE   156

Chapter 16.   APPLIED POLLINATION ECOLOGY   159

Chapter 17.   CASE HISTORIES   164

17.1. Pollen Presentation Types   164

17.2. Extra-floral Nectar   168

17.3. Hydrophily   169

17.4. Trap Blossoms   170

17.5. Brood-place Blossoms   173

17.6. Chiropterophily   178

17.7. Pollination Syndromes within the Papilionaceae   180
17.8. Pollination Syndromes within Labiatae—Scrophulariaceae   188
17.9. Pollination Syndromes in Orchids   199

Chapter 18.   EPILOGUE   205

References   207

Index of Plant Names   227

Index of Animal Names   235

Subject Index   239

PREFACE TO THE THIRD EDITION

The object of this book is stated in the title: *principles* of pollination ecology. It is not a manual. A comprehensive manual of pollination ecology would run into many volumes; indeed, Knuth's handbook dating back to the turn of the century covers 2972 pages, and very much has been added to our knowledge since then. Besides, such a manual would chiefly contain redundant or irrelevant matter, dealing with plants of little or no direct interest except for the few scientists specializing in pollination ecology. What might be desirable is a set of regional manuals, geared to the local floras or treating families or other major taxa in a systematic fashion. One or two such manuals might be written today, and might also count on reader interest, sufficiently widespread to warrant publication. However, that is not the task we have set ourselves, and when we quote specific examples it is simply to give concrete demonstrations of how pollination functions in practice. We make no attempt at being complete: for each example quoted, probably dozens of others might just as well have been used. We may also neglect aberrant types: we quote the genus *Cornus* as an example of plants in which the bracts are enlarged and form visual attractants – notwithstanding the fact that in *C. mas* the flowers are equally important in that respect. For examples that belong to general botanical knowledge, we do not, as a rule, quote any authority; such quotations would have inflated the size of the book, and most of them can be found with details in Knuth's handbook (of which the original third volume was unfortunately never translated, vol. 3 of the translation being vol. II, 2 of the original.) Personal observations are sometimes initialled, especially if unpublished.

We have tried to formulate *general principles* of pollination ecology, applicable anywhere, and it is our hope that the reader will acquire a sufficient measure of these general principles to make his own observations within the confines of the flora of the area in which he is living.

We cannot here describe all the different ways in which, for example, stamens are modified in response to pollination: what we can do, and what we hope to achieve, is to make the reader aware of the problem of, and the existence of, variations in the androecium and the possibility that these variations may, in one way or another, be functionally explicable in terms of the pollination ecology of the flower in question.

Thus, it is our hope that the book will help first the college or university teacher, giving him the necessary background for teaching pollination ecology to his students, and secondly, those students when they go back to schools (as many of them will) and in their turn take up teaching, whether this happens in an area with a similar or a different flora. Further, it is hoped that the book will serve the college or university student as a frame of reference to which the concrete data of the courses can be referred, and where he can find the general principles behind the individual observations he is able to make in the field or laboratory.

This book should therefore not be studied in isolation. Metaphorically speaking, one should hold it in one hand, the living plant in the other, correlating the concrete facts of the latter with the general principles of the former. The aim of the book is to give the reader the potential for a more profound understanding of what he may have seen many times, but has never observed. Perhaps it is not superfluous to stress the importance of observations of what actually happens in nature. Too many erroneous conclusions about pollination ecology have in the past been made from morphological observations, in gardens, or by laboratory experiments alone. The only phenomenon that is really valid is what happens where the plant is growing spontaneously. Ecology deals with the interaction between organisms and their habitats, and while it is very interesting to observe pollination in plants that have been transplanted outside their proper area, this only gives information about the adaptability of pollination agents, not about the mutual adaptation of flower and agent (see van der Pijl 1937a).

"Modern" biology has made spectacular advances during the last few decennia. The results of genetical, biochemical, and other investigations have given us a new insight into phenomena of life never before understood, hardly even perceived, but to a majority of biologists many of the spectacular achievements of modern biological research can be but a theoretical knowledge, to be learnt from textbooks, theses, or in the lecture hall. Many phases of "old-fashioned" biology can be studied, albeit primitively, by and brought home to every schoolchild. To deprive him or her of the opportunity of this direct contact with nature is, we think, wilfully to take away the bread and offer stones instead, be these stones full of precious minerals. Pollination ecology is therefore a subject of the greatest value in teaching, just because the teacher who knows the basic principles can make them come alive everywhere, to all his students (Webb 1957).

The three R's may seem trivial, but they are fundamental. Too many "modern" biologists seem to forget the importance, not only of those three R's, but of a long series of other ones, leading up to the summits of today's science. "Nature study" (including pollination ecology) may perhaps seem rather far down the ladder, but it is our conviction, unfortunately confirmed by *bêtises* on the part of "modern" biologists, that a biology neglecting these apparently simple branches is doomed to sterility, and that the functional aspect is all too frequently neglected.

This does not, of course, mean that pollination ecology should not benefit by the achievements of modern biological research. Not entirely without reason pollination ecologists have in the past been accused of neglecting the results of research in other branches of biology. However, this situation has changed materially in recent years, and it changes from day to day, not only in the use of sophisticated equipment, but also in the integration with ideas from evolution, ethology, etc.

To illustrate the principles of pollination ecology we have described some case histories as examples. For practical reasons the nomenclature of the original publication is usually retained when cases have been quoted from the literature.

Using this book in the field a reader may feel that too many of the examples have been taken from too few floristic regions. The explanation is simple enough: pollination ecology has chiefly been studied in a few areas, and so we have had to take our examples from them. We have a small, but fervent, hope that this book may initiate studies in pollination ecology in other regions. Should this be achieved we are in no doubt that not only would it be possible in future books on this subject to supplement our examples by better ones from other floras, but also that completely new principles may come to light, corresponding to the discovery of the sexual attraction, or the pollinating marsupial. If this book rapidly becomes outdated for that reason, nobody would be happier about it than the authors.

So far we have mainly quoted from the preface to the first edition. Since its publication, the position of pollination ecology has changed from being a backwater of biology to become an important field of research, as witnessed by the multitude of research papers — far too many to be adequately covered — and several textbooks aimed at various levels. The increased activity, for which this book may be partly responsible, has many causes; one of them is the practical importance of understanding the pollination ecology of crop plants for high yield and for purity of strains. Also, modern research in pollination has revealed many important facets of biological evolution. What some twenty years ago seemed a rather static subject, fixed in its conceptual foundation, is today in the process of reorienting itself towards a higher level of understanding of basic principles. Also taxonomists are increasingly aware of the importance of pollination ecology for speciation, a subject completely neglected earlier, and modern taxonomic monographs tend to include pollination data.

We have chosen the term pollination ecology for precision. We understand "floral biology" to comprise all manifestations of the life of the floral region, also those not directly connected with pollen transfer, even if they interact with the phenomena that are treated in this book. Our aim has been to discuss only the interaction of plants and their pollination vectors, the ecological problem.

L. van der Pijl
K. Fægri

CHAPTER 1

A SHORT HISTORY OF THE STUDY OF POLLINATION ECOLOGY

1.1. THE SEXUALITY OF PLANTS

Certain agricultural practices with the aim of achieving or furthering the production of fruits or seeds by cross-pollination have been known, perhaps, as long as organized agriculture has existed (see Chapter 16). Some of these practices are recorded in documents from classical and pre-classical ages; we may assume that there have been others.* Even so, the scientific concept of what pollination represents took a remarkably long time to establish itself, and the idea of sexuality in plants, whether it existed and how it manifested itself, was a standard problem during the Classical Age, the Scholastic Period, and even during the first centuries of more modern science. The question was frequently settled by criteria for maleness and femaleness which we cannot now accept as valid, for example those expressing themselves in still existing names like *Athyrium filix-femina* and *Dryopteris filix-mas*. As can be surmised, the scholastic argument in this case makes a rather amusing lecture, but does not contribute much to a solution of relevant problems. Perhaps it is more than a coincidence that the sexually repressed Middle Ages could not grasp an idea which presented no difficulty to the more licentious eighteenth century (see Taylor 1954).

It was left to the age of Linnaeus to solve the problem. He himself realized the importance of the so-called sexual organs of the flower, and based his division of the plant kingdom on them. However, his own contributions to pollination ecology are insignificant, and there were other botanists of that era to whom we are more indebted.

To Rudolf Jakob Camerarius (1665–1721) we owe the recognition of the sexuality of plants and the understanding of the function of the sexual parts of the flower. In a published letter *de sexu plantarum* (1694), based upon his own experiments, he states that two different parts of the flower, viz. stamen and pistil, must work together to produce ripe seed. He concluded, therefore, that these two parts represented the sexual organs, a concept still generally maintained, even though it is not quite to the point. The flowering plant represents the asexual diploid generation, therefore it is not strictly accurate to call its organs sexual.

1.2. CROSS-POLLINATION.
THE FOUNDATION OF POLLINATION ECOLOGY

Self-pollination of the bisexual flower was apparently so obvious to Camerarius that he never thought about other alternatives, and the first discoveries of cross-pollination are said to be due to A. Dobbs (1750) and H. Müller (1751) who also discovered the rôle played by insects in pollination (Sachs 1875, V. Grant 1949a). However, neither Dobbs nor Müller

---

*For the history of pollination ecology with references, see Loew 1895; Lorch 1966; Schmid 1975.

1

pursued their studies, and two other botanists, Koelreuter and Sprengel, are generally accepted as the fathers of pollination ecology.

Joseph Gottlieb Koelreuter (1733–1806) was the first botanist to carry out hybridization on a large scale for scientific purposes, thereby definitely proving the sexual nature of pistil and anther; he described the pollen grain and its function in some (not always correct) detail, and defined the stigmatic surface as a separate part of the surface of the pistil. He recognized autogamous, anemophilous, and entomophilous pollination, and the rôle of the nectar. He also recognized dichogamy and the function of certain movements of floral parts. Koelreuter's main work on pollination was published 1761–6.

The next major publication was Christian Konrad Sprengel's (1750–1816) famous work published in 1793. Without knowing about Koelreuter's paper he rediscovered entomophily and dichogamy and interpreted their importance more correctly. In the same way, he found that cross-pollination was obligatory in some plants. Among his other discoveries, based upon innumerable individual observations, were nectar protection, nectar guides, trap flowers, the function of the corolla as an organ of advertisement, and the differences between day and night flowers. His main discovery was that of the mutual adaptation of flowers and their pollinators. Pollination ecology before and after the publication of his book were two different things.

1.3. THE PRE-DARWINIAN PERIOD

The period between Sprengel and Darwin is one of almost complete neglect of pollination ecology, and prevailing concepts were generally much less advanced even than those of Koelreuter. However, the studies of pollination ecology of Darwin and his contemporaries presumed the understanding of floral morphology and of the fertilization process gained during this period.

For pollination ecology itself only one name may be mentioned from this period: Thomas Andrew Knight, who in 1799 formulated the principle that no plant can without detrimental effects pollinate itself for a sequence of generations. This principle was based upon observations of the behaviour of obligate outbreeders and was later, under the heading of the Darwin–Knight law, to play a rather unfortunate part in pollination ecology, leading less critically minded followers of Darwin to neglect the phenomenon of self-fertilization, and to search, sometimes rather frantically, for cross-pollination mechanisms and adaptations even in habitual inbreeders.

1.4. DARWIN AND THE POST-DARWINIAN PERIOD

Charles Darwin influenced pollination ecology more deeply than anybody else during the nineteenth century. Within this field he published a number of papers that gave evidence of patience and acuity of observation. Pollination ecology had not made such progress since Sprengel's time. His methods of observation marked the end of the deductive, philosophical approach of the previous half of the century.

Even so, Darwin's most important contribution to pollination ecology was of a more profound nature. To Sprengel the wonderful adaptations he observed in his flowers were witnesses of the Creator's wisdom and of the beauty of His creation. This admirable view point could hardly appeal to scientists at large, and pollination ecology would therefore tend to assume the character of a mass of uncorrelated facts. The acceptance of the theory of natural selection immediately changed this situation: adaptation, not to speak of mutual adaptation, was one of the phenomena, indeed *the phenomenon*, at the centre of interest of early selectionists; the study of pollination ecology with its remarkable adaptations became scientifically fashionable, and, more important, acquired a new philosophical dimension, and a unifying theory.

Whereas it is almost impossible to find a single botanist worth mentioning in pollination ecology during the first half of the nineteenth century, their names abound later to such an extent as to make any attempt at a complete enumeration futile. The two Müllers, Hermann and Fritz, Federico Delpino (to whom we owe many modern terms and concepts), Friedrich Hildebrand, and many others (e.g. Trelease and Robertson in the U.S.A.) were busily collecting data on pollination, sometimes with more eager than critical judgement. This period culminated in Knuth's great work from 1898 to 1905, which was completed by Loew. It contains all data known up to then and is still a most useful work of reference, though usually no more critical than the individual papers on which it is based.

1.5. THE MODERN PERIOD

Classical pollination ecology had apparently spent itself with Knuth's handbook. Its philosophical approach and methods were obsolescent, and the study of pollination more or less fell into disrepute during the first decennia of our century as an occupation for retired schoolteachers, but unworthy of a real scientist, even if noteworthy contributions were published during this period, e.g. by Clements and Long. The impact restoring it came from outside, from the study of animal behaviour. The studies of Frisch, Knoll, Kugler, and others on the behaviour of different groups of pollinating insects had important repercussions on pollination ecology because the results of these studies gradually restored confidence in many of Sprengel's acute observations and deductions which had been discredited by a previous generation.

Modern trends in pollination ecology can be summarized under the following headings: (1) Observations in countries outside Europe. Most of the work done by previous generations had been carried out in Europe, even exotics being studied in European botanic gardens. As important classes of pollinators are absent from the European fauna this led to grave errors. (2) The experimental approach to verify deductions and to test the reactions of pollinators. (3) A genetic and phylogenetic (also population genetic) approach to the problems underlying phenomena of pollination ecology, and a more profound genetical understanding of the implications and importance of cross-pollination. (4) As a consequence of this, pollination data ("breeding system data") are now gradually being incorporated into biosystematic reasoning. (5) Quantitative data about the energy flow in plant/pollinator interaction. (6) The realization that pollination ecology must be seen in a total community context, even if there are few quantitative data in literature as yet. Until now, the evolution of community pollination spectra has only been touched upon.

CHAPTER 2

TECHNIQUES IN POLLINATION ECOLOGY

The primary technique of pollination ecology (cf. Porsch 1922) is the same today as in Sprengel's or Darwin's days: consistent observation of what really happens in nature, in the original, natural habitat of the plant under investigation. A plant growing in strange surroundings is exposed to other pollinators which may work the pollination mechanism "correctly" or may not. Even wind conditions may differ in a foreign habitat. An "incorrect" pollination mechanism may be of great interest in itself, but, as previously mentioned, it will not give information about the existence of mutual adaptations between that particular plant and the fauna of the region. The fact that, for example, American ornithophilous flowers in the Old World are pollinated by local birds belonging to taxonomic units completely different from the American ones, proves the general character of ornithophily, but gives no information about the specific adaptation mechanisms.

Straightforward observations as to how pollination is carried out are not always made without difficulty. Some pollinators work at inconvenient times of the day, or out of the reach of ordinary visual observation, e.g. in tree-tops. In other cases it may be desirable to watch the behaviour of individual pollinators, which necessitates marking and other complicated techniques. Or pollinators may have to be captured and investigated for their nectar and pollen loads, both on the outside and in the alimentary canal. All classes of visitors should be recorded. One sometimes suspects that some of the classical investigators have discarded observations or overlooked insects because of a preconceived notion that they did not "belong", or because they were too small, etc. That both plants and animals should be properly identified and voucher specimens kept is another obvious admonition. Periodicity in the appearance of attractants and of suspected pollinators is an important fact that should be closely observed.

The movements of pollinators in flowers are frequently very swift, and often photographic techniques are desirable to analyse the behaviour of animals, sometimes also the function of floral parts. Modern photographic techniques — powerful flashlights, stroboscopic light systems, rapid colour film, on-the-spot developing — have immensely increased the potentialities of photographic methods. Photomultiplying devices are important in the study of pollination. Ayensu (1974) gives very good night photographs of pollinators and fruit-eating bats taken without additional light. Cinematographic techniques are even more valuable than still-photography. Pollination observations in the field are always extremely time-consuming, and days may pass without anything happening. Photographic techniques may take even longer, because it usually does not pay to run after the pollinator with a camera; it is better to focus one or a few blossoms and wait until the pollinator arrives.

In some respects television techniques are superior to photographic techniques and visual observation: the potential wavelength spectrum is wider and results can be obtained with

4

lower light intensities. Combined with video techniques they give an instantaneous permanent record (Eisner *et al*. 1969).

The wing-beat of insects, especially bumblebees, changes with different activities. Tape recordings have been used to analyse this. Some of Macior's recent papers give examples of very refined field techniques.

In the field there are also negative observations to be made. Sometimes it is as important to note which animals do *not* visit, though present in the locality. And, similarly, it is important to note which animals visit other blossoms in the neighbourhood instead of, or together with, the species under observation.

Hagerup (1951) has rightly pointed out that insect visits, even regular insect visits, to a flower do not always mean that the particular insect is the pollinator of that flower: there is always a chance that autogamy may have occurred earlier, even before the flower opened, or that some other less conspicuous insect may have carried out the pollination. The state of the stigma should therefore also be checked before the pollination act under observation.

Very simple equipment may help to supplement the field observations. Odour-producing organs within the flowers (osmophores) can be localized by cutting out the suspected parts and keeping them for some time in a closed vial. The accumulated odoriferous gases are easily smelled when the vial is opened. The presence or absence of sticky or oily substances on pollen grains or on animals can be checked by bringing the organ in question into contact with a clean glass plate on which traces of oil are easily seen (Daumann 1966). Self-pollination is checked by enclosing flowers in a bag sufficiently tight not to let insects (or wind-dispersed pollen) in, taking care that temperature and humidity inside the bag do not rise to dangerous levels.

Chemical analyses of varying degrees of sophistication are necessary for the identification of nectar constituents or active liquid or gaseous substances in or emanating from the flowers. Thin-layer and gas chromatography are obvious methods for the solution of many of these problems.

Physical techniques used are spectral analysis of flower colour, both inside and outside of the visual spectrum, or electrophysiological techniques, especially for the study of sense reactions, e.g. electroantennograms for the study of odour perception, Kullenberg and Stenhagen (1973) give many examples of the use of refined chemical and sense-physiological techniques in the study of pollination, above all various phenomena of sexual attraction.

Pollen adhering to pollinating insects indicates which species have been visited. The pertinent palynological techniques are presented elsewhere (Fægri and Iversen 1975).

The next question is why pollinators behave as they do. Conclusions by analogy may be of considerable help, but in many cases experiments are necessary to decide between hypotheses. It is not easy to carry out conclusive experiments on ecological questions, and all too frequently "no effect registered" has been interpreted as "no interrelationship extant", which is not always the case. Some of the more common errors are abnormal behaviour of many insects when caged, different colour perception, due to the effect of radiation invisible to the human eye and different behaviour of conditioned and "virgin" pollinators (a nectar guide may greatly help an inexperienced pollinator, but be of no value or have a different meaning to one that has previously visited many flowers of the kind, cf. Kugler's different evaluation of the function of nectar guides in 1930 and in 1936). In reality the "why" of pollination ecology is largely animal psychology, and experiments in pollination ecology without a thorough background in animal psychology are generally less valuable than simple field observations as to "how".

Earlier observers presumed that the senses of pollinators, which then meant pollinating insects, corresponded to those of man. Today we know that they do not necessarily do so, but in modern experiments it is often astonishing to see how similar are the effects of sensory activity in insects and in man (see Ribbands 1955). However, there are also obvious exceptions, like different ranges of the visual spectrum (Kugler 1962), the presence of chemical contact receptors on extremities, or special smells connected with sexual activity (see Kullenberg 1956b and later). One rather specific, but very serious source of errors in experiments of this kind is that both the experimenter and his animals ultimately become too clever, and the latter may be trained to perform tricks never performed under natural conditions.

However important, and often indispensable, physiological or morphological analyses are for understanding ecological adaptation, they cannot replace the ecological, functional aspect, any more than the chemical fact that the blue paint is ultramarine explains why the artist has decided to use blue at that place in the painting.

Pollination processes are particularly well adapted for cinematographic representation. For example, the films issued by the Institut für den wissenschaftlichen Film, with comments by Vogel, or the Uppsala *Ophrys* films (Kullenberg and Bergström 1976) are of very high scientific and educational quality.

CHAPTER 3

POLLINATION AS SPORE DISPERSAL

Although pollination exists only in higher plants, i.e. those possessing pollen, it is, in fact, a specialized type of a phenomenon that occurs throughout a major part of the vegetable kingdom. To forget this tends to confuse the issues and to present pollination as something more complicated than it really is. For a proper understanding, pollination must be referred back to related phenomena in lower plants. As we see it in nature, the phanerogamic plant represents the diploid generation, the sporophyte. As the name implies, this generation produces spores, generally accompanied by chromosome reduction. These spores germinate into the haploid generation, the gametophytes, which are male or female, and produce sexual cells that fuse again at fertilization, thus reconstituting the diploid number and producing a new sporophytic generation.

Schematically, this can be represented, e.g. in the green alga *Ulva*, as follows:

sporophyte → reduction division → spore →
               gametophyte   → gametes → fertilization → sporophyte

In *Ulva*, all specimens have the same sexual potentiality; in others, e.g. *Selaginella*, there are two types of spores, producing two different types of gametophytes:

sporophyte → reduction division ↗ microspores →
                                 ↘ macrospores →

male gametophyte   → spermatozoid ↘
                                   fertilization → sporophyte
female gametophyte → egg cell   ↗

A corollary to the existence of two types of spores is the need for two different dispersal strategies. Pollination is related to microspore dispersal strategy only.

Phanerogams also produce macro- and microspores, but the macrospore is included in the pistil, and never becomes a discrete unit. As its presence is therefore generally not realized, this makes the microspore, i.e. the pollen grain, appear more exceptional than it really is, because of the apparent absence of a female counterpart. Actually, apart from the existence of a few gametophytic cells (mostly abortive), there is little to distinguish between a young pollen grain and a typical microspore. In a wider biological sense, pollination is simply a specialized case of microspore dispersal, and pollination ecology a special aspect of spore dispersal ecology. The nuclear divisions taking place in the pollen grain later do not influence its dispersal strategy and can therefore be neglected in this comparison.

Apart from the most primitive ones, plants have two sedentary stages, sporophyte and gametophyte, and two mobile or motile ones, spores and gametes. Originally there was not much difference between (zoo-)spores and gametes, but gradually spores changed as a response to ecological conditions. During phylogenetic development gametes did not change very much, except that ultimately they were so reduced as almost to lose their individuality. The uniformity of gametes entails that the fertilization process is also rather uniform throughout the plant kingdom. Even in angiosperms the main features of fertilization are not very different from those of the more primitive process in lower plants.

With gametes being confined to the original aquatic (or at any rate moist) conditions of primitive plants, spores are the stage adaptable to dispersal under dry conditions, and the existence of terrestrial plant life presupposes the existence of spores resistant to the adverse forces of land life. We see, therefore, that spores even of the most primitive land plants have a resistant outer cover that is chemically related to that of pollen grains. For their pollination higher plants were thus able to take over an already existing spore type and a mechanism already developed for its dispersal.

Yet in one respect there is a very great difference between spore dispersal and pollination. An ordinary spore has a comparatively wide ecological niche; even if a gametophyte may be rather exacting in its demands (not all of them are), there are many places and many types of habitats in which the spore may germinate. Further, being motile, the gamete(s) can to some extent compensate for an unfortunate mutual position of the male and female apparatus. The pollen grain, on the other hand, can germinate and the gametophyte grow successfully in one single, very restricted place only, viz. on the stigmatic surface of a compatible flower. If germination exceptionally takes place elsewhere, the resultant plant, the pollen tube, cannot fulfil its biological function, or even develop properly.

This enormously restricts the germination potential of pollen grains and calls for a much greater *precision in the transfer* than in the dispersal of ordinary spores. The many remarkable adaptations (Section 5.1) observed in pollination ecology can only be understood against the background of this demand for precision, and the evolution of pollination mechanisms in entomophilous blossoms is, on the whole, an evolution towards increasingly higher precision.

With the evolution of heterospory in many lower plants, a certain demand for precision in the transfer of the microspore will establish itself in these groups as well, modified by greater or lesser motility of the "male" cells themselves.

In higher plants, one of the primary ecological functions of the spore in lower plants, viz. dissemination, has been taken over by an entirely new dispersal unit, the seed, an arrested developmental stage of the new sporophyte. Ecologically the seed may be compared to the resting zygote of lower plants. Seeds have a much wider ecological range than pollen, to which corresponds the fact that seed dispersal mechanisms are generally less evolved (but not always less varied) and function less precisely than pollination mechanisms.

For a proper understanding and evaluation of pollination mechanisms, we shall first recapitulate the main features of spore dispersal in lower plants.

CHAPTER 4

SPORE AND POLLEN DISPERSAL
IN LOWER PLANTS AND GYMNOSPERMS

4.1. SPORE DISPERSAL IN THALLOPHYTES

In algae spore dispersal depends on water. Some resting stages are dispersed by wind, and so are some vegetative stages of aerophilic algae, but generally algae are too primitive to be of interest in this respect. The same holds for bacteria. Myxomycetes on the whole seem to be wind-dispersed and have developed special mechanisms for that purpose, including an arresting mechanism, the capillitium.

In fungi, the picture is entirely different (see Ingold 1971). Primitive types depend on water, whereas the more evolved ones have gradually become independent of it, as seen in the series from the water-dependent *Saprolegnia*, via *Peronospora* (in which one dispersal stage depends on water) to *Mucor* (completely independent). An astonishing variety of spore dispersal mechanisms are met with within terrestrial fungi. Wind dispersal of small, more or less ellipsoidal spores is most frequent. This dispersal may be entirely passive, as in *Ustilago*, or combined with an active phase as in the majority of higher fungi, which shoot their spores from asci or basidia. More spectacular active dispersal is found in *Empusa* (individual conidia), *Pilobolus*, or *Sphaerobolus* (larger units). A rain-splash mechanism with or without subsequent wind dispersal is found in *Nidularia* (Brodie 1951) and the puff balls (Bovistaceae). Dispersal by carrion insects, attracted by scent imitating that of decaying protein, is found in Phallaceae, whereas sugar is the insect attractant in the sphacelia stage of *Claviceps*, or in the preaecidium of Uredinales. According to Schremmer (1963) the dispersal in *Phallus* is endochorous; the exact mechanism is unknown in the other cases. Sugar, in this case secreted by the host plant, is also the attractant in *Ustilago violacea*. Endozoic spore dispersal is found in hypogeic fungi like *Tuber*, and the dispersal mechanism of *Tilletia tritici* may even be a secondary phenomenon adapted to the conditions of cultivation.

4.2. SPORE DISPERSAL IN MOSSES

Most mosses have a rather simple spore dispersal mechanism: small units dispersed by wind. Refinements are chiefly represented by the arresting and regulating action of the capillitium in hepatics and the peristome in leafmosses. The spores of a few aquatic species are evidently spread by water; this has in some cases lead to reduction of the dispersal mechanism. *Sphagnum* has an explosion mechanism followed by wind dispersal (Grout 1926).

More interesting in this respect are those Splachnaceae that live on dung or carrion and attract coprophilous or carrion insects by means of smell resembling that of decaying protein, and which have developed separate morphological devices for that purpose (Bryhn

1897). It is, perhaps, not unreasonable to think that future investigations may reveal that spore dispersal strategies in mosses may be more varied than it seems today.

4.3. SPORE DISPERSAL IN PTERIDOPHYTES. CONCLUSION

Compared with those of the groups already mentioned, spore dispersal of pteridophytes is comparatively monotonous, being dependent on wind, or water for aquatic species, with few special adaptations for it. In some families spores are actively dispersed by violent movements of the bursting sporangium walls. Symbiotic ants disperse the spores of some tropical ferns (Docters van Leeuwen 1929).

One very important conclusion can be drawn from this list of spore-dispersal mechanisms (the main references for which may be found in any textbook), viz. that pollination, in coming into existence with the appearance of higher plants, evolved on the background of an extremely varied range of spore-dispersal mechanisms in the ancestral forms. Even if most variation of such mechanisms is found in the fungi, a group which definitely does not include the ancestors of higher plants, spore-dispersal mechanisms in the ancestors of higher plants may have been any of a series of different types, and it is very far from being *a priori* certain that it was wind dispersal. With this background the problem as to which is the more primitive: wind or insect pollination, loses a great deal both in meaning and in significance. It should be replaced by another question: Which, if any, changes took place in the microspore dispersal mechanism of the ancestral forms concomitant with the evolution of angiospermy? Pollination is not new; it developed out of something more primitive, and any discussion must start, not with pollination as such, but with the corresponding mechanism in ancestral forms. Nor should the fact be neglected that all the dispersal and attraction principles met with in pollination are also realized in some lower plant group (where the term "lower" does not imply direct relationship). Indeed, spore dispersal mechanisms in lower plants are at least as varied as the pollen dispersal mechanisms known today.

4.4. POLLINATION IN GYMNOSPERMS

Pollination in gymnosperms occupies an intermediate position between the simple (micro-)spore dispersal of lower plants and pollination proper. As in the latter, the germination possibility is restricted: unless the pollen grain reaches a micropyle in a compatible female blossom it has no chance of germinating and producing offspring (spermatozoids or male nuclei).

In the absence of a stigma the receptive surface of a "female" gymnosperm flower is the micropyle or an adjacent cone scale. It is not known that these receptive surfaces can react selectively with the germinating pollen grain or the pollen tube, which implies that self-incompatibility in gymnosperms, if present, must be due to incompatibility in the haploid stage: between nuclei or between the male nucleus and the female endosperm, but not the haploid—diploid interactions, which are so important in angiosperm incompatibility systems.

If we define pollination so widely as to cover that of gymnosperms as well as of angiosperms,* pollination is probably polyphyletic: the transition from microspore to pollen

---

*This we consider permissible, but reserve ourselves against an extension of the concept to comprise also the transfer of microspores in vascular cryptogams. Even if some of the structures met with in the macrosporangia are remarkably similar to stigmas, etc., their origin is completely different (see Maheshwari 1960). This does not, of course, preclude that similar transfer mechanisms may have been at work.

took place independently in various developing lines of vascular cryptogams, some of which succeeded in the present-day higher plants: gymnosperms and angiosperms.

In conifers, wind pollination is prevalent, perhaps exclusively so, and with the quantitatively great rôle played by conifers among present-day gymnosperms, one is easily led to premature generalizations. In view of the possibility that gymnosperms are polyphyletic, the other orders may be as significant, or even more significant, for a phylogenetic discussion of pollination. *Ginkgo* seems to conform to the pollination pattern of conifers though fertilization may take place even after detachment of the macrosporangium.

In Cycadales, there are scattered observations indicating that insects pollinate at least some species; whether this is accidental or is the regular pollination pattern of the particular species is not definitely known. Pollen-eating beetles are known to visit male cones of *Encephalartos* and of *Zamia integrifolia* (Wester 1910), and small bees are reported to collect pollen in *Macrozamia* (a variety of *M. tridentata* [Willd.] Regel according to Schuster, 1932). As these plants are dioecious, such activity alone does not constitute pollination, and the question remains as to why insects visit female cones. Hemipters have been observed in *Macrozamia*. The pollination drop may serve as an attractant for small insects; on the other hand, it is questionable if the rise in temperature observed (in male inflorescences only) serves as an attractant in a tropical climate. In some cycads both male and female cones emit a very strong odour. This is unpleasant to man in some species, whereas it is said to be pleasant in *Macrozamia* (cf. visits by bees in this genus). According to Rattray (1913), beetles visit male cones of *Encephalartos*, attracted by the smell, and also female cones to deposit eggs. We may perhaps here perceive a pollination mechanism corresponding to the one operated by semi-destructive visitors like *Hadena* and *Chiastochaeta* in some angiosperm flowers (Section 8.5). Finally, it should be mentioned that Baird (1938), although believing that cycad pollen is windborne from one inflorescence to another, maintains that transfer within the female cone (of *Macrozamia*, cf. above) is dependent on insects, as the micropyles are situated too far in for pollen grains to reach them by wind transport only. Insects are found within cones of both sexes. Thus there is sufficient indication that insects play a part in pollinating cycads; but there is not much conclusive evidence, and direct observations of pollen transfer from male to female inflorescence by insects have not been made, as far as we are aware.

In Gnetales (*s.l.*), wind pollination is generally presumed prevalent. There are, however, the well-known instances of some species, e.g. *Ephedra campylopoda* and *Welwitschia* (cf. Porsch 1958; we do not accept Bornmann's 1972 assertion that *Welwitschia* is wind-pollinated), in the male inflorescence or flower of which there is a single sterile female flower, respectively pistil or rather ovule. Apparently, this has only one function: to exude a (sweet) pollination drop attracting insects. In these plants, the situation is the opposite of that of the cycads: *there* we found a possible attractant in the male flower, viz. pollen for pollen-eating beetles, but had difficulty in explaining the attraction to female cones. *Here* we find a possible attractant in the female flower, viz. the pollination drop, but there is apparently nothing in the ordinary male flower to attract insects. This is remedied by the existence of the rudimentary female flower in the plants mentioned. If we accept the general phylogenetic rule that organs never start as rudiments, the conclusion would be that *Ephedra* and *Welwitschia* must descend from plants with hermaphroditic inflorescences or flowers, which have later become unisexual, like those of many other wind-pollinated plants. Some *Ephedra* species (see Mehra 1950) and *Welwitschia* have remained more primitive due to the fact that they are still insect-pollinated (*Ephedra* honey is mentioned by Ordetz 1952). Male flowers of *E. campylopoda* possess a fruity odour. This again would mean that insect pollination must be older than sexual differentiation between individuals, and perhaps also that the latter has come about in response to secondary wind pollination like it has in many higher plants. In *Gnetum*, too, insects have been observed to collect sticky pollen in the showy, sweet-scented nocturnal male flowers. The female ones possess a sweetish pollination droplet (cf. Hendryck 1953). Whereas it is highly interesting to find such indications of primary hermaphroditism and insect pollination in the flowers within these groups of gymnosperms, care must be taken not to over-generalize; conifers are but distantly related to Gnetales, and their unisexuality may be primary, i.e. the character of unisexual strobili developed in their ancestors before the characters qualifying these plants as gymnosperms. Nor does the possible existence of primary entomophily in cycads and Gnetales preclude that there may exist primary anemophily in *Ginkgo* and conifers, perhaps the only primary anemophily in pollination ecology. Finally, the phylogenetic relationships between these groups and angiosperms are so obscure that with regard to the latter group conclusions can only be drawn by analogy, but there may be reason to suspect that nectar from nectaries was not the primary attractant when the angiosperm flower developed.

At any rate it is obvious that the problem of the pollination drop is not as simple as it may seem. To quote McWilliams (1958: 115): "the absorbtion of the micropylar fluid under the influence of pollen is not really explainable." Doyle (1945) maintains that the pollination drop is present in primitive conifers, whereas its receiving function has been taken over by (dry) stigmatic flanges of the micropyle in more advanced types (*Abies*, *Tsuga*). This might mean that anemophily becomes more fixed during the phylogenetic development of the group and that the potential for insect pollination, originally present, is being lost. However, Doyle and Kam (1944) maintain that the nucellus is inaccessible to airborne pollen in *A. homolepis*. Fertilization presumes the intermediary of a pollen tube, growing on the surface of the seed scale.

CHAPTER 5

POLLINATION IN ANGIOSPERMS

Only in angiosperms is pollination typically developed in its three phases: (1) release of pollen from the "male" part of a flower, (2) transfer from the paternal to the maternal part, and (3) successful placing of pollen on the recipient surface of the latter, followed by germination of the pollen grain. This introduces the next phase, which is fertilization. Whereas the germination process as such is no concern of pollination ecology, the germination of the pollen grain on the stigma and the subsequent fate of the pollen tube and the male nuclei indicate whether the surface upon which the grain was deposited, was a "correct" one or not. In self-incompatible plants a successful transfer of pollen is not in itself a successful pollination. Each of the three phases of pollination shows great diversity.

The angiosperm flower can be regarded as a collection of sporophylls* developed in response to microspore transfer to a much higher degree than the fructification organs, the sporangiophores, of any other group taken as a whole, and its structure can only be fully understood when considered in relation to pollination ecology, i.e. as a functional unit. Without a functional background, a flower becomes meaningless morphological play, and the development of the flower incomprehensible and only to be described by referring vaguely to orthogenesis, *Gestalt*, and similar theoretical concepts.

We shall discuss the flower in relation to the function of its parts, without too much consideration of the basic morphological value and evolution of its different members. Our morphological starting point is therefore the already complete, simple flower, as represented in the diagram, Fig. 1, which defines the main concepts and terms.

We define anthesis as beginning when anthers and stigmas are exposed to the pollinating agent(s), either because the flower opens or because the organs protrude from a closed flower in such a manner as to expose themselves to the same agents. All organs may not be continually receptive during anthesis. Similarly, anthesis ends when the same parts are no longer available to the pollinating agents. There may be "sterile" periods both at the beginning and at the end of flowering periods when anthers and/or stigmas are exposed, but not functional. In cleistogamous flowers there is, by this definition, no anthesis. We consider anthesis as a function of the flower or the blossom as such and are not in favour of using the term when describing the state of single parts, e.g. "anthesis of a stamen". Flowers may close temporarily during anthesis, even to such an extent as to be indistinguishable from withered ones (*Ipomea* spp.).

5.1. POLLINATION AND ADAPTATION

In dealing with a theme like pollination ecology it is hardly possible to avoid a certain finalistic way of expression, and even if it were, the result would be an artificial and laboured language. We therefore stress once and for all that using the word "adaptation",

*Sporophylls are in principle asexual, even if in this special case the connection between sporophylls and the sexual function is very close.

13

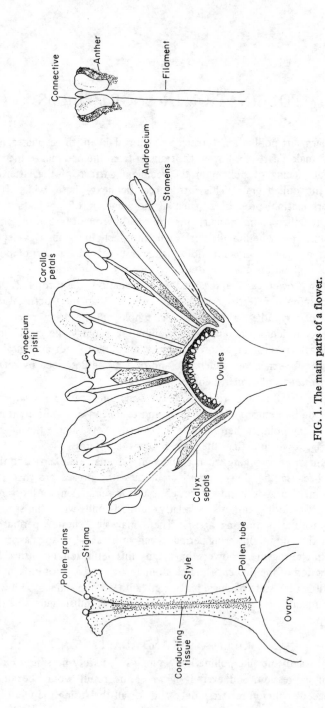

FIG. 1. The main parts of a flower.

which will frequently be used, we have in mind a statistic relation, viz. the observed (or *ex analogia* inferred) fact that in most instances a certain structure functions in a certain manner, which is made possible by the morphology (and ethology) of the organism(s) concerned. Such adaptation is a fact that can be objectively registered, independently of any possible previous erroneous statements by careless investigators. In this connection the question of "adaptation" versus "utilization" (Goebel, Troll) is immaterial; we only register here our disagreement with the "anti-adaptation" point of view. It is possible to give a description of an eye or of a motorcar without taking into account its function; but there is hardly any doubt that an understanding of the description would be greatly furthered by the introduction of a functional viewpoint. To quote Moll: "No hypothetical element is present when we simply establish the fact that eye and ear are useful to man . . . . Usefulness of organs . . . can not be denied simply because we are ignorant concerning its causes." The question whether this should be called adaptation or utilization is more semantic than real. On the other hand, it would take extraordinary credulity to believe that the many extremely exact and complicated interactions between animals and plants found in pollination have arisen fortuitously as a result of non-directional variability (Wanndorp 1974).

This should not be interpreted to mean that accidental utilization of blossoms by insects with concomitant pollination does not take place. Few blossoms are so specialized that pollination can only be carried out by one vector. When plants are moved to other geographic regions with a different fauna, such plants like *Yucca* or *Ficus* remain unpollinated, except for a few, irregular, and accidental visits. Unspecialized flowers are usually utilized in one way or the other. A good example in point is the myophilous flower of the cocoa plant, which is utilized by different insects in different parts of its present, great cultivation area. There is no direct adaptation of an American plant to African insects, but these utilize — adapt to — the imported source of food.

The term *co-evolution* has often been used for adaptations between blossoms and insects. Various authors have pointed out that this must be conceived of as a leap-frog process: a small change in the blossom starts reactions in a pollinator, which again has an effect on the blossom, etc. The term should not be used indiscriminately for any adaptation between two organisms, but should be reserved for directed mutual adaptations.

Attempts to "explain away" functional interpretations by reference to morphological or physiological characteristics of the organ(s) in question are fallacious and severe breaches of the simplest rules of logic. The demonstration that cleistogamous flowers are flowers in which development is inhibited at an early stage is perfectly valid, but has no direct bearing on the functional interpretation of cleistogamy.

Adaptation has two sides: a positive, adaptation *to* something, and a negative, adaptation *against*. The one is frequently an automatic function of the other. Thus, one may ask whether the long, narrow tubes or spurs found in butterfly blossoms are "primarily" an adaptation *to* the corresponding prosbosces of their "adapted" visitors, and, incidentally also to their instinct for penetrating into narrow openings with their prosbosces, or if it is an adaptation *against* other insects getting at the nectar. The answer probably is that both developed simultaneously, and the reaction to selection pressure took the same direction for both phenomena; the exclusion of, for example, bees leaves more nectar for butterflies and thus gives the blossoms a greater value to them. The absence of a landing-place in hummingbird and moth blossoms, the closed corollas of bumblebee blossoms, etc., may all be interpreted in a similar manner. All these "adaptations" have a positive and a negative side: they encourage visits from some animals and exclude others.

The general evolutionary tendency in pollination is directed towards greater refinement, locking blossom and pollinator more closely together in mutual interdependence. Whereas this has advantages for both parts as long as things go well, Baker and Hurd (1968) have pointed out that this close interdependence also constitutes a potential weakness. If one of the partners should fail, the other one is also doomed unless it can save itself by a compensatory mechanism. The dispersal of either partner to new areas is dependent on that of the other, and it is significant in this context that the most successful colonizers (weeds in new, unstable habitats) in the plant world have very unspecialized breeding (including pollination) systems ("Baker's Law").

5.2. STRUCTURE AND FUNCTION OF THE ANGIOSPERM FLOWER

The simple flower structures represented in Fig. 1 have their very distinctive functions which may be inferred statistically by analysis of how they really do function without any reference to "adaptation". Thus the most frequent function of the calyx is that of protecting the young flower before anthesis. In some plants, this function is so exclusive that the calyx drops off or withers immediately anthesis has begun, e.g. in *Papaver*. Similarly, the corolla generally attracts pollinators by long-distance advertising of the flower and its attractants. Anthers produce and present the pollen, and the pistil contains the ovule; more specifically, the stigma constitutes the receptive surface, and the style guides the growth of the pollen tube towards the ovules.

This short presentation serves to link certain morphological units with their most common function, but any of these functions may be missing in the flower under examination; and some of them may be taken over by other organs, which more rarely perform them. In such instances the organ in question is morphologically different from its "typical" state; it shows "adaptations" to its new function. Thus, the function of protection may be afforded by bracts (Compositae) or other appendages outside (below) the calyx as well as by corolla segments (Umbelliferae). For advertising, the corolla may be supplemented by calyx (*Calluna*), stamens (*Canna*), or pistil (*Iris*), either alone or in conjunction with each other; extrafloral parts may come into the picture here too (*Cornus*). The production of pollen is exclusive for anthers, but the presentation may be taken over by other organs. All these modifications of function are better understood in relation to the specific mode of pollination — they form part of the *syndrome* of the pollination in question.

Whereas it is redundant to describe the normal morphology of the outer members of the flower, the functional structure of stamens and pollen may need some elaboration.* In supposedly very primitive flowers, e.g. in *Degeneria* (A. C. Smith 1949), the stamen very much resembles the (micro)sporophyll of a cryptogam or a cycad, with the thecae disposed as individual sporangia on or immersed in a small sporophyll. In more evolved flowers, the

---

*We shall not take sides in the controversial question on the morphological nature of the "sporophyll" and the original position of the sporangia. For a discussion of modern theories of the morphological nature and the origin of the angiosperm flower we refer to papers by Gottsberger (1974) and A. Meeuse (1975), with whom we are, however, not always in agreement. We do not find Leppik's differentiation between sporophylls and semaphylls (1961: 4) particularly useful in pollination ecology, also not after Meeuse's redefinition (1974: 88): "all laminiform elements functioning as optical attractants in the reproductive region of zoophilous plants." It fails to account for the attraction function of sporophylls, which is very important in some flowers, nor does it take into account olfactoric attraction, which especially in primitive blossoms may be more important than optical.

laminar part of the stamen is very much reduced, but the thecae still distinctly retain their character of sporangia with an opening mechanism. Openings are most often slits along the total length of the thecae, and often the wall curves back so much that it turns the thecae inside out. The pollen of insect pollinated flowers adheres to the outer surface of the anther until removed by a pollinator. The dry pollen of wind-pollinated plants generally falls or blows out of the anther at once; but there are exceptions to both rules and they are modified in different ways.

Anthers may also open by means of lids or valves as in Berberidaceae, or empty themselves through pores (Solanaceae, *Cassia*), sometimes elongated into tubes, as in Ericaceae; or parts of the theca wall may be detachable and be removed by the pollinator as in some *Garcinia* species. Many of these aberrant types of opening mechanisms are part of the syndrome of a specific pollination technique, e.g. in the pore-opening anthers of *Rhododendron*, *Exacum*, and many Melastomataceae the pollen is squeezed or thrown out (van der Pijl 1939). Sometimes appendages to the stamens have to be struck or moved to effect discharge of pollen (*Vaccinium*, *Thunbergia*, cf. Section 17). On top of the anther in many Mimosoideae there is a gland, the function of which is still unexplained.

In *Salvia*, e.g. *S. pratensis* (Section 13.2) the sterile middle part of the anther, the *connective*, is very long (more correctly broad), and forms a pivot link with the stiff, short filament; and in addition one theca is sterile. This complicated structure, which with certain modifications is also found in *Calceolaria* (Vogel 1974: 140) and *Roscoea* (Nordhagen 1932: 32), has a definite function during insect visits. The connective may be greatly developed in other families too, e.g. Melastomataceae. In *Canna* most stamens are sterile and supplement the perianth as advertising organs; only one remains fertile, and even that one has only one fertile theca, the rest being petaloid.

Secondary pollen presentation (not to be confused with accidental dislocation) is found when pollen is emptied out of the thecae, generally before or at the outset of anthesis, and deposited at some other place in the flower, generally concomitant with the withering and more or less complete disappearance of the anther. In Proteaceae pollen is deposited on a region that later develops into the mature stigma. A classical example is that of *Campanula* (Section 17.1) and many Compositae, the individual florets of which behave like *Campanula* flowers. In the related family Goodeniaceae the morphological development has gone further, inasmuch as there is a definite cup on top of the style in the genus *Scaevola*. Pollen collects in the cup before anthesis and is pushed out by the stigma (Kugler 1973). A more evolved mechanism is found in some plants, e.g. species of *Centaurea*, in which exposure of the pollen deposit does not take place until the filaments are irritated by the visit of a pollinator. They then contract, exposing the pollen. Secondary pollen presentation by means of parts of the pistil occurs also outside the Campanulales kinship, e.g. in Rubiaceae, tending towards dioecy; the style is maintained for pollen presentation in staminate flowers. Similar structures are found inside the inflorescences of various Compositae in which florets that have lost their pistillate function, nevertheless maintain the style as a presentation organ for pollen (van der Pijl, unpubl.; Stuessy 1972).

The mechanism in *Centaurea* is an example of a more general phenomenon, viz. that of pollen protection, in this case carried out by concealment. Pollen may need protection against theft by insects that do not pollinate, against being blown away from the flower by wind (except in anemophilous flowers), against being washed away by rain, and, above all, against moisture which would kill the pollen of some species either directly (cf. Kerner 1873) or, for example, by causing premature germination (Eisikowitch and Woodell 1975).

A very effective and simple protection against wetting is furnished by the hanging position of many flowers, though this position has other effects, too (see Hallermeier 1922). Total or partial closing of the flower also excludes water, see the discussion of the *Pedicularis* flower (Section 17.8). In other cases, pollen is concealed in narrow tubes into which water does not easily penetrate, or protected by a cover of hairs or scales that prevents the water from running into the flower. Hairs in flowers do not always seem to have this function, though, and there are also pollen grains that do not seem to be damaged by moisture, at least for a short period. More elaborate protective measures are the closing movements carried out by many flowers at the approach of inclement weather. Such movements have the great advantage that neither pollen nor nectar, if present, are exposed at times when the pollinating agent (wind or animal) does not function. The periodic opening of flowers, day- and night-flowers, etc., may be evaluated in the same way: as a measure of protecting the valuable products of the flower at times when their exposure would serve no "proper" function. And the example of *Centaurea*, mentioned above, represents one more step in the refinement: pollen is exposed only at the very moment the pollinator is active in the particular floret.

Whereas some of the movements carried out by stamens are clearly adaptive, others apparently are not. However, many floral structures previously considered non-functional have, on closer scrutiny, been found to occupy an important place in a pollination syndrome, so we do not want to state categorically that stamens carry out non-functional movements, even if there are cases that cannot be interpreted at present.

Pollen may live for a shorter or longer time after ripening. Generally, the order of magnitude is a few days, but both shorter and longer periods are known (see Visser 1955; Brewbaker 1959, Pruczinsky 1960). After having been packed in the corbiculae of pollen-collecting hymenopters, pollen rapidly loses its power of germination (Maurizio 1959). According to Lokoschus and Keulart (1968) this is due to the effects of a fatty (10-hydroxy-2-decenoic) acid from the mandibular glands. Under experimental conditions the longevity of pollen can be considerable. Fechner and Funsch (1966) report *Pinus ponderosa* pollen that germinated after 11 years' storage at low temperature (0–4°C) and medium humidity (25–50 per cent). Hanson and Campbell (1972) report the same result after vacuum-drying, and Collins *et al.* (1973) report positive results by storage in liquid air.

In nature, the life expectancy of pollen grains depends on their metabolic rate; the metabolically more active trinucleate grains are generally short-lived. In *Digitalis* the pollen may have lost its germinative power before the end of anthesis (Daumann 1970b). On the other hand, a *Cycas circinalis* in the Botanical Garden in Copenhagen was fertilized by pollen that had been stored at ordinary temperature for 11 months (Sørensen 1970). Mention should also be made of the extremely slow development of the fertilization process in many spring-flowering trees, e.g. *Corylus*. The first stages of germination seem to follow rather soon; then the young pollen tube may rest for a long period before resuming growth.

After having been wetted, pollen is frequently unable to germinate. This may be due to the leaching out of the exine of easily soluble proteins (enzymes) from the pollen coat (Knox and Heslop-Harrison 1971), although the role of these proteins in the germination process, if any, is still conjectural, cf. p. 29.

The gynoecium is generally more conservative than the outer parts of the flower, and largely conforms to the well-known, bottle-shaped pattern. In some families (Orchidaceae, Asclepiadaceae) the upper parts of the gynoecium and the androecium fuse to form a single, central structure, the *gynostemium*, in which the thecae form pockets.

The stigma is the most variable part of the gynoecium. Usually, it forms a small, viscid knob or cleft at the tip of the style; in *Iris* and *Sarracenia* the active stigmatic surface forms a very small, localized, part of the total. The extent of the functional part of the stigma can be demonstrated by chemical reactions, e.g. the hydrogen peroxide reaction described by Zeisler (1938). In some genera, style and stigma are more extravagantly developed, although the functional reason for this is not always quite clear. The pollen grains adhere to the stigma due partly to their own stickiness, partly to the glutinous nature of the stigmatic surface, and partly to its roughness. The relative importance of these three factors varies. The stigma is sometimes so placed as to comb pollen off the pollinator as in *Iris*.

Finally, the stigma must offer a suitable germination bed for the pollen grains. The specific demands vary: some pollen grains also germinate easily on artificial media, others need a specific chemical substance to germinate on such media, and some are very restricted, e.g. the pollen grains of *Pavetta javanica*, said to germinate only on stigmas of the same species, and of *P. fulgens*. The interaction between stigma and pollen grain is further discussed in Section 5.3.

A primary problem in the study of the angiosperm flower is that of angiospermy itself. Geologic evidence suggests that angiospermy is an advantageous condition, angiospermous plants dominating and out-numbering all other groups since their appearance. Whereas this might be due to other features concomitant with angiospermy (Fægri 1974), it certainly shows that the latter condition cannot be disadvantageous. The term angiospermy is in itself a misnomer: ovules (not seeds) are the organs primarily enclosed. *Angiovuly* would be better, and is to a certain extent replacing the customary one.

Some effects of angiospermy can easily be envisaged. On the negative side there is the more difficult process of pollination and fertilization, necessitating the establishment of a whole set of new organs: stigma, style, and conductive tissues on the part of the female apparatus, and long pollen tubes on the part of the male gametophyte. It, incidentally, also involves changing the function of what remains of the gametophyte from that of a rhizoid-like organ of nourishment (not too unlike the microgametophyte of, for example, *Laminaria*) to a very specialized organ for transporting the fertilizing nuclei. Tendencies in this direction are also found, rather incompletely, in some conifers, e.g. *Tsuga* (see Doyle 1945).

On the positive side there is primarily the effect of protecting the ovule against desiccation. The fact that completely unprotected ovules exist also in land plants, e.g. *Taxus*, does not in itself disprove the usefulness of this function, especially in the advancement of plants towards drier regions; but it would indicate that external, mechanical protection against desiccation to a great extent may be replaced by an internal, physiological mechanism. It should be noted also in connection with angiospermy that the fertilization process in angiosperms is less dependent on a liquid phase than in any other plant group except the higher fungi. All told, the effect of angiospermy on moisture conditions in the gynoeceum is hardly sufficient to explain its universal success.

A second point to consider is the possibility that the stigmatic surface and the tissue of the style exert a selective effect on the germination of the pollen grain and the growth of the tube, as discussed in Section 5.3. The possibility of a sieve effect against incompatible pollen gives greater economy and efficiency in establishing cross-fertilization as the preferred mode in higher plants (see Whitehouse 1950). Without this effect, incompatibility between genomes in the would-be fusing nuclei is the only possible guard against self-fertilization. Thus, the sieve effect is of the highest selective value, and may certainly have contributed very much towards making angiospermy so successful. Whether this is enough to explain the present-day preponderance of angiosperms is questionable.

A third effect of angiospermy is protection of the vulnerable ovules against mechanical injury. This is especially important if we accept the view that the ancestors of angiosperms were animal-pollinated, probably by awkward, pollen-eating beetles (see Section 11.1.1). Inclusion of the ovules to protect them from rude handling would probably in that case have great selective advantages, as maintained independently by several botanists approaching the problems from different angles. This viewpoint also places the states of perigyny and epigyny in a wider biological context:

Ovules:
unprotected → protected by   → protected by gynoecium wall plus tissues derived
                 gynoecium wall   from the outer members of the flower or the pedicel
(gymnospermy)   (angiospermy)   (epigyny)

The origin of angiospermy can hardly ever be demonstrated palaeontologically, and must be the subject of conjectural and speculative hypotheses. However, if we consider the flower as a functional unit in relation to pollination, its various features, including angiospermy, form a consistent pattern. The fact that the morphological elements involved in the creation of peri- and epigyny vary widely in different plant families (Douglas 1957), may be taken as indicating that these states arose in response to a functional demand, not as some kind of morphological orthogenesis (see V. Grant 1950a).*

For evaluation of the problem of how angiospermy originated, it is, perhaps, not without significance that Harris (1956) has found faecal pellets showing that some small reptile lived on microspores (and "fruits") of Caytoniales, one of the few plant groups which independently reached a kind of angiospermy.

The pistil has two main functions, namely (1) protection of ovules against unfavourable influences, and (2) reception of pollen. The first function is common to all (non-male) flowers, with small modifications. The second function is highly modifiable. In habitual self-pollinators, it is weakly developed. In the case of abiotic pollination the protective function of the perianth may cause it to obstruct the path of pollen transport. The perianth is therefore a negative factor in the pollination ecology of such plants. Finally, in animal-pollinated plants the second function of the flower also entails the positive attraction of pollinators, generally by means of the perianth. These three cases are fundamentally different, and must be considered separately.

In dealing with pollination, it will be necessary to introduce the functional concept of a *pollination unit* (see Berg 1959) for which we propose as a scientific term *anthium*. In cases like *Tulipa* or *Paeonia* the individual flower is obviously also the unit of pollination: it attracts visitors even if completely alone. Pollinators land on the flower, i.e. they stop travelling and remain immobile except for the secondary movements necessary to utilize the sources of attraction: pollen, nectar, etc. Equally obvious, the pollination unit in *Chrysanthemum leucanthemum*, *Trifolium pratense*, *Cornus*, or *Zanthedeschia* is the whole

---

*Whereas there is no doubt that bird pollinators also handle the flower very rudely, and that additional protection is advantageous, we hesitate to accept Grant's idea that bird pollination is old and precedes bee pollination. On the other hand, there is no doubt that the more highly evolved pollinators, hymenopters and lepidopters, handle flowers more carefully, and that very little protection would be necessary against them.

inflorescence. Once landed, the pollinator has stopped travelling (and using energy for that purpose), and until it again leaves the inflorescence the subsequent movements are in principle not different from the secondary movements performed by an insect utilizing separate nectaries in an individual flower. The morphological difference between flowers and inflorescences is in itself irrelevant in pollination ecology; the pollination unit may be one or the other. There are many intermediate types, less obvious than those referred to above. Particularly, inflorescences may appear as one unit at a distance, but dissolve into separate units at closer range. The *Linaria vulgaris* inflorescence can be defined as an *attraction unit*, but in this case the movements necessary to utilize the nectar of individual flowers are so completely different from those taking the pollinator from one nectary to another that they must be included under the term travelling, even if this may sometimes be crawling instead of flying. One frequently sees that in such cases pollinators travel to the next inflorescence very freely, returning perhaps to the first one later only. Incidentally, this "unsystematic" behaviour of pollinators increases the chances of cross-pollination. The *Iris* flower constitutes one attraction unit, but contains three separate pollination units, as distinctly separated as the flowers of *Linaria*, and in *Gloriosa rothschildiana* one flower even forms six pollination units (five male and one hermaphroditic) with a rather loose connection between them as an attraction unit. Sometimes the attraction unit may include organs that form no part of the pollination unit, like the standards of *Iris*, the bracts of *Castilleja*, or the pedicels of *Mesadenia*. These exclusive attraction organs are derived both from that part of the flower which has attraction as its "normal" function, viz. the perianth, and from other organs which have taken over all or part of that function, as in the last two genera mentioned above. The *Castilleja* corolla is almost invisible from without, and the *Mesadenia* inflorescences have relatively conspicuous bracts and corollas as well as pedicels.

For the current text we shall use *flower* as a morphological term, and *blossom* ("flower form" in Percival 1955) for the pollination unit as an ecological term. This corresponds to the differentiation, in German, between *Blüte* and *Blume* (see Knoll 1926) except that we include in "blossom" also the partial flowers of *Iris*, etc., not mentioned by Knoll, *loc. cit.*, and do not accept Knoll's later (1956) exclusion of anemophilous pollination units. Thus a "blossom" may be either an inflorescence (*pseudanthium, Überblüte*), a flower (*euanthium, Blüte*), or part of a flower (*meranthium, Teilblüte* cf. Fig. 2).

The blossom develops in relation to pollination. The primitive pollination unit is an open, undifferentiated flower, which presents pollen (and nectar?) and stigmas freely, but passively, to the pollinating agent; it is terminal on main or side axes. In wind-pollinated plants development goes towards reduction of perianth and increase of pollen production and size of stigma together with more effective presentation of both. Active pollen presentation by explosion develops in some species.

In animal-pollinated plants the developmental trend leads to higher integration and precision in the presentation, transfer, and deposition of pollen.

We may assume that loose microspores were originally presented by primitive, unspecialized structures, later followed by typical stamens which originally occurred in large number. In the later development the number of stamens decreased and their position became better defined. This line culminates in the coherence of pollen grains in pollinia, which are in Asclepiadaceae integrated to such an extent as to have one common germination pore (cf. Linskens and Suren 1969).

For the deposition function a similar line can be established: (a) free, open carpels with a marginal stigmatic area, but no style (*Drimys, Nypa, Akebia*); (b) free, closed carpels with

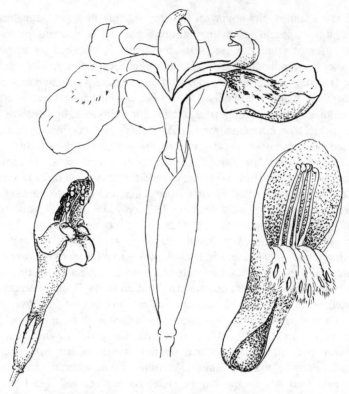

FIG. 2. The same type of blossom, viz. a gullet, realized in three different ways. Left: As a whole flower in *Galeopsis*. Middle: As part of a flower, between stigma and one perianth member in *Iris pseudacorus*. Right: As an inflorescence, between a bract (upper) and perianth members of the individual flowers in *Mimetes hartogii*.

style and stigma; (c) carpels temporarily joined by a stigma (Rutaceae, the synstigma in *Ficus*); (d) the integration to syncarpic ovaries. Also in this development there is a trend from a diffuse, central placing towards a well-defined, often asymmetric placing, of a few carpels.

If both functions are present in the same blossom, the end result may be a total integration as in Orchidaceae, Asclepiadaceae, or in *Pedilanthus*, enclosing nectaries and anthers in deep narrow parts of the flower, but it can also be effected by movements, so that flowers are open only when insect visits are likely to occur. A special case of the latter are the ephemeral flowers. Further refinements are development of traps, closed flowers, and zygomorphy, differentiation between parts of the blossom, introduction of conducting structures, either mechanical, like rows of hairs or spines, or optical or olfactorial as nectar guides. Another refinement is the introduction of actively moving parts in the flowers, like the stigma in *Mimulus*, the stamens of *Centaurea* or *Cytisus*, or almost grotesquely, in *Catasetum*.

Finally, there is, in both groups, a tendency for flowers to collaborate in inflorescences. There is also among the blossoms a gradual development from allophilic organization (available for pollination by any visitor) to euphilic (restricted to pollination by highly specialized, sometimes unique agents). This parallels the corresponding development among pollinators from dys- and allotropic to eutropic, a development found both between the large classes, from beetles to lepidopters, and at the same time within each class. Similar developments are found in classes of vertebrates (cf. Table 2 on p. 48)

The motions exhibited by plants under the influence of wind (and water for hydrophytes) may also be of importance in pollination ecology, and so may the more or less accidental movements most flowers perform under the impact of a visitor. However, there are also some specific movements of the whole flower, to be dealt with later. Spontaneous and induced movements may change the total architecture of the blossoms, in addition to closing them under unfavourable circumstances. Some blossoms change more or less completely in the course of anthesis, presenting their pollinators with varying structures, generally trapping them (*Calycanthus*, *Nymphaea*, *Victoria*). In some flowers, parts may move after a pollinator has landed bringing it into a position in which pollination is better ensured. The labellum in many orchid flowers performs such movements; very simple, passive ones under the influence of the weight of the pollinator (*Epipactis palustris*, *Bulbophyllum*), more complicated ones due to the release of a built-in tension (*Plocoglottis*) or to an active (seismonastic) movement, triggered by the touch of the visitor on some sensitive part of the flower (*Pterostylis*).

Each part of each blossom functions in its own way, but functions of various parts within a blossom are correlated. Experience shows that certain combinations, certain patterns, return more frequently than others, thus producing definite blossom types. Further, it will be found that these blossom types are also characterized by definite pollination mechanisms. However, this does not mean that all the "typical" features are present in all cases; there is a certain *syndrome* of blossom characteristics corresponding to each pollination type, but in any given case any feature belonging to the syndrome may be missing just in the blossom under consideration. Thus if it is stated that fly-pollinated blossoms have dull colours, this only means that, statistically, most fly-pollinated flowers have such colours; but it does not preclude some bright-coloured flower from being more or less exclusively pollinated by flies. Similarly, the long, narrow beak belonging to the syndrome of flower-birds is in the flower-parakeets (trichoglossids) replaced by the ordinary, short, thick, curved parrot-bill (compensated for by the extremely specialized tongue). Syndromes may be weakly developed in hemitropic blossoms, but the loose characters may assist in speciation and in the building up of adaptations. On the other hand, the syndrome is no exclusive prerequisite; after all, a hummingbird may be fed from a bottle, and innumerable bees are fed from coloured squares and triangles in experiments. However, the constant occurrence together in nature shows that the combinations of characters involved in a syndrome are far from being accidental or redundant. On the other hand, a hasty observer should not be content with identifying one single character of a syndrome. Attacks on the syndrome concept because of such premature conclusions based upon too slender evidence completely miss the target.

5.3. CROSS- AND SELF-POLLINATION

After Koelreuter's and Sprengel's establishment of cross-pollination as a general principle, a period followed during which there was a tendency on the part of botanists to disregard self-pollination altogether, giving us the so-called "law" of Knight–Darwin (cf. C. Darwin 1876). This "law" undoubtedly lead to an unwarranted search for crossing mechanisms even where they did not occur, and to the stretching of the evidence for preventive measures against self-pollination.

Contributing to this situation was the fact that at that time the fundamental difference between self-fertilization as a genetic phenomenon and self-pollination (see below) was not understood.

Phylogenetic experience indicates that cross-*fertilization*, outbreeding,* has some positive selective value, and there seems to be a general consensus of opinion that this positive value is based upon the greater genetic variability produced by constant change of genetic recombination in heterozygotic individuals by crossing on the species level (V. Grant 1955; Stebbins 1957). Mutations of different origin are brought together and multiplied by intercrossing and may, in favourable cases, interact in such a way as to make the new individual more successful in competition, or even make it able to occupy another ecological niche. Outbreeding thus seems to be a necessary condition if the sexual process and the diploid state are to have a positive effect in selection (for this, and the following, cf. Whitehouse 1950 and references). The inherent variability of individuals produced by inbreeding is necessarily much smaller (Chapter 12). In genetics the concepts of in- or outbreeding relate to the genotype, namely the individual plant including all clonal subdivisions.

On the other hand, in classical pollination ecology, the concepts of self- and cross-pollination were related to the *flower*. The two opposites are *autogamy*: pollination taking place within one flower, and *allogamy* pollen from one flower being carried to the stigma of another one. Autogamy left aside, allogamy may be further divided according to whether the two flowers are on the same plant (*geitonogamy*) or on different plants (*xenogamy*). Whereas in a genetic sense geitonogamy is equivalent to autogamy, it involves the same kind of work on the part of the pollinator as xenogamy. Xenogamy may even be genetically equivalent to autogamy if the plants involved belong to the same clone. Self-incompatibility† may explain why many garden varieties generally do not set seed even after xenogamy, and probably also why many plants in the wild do not, e.g. *Butomus umbellatus* in many small lakes, because such populations are clones, i.e. genetically the same individual. The confusion of geitonogamy and xenogamy in the older literature has made it very difficult to evaluate properly the occurrence of self-incompatibility.

Barriers in reproduction are important features in the study of pollination ecology (cf. Baker 1960), even if many of these barriers fall within the domain of other branches of biology. The list below includes many features which will not be discussed further.

If we assume outbreeding to be a process favoured by selection, there are two types of offspring production which are apparently disadvantageous. On the one hand, there is chaotic, promiscuous pollen dispersal and indiscriminate crossing in mixed plant communities. A successful interspecific cross may lead to the formation of new species, but most results of this kind of pollination fail. On the other hand there is inbreeding which, if

---

*Terminology at this point is confused. As used here, outbreeding designates a sexual process through which seeds are produced by the union of gametes from two genetically different individuals (genotypes). At the other extreme, inbreeding, seeds are produced by the union of gametes from one individual, i.e. by self-fertilization. Other terms frequently used, like outcrossing, natural crossing, cross-breeding, etc., are too vague to be of any use and merely add to the confusion especially prevalent in plant-breeding literature, where geitonogamy (also within a clone) may cause conceptual difficulties, but is no excuse for confused terminology, which even brings in the term hybridization. The latter term should be reserved for fertilization between gametes representing two different taxa, i.e. different genomes.

†The usual term self-sterility is logically untenable; sterility, like fertility, is a condition *per se* and has nothing to do with pollination effects. A sterile individual cannot produce offspring under any circumstance.

successful, will — over the generations — lead either to genetical impoverishment and limited possibilities for adaptation to new situations, or become the starting point for the formation of successful inbreeding species.

Reproductive barriers for isolation, i.e. against hybridization, may be divided into two groups:

(a) Vegetative isolation:
    (1) Geographic (historical) barriers.
    (2) Habitat specificity: species are separated owing to diversity of habitats.
    (3) Hybrid habitat specificity: if hybrids are formed, their habitat demands may not be satisfied.
(b) Reproductive isolation:
    (4) Seasonal flowering.
    (5) Pollinator specificity including constancy and monotropy of visitors. Isolation may be mechanical because of different construction of the blossom including differences of pollen deposition on the pollinator, or ethologic because of differences of signals (attractants, cf. V. Grant 1949b).
    (6) Statistical effect of back-crossing: as long as the hybrid population represents a minority in relation to the populations of the parent species, and there is no pollination or fertilization bias, there is a higher probability that the hybrid will be fortuitously back-crossed to the parents than pollinated by itself. The hybrid genome therefore dilutes rapidly and changes back to those of the parents.
    (7) Interspecific incompatibility of pollination and fertilization: the pollen grain does not germinate on the stigma or the developing pollen tube interacts with the style in such a way that the male nuclei do not reach the egg cell.
    (8) Gametic incompatibility.
    (9) Endosperm incompatibility.
    (10) Hybrids inviable (under natural conditions!) or sterile.

In order to be effective, selfing, i.e. seed-setting by means of the plant's own pollen, presumes two independent conditions, both of which must be fulfilled, viz. self-pollination, pollen transfer, and self-fertilization, i.e. self-compatibility. The barriers against selfing are partly the same as against hybridization, mentioned above, viz. items (7) to (10). In addition there are external barriers, which are partly negative, preventing self-pollination, partly positive, at the same time furthering out-crossing.

Pollination ecology has been less concerned with genetic phenomena than with external devices to further allogamy. These devices can be defined in relation to their effects on the two functions (often, but erroneously called "sexes") in the blossoms affected:

(a) Dichogamy — all blossoms identical; presentation and reception of pollen in the individual blossom separated in time.
(b) Herkogamy — all blossoms identical; functions separated in space within each blossom.
(c) Heteromorphy — both functions present at the same time in each blossom, but different, functionally complementary and interdependent blossom types exist.

(d) Dicliny ("unisexuality") — the two functions are separated in different blossoms: d_1 on the same plant: monoecy; d_2 on different plants: dioecy.

These devices are not mutually exclusive and may occur together, and also together with the internal barriers mentioned in the list. Such apparent redundancy has been misunderstood; it may represent relics from earlier pollination syndromes, left in the course of further evolution, but it is frequently functional in its present form. To understand properly the avoidance of self-pollination one should take into account not only the fact that (in many species) the resulting seed is poor, but also that the wrong pollen may have immediate negative effects, in extreme cases resembling a poisoning. In the evolution of pollination, economy plays a great part. Prevention of wastage of pollen and eggs seems to be a legitimate consideration; it is also accepted as a background for corresponding pre-mating behaviour in zoology. The devices listed above therefore do not only affect the pollination process proper, but even more prominently the economy and prevention of nonsense or even deleterious pollination.

Usually, visiting pollinators exhaust a blossom and leave it, not to return. However, in some blossoms the visiting urge is so strong that they will try to enter the same blossom again immediately (*Stanhopea*, van der Pijl and Dodson 1966: 67). In such a blossom dichogamy obviously serves an important function by preventing nonsense pollination and loss of pollen.

In certain pollination syndromes the morphological demands of the two functions are contradictory, e.g. in wind-pollinated blossoms, where it is important that both stigma and anthers are freely exposed and do not get into each other's way. Unisexuality is the answer to this. However, the same may be the case in some of the most evolved pollination syndromes known: the catapult mechanism of *Catasetum* and similar orchids (van der Pijl and Dodson 1966: 63). Pollinia cannot be catapulted back against the direction of gravity, and so the receiving ("female') flower is inverted, and pollinia can swing downwards onto the stigma.

Separation of the functions, on different plants, *dioecy*, is one safe way of avoiding self-fertilization, though it should be noted that principles of sex distribution in plants are very rarely absolutely consistent. Dioecy in conifers* may be primary, but probably always secondary in angiosperms. The effectivity of the mechanism is acquired at the expense of half the population not bearing seeds, and creates dangers of non-pollination. The more complicated forms of gynoedioecy (the existence of separate pistillate and hermaphrodite flowers) can be considered transitional stages, in some of which self-fertilization is not excluded. In others, the morphologically hermaphrodite flowers are functionally male, as their seed-set is very bad, if any (Arroyo and Raven 1975). Whereas monoecy obviously does prevent autogamy, it is no guarantee against the genetically equivalent geitonogamy. As a preventive measure against self-fertilization, monoecy is therefore of little value unless combined with a second-order dichogamy (see below). In some cases a functional dioecy may be created inasmuch as all male flowers may have finished anthesis before the female ones open or vice versa: "heterodichogamy", e.g. in *Acer* spp. (cf. Stout 1933, Gabriel 1968). Monoecy has the advantage that all individuals are potentially seed-bearing.

---

*Dioecy as such is certainly not primary; the expression only implies that the ancestors of conifers may also have been dioecious. The terms dicliny and monocliny refer to the individual flowers, or blossoms not to the plants. For a discussion, cf. Heslop-Harrison (1963) who, however, maintains that there may also be primary "unisexuality" in some groups of angiosperms.

The morphological development in unisexual blossoms is strongly influenced by the contrasting demands of the two functions, especially when the less-influenced attraction apparatus is reduced to unimportance. Anemophilous unisexual blossoms are extreme exponents of their functions, and are, accordingly, completely dissimilar. On the other hand, (primitive?) cases of dioecy are known which are functional but not morphologically expressed. In *Pimenta dioica* the flowers of "barren", i.e. male, trees are hardly distinguishable from those of "fertile", i.e. female specimens (Chapman 1966). Where nectar is the attractant, there are no difficulties, but in blossoms with pollen attraction, like *Tetracera* (Kubitzky 1969) or *Vitis,* the "female" blossoms must use sterile food-pollen for an attractant (Brantjes, unpubl.).

Herkogamy, the spatial separation of anthers and stigma, is so much the rule as to be mostly taken for granted and passed by unnoticed like M. Jourdain's talking prose. The more spectacular phases include phenomena like the very strong fixation of relative position met with in orchids or in asclepiads. In general, however, one is more likely to notice the *absence* of herkogamy in self-pollinating flowers or in self-pollinating phases of flowers (Section 12.2). Being so widespread, herkogamy is frequently found together with dichogamy, but may also occur independently, as in *Tecoma radicans.*

Dichogamy is the separation of reception and deposition functions in time, allowing, however, seed-setting in all flowers. It includes the phenomena of protandry and protogyny. Its opposite would be called *homogamy*, with both functions taking place at the same time.*

Self-pollination is precluded if anthers dehisce and shed their pollen at a time when the stigma is not receptive, yet the mechanism will allow for seed-setting in all the flowers. *Protandry* exists where pollen is available before the stigma is receptive. If the stigma is receptive before pollen is liberated, this is *protogyny*. The classical example of protandry is *Chamaenerium angustifolium*; the condition is widespread in Compositae, Campanulaceae, Labiatae, Malvaceae, Caryophyllaceae, Leguminosae, and others. Extreme cases with anthers shedding before the female phase occur in *Saxifraga* and *Impatiens.* Protogyny occurs in Cruciferae, Rosaceae, Berberidaceae, Thymelaeaceae, Caprifoliaceae, and Juncaceae; shedding of the style before the male phase in *Parietaria* and Annonaceae. In many instances dichogamy is partial, the second, e.g. the female, phase commencing before the first one has terminated, here before the anthers have ceased to function. Thus the flowers are functionally unisexual at first and hermaphrodite later, whereas in complete dichogamy there should be two separate unisexual phases.

Essentially, the flower represents the end of a stem, and the "natural" development of appendages is centripetal. This means that lower leaves form before the higher, leaves before bracts, bracts before perianth, and, continuing the sequence: calyx, corolla, androecium, gynoecium. Protandry is thus seen to be nothing but the normal sequence of development of the floral appendages, the only abnormality being that a telescoped development sequence of calyx—androecium suddenly comes to a temporary halt before the next stage. It is remarkable that protandry is so dominant in derived families.

Whereas simultaneous ripening of pollen and stigma is nothing but consistent telescoping of the development of appendages in the floral region, protogyny is rather remarkable, as it constitutes the reversal of a regular order of events. It is difficult not to relate this reversal to

*The term homogamy has in some Italian publications been used for autogamy. Zoologists have used it for "preferential mating between the most similar individuals" and some botanists have followed this usage in population dynamics (Straw 1955).

the fact that protogyny is more effective than protandry as a measure against self-pollination. After pollen has germinated on the stigma, there is a virtual competition for ovules. Under these circumstances, it is obvious that receptivity of the stigma, even a few hours before the anthers open, will mean that any allogamous pollen received during these hours will, as a rule, carry out the fertilization. By contrast, in order to be effective, protandry must be total, all the self-pollen being swept out of the flower before the stigma opens. Only if anthers are shed completely before ripening of the stigma, is the danger of contamination with the self-pollen almost eliminated, and even then, some of this pollen may have been left somewhere in the flower and may contaminate. It is remarkable that protogyny is found in some otherwise very primitive families, e.g. within Polycarpicae, within which protandry is found in the relatively derived family Ranunculaceae. In trap blossoms protogyny is an essential part of the pollination syndrome, which "explains" its occurrence, for example, in Calycanthaceae.

Whereas the adaptive character of protogyny is thus more distinct, both from the point of view of function and of developmental physiology, protandry is also an adaptive measure, especially in its more complicated form, as seen, for example, in *Saxifraga* or *Geranium*, where anthers ripen, move into and out of position, and are shed one by one in sequence. The effect is that pollen is presented during a very long period of anthesis, but only part of the total pollen production is exposed at any one time. It is a case of not putting all eggs in one basket, easily correlated with the fact that the flowers of the genera mentioned are open and their anthers unprotected. Only after the last anther is shed do the stigmas exhibit external signs of being receptive. In the dense inflorescences of many *Saxifraga* species, which almost form one pollination unit, there is a curious discrepancy between the very complicated mechanism preventing autogamy and the rich possibilities for geitonogamy. The same applies to the dense heads of some Compositae, whereas in others second-order dichogamy counteracts geitonogamy. Also, differences between various zones of the inflorescence in the production and presentation of attractants (nectar, pollen) may serve to channel visits to flowers in a definite phase (*Helianthus*: Free 1970). In racemose inflorescences dichogamy causes a second-order herkogamy in as much as the centri- (acro-) petal development of flowers ensures that those belonging to a certain zone within the inflorescence are in the same state — male or female. This is readily seen in *Plantago* inflorescences; e.g. *P. major* is distinctly protogynous, and the style projects from the still closed corolla; only later do corollas open and stamens develop. The frequent occurrence of second-order herkogamy caused by protandry has been correlated with the habit of bumblebees of alighting at the bottom of an inflorescence and climbing upwards, which will generally give the pollinator a better starting position in relation to the individual flower than climbing downwards. The pollinators then first visit flowers in female stage with receptive stigma, if any are available, and will only later be covered by pollen from younger, higher flowers in male stage with open anthers. Conversely, since Sprengel's days it has been maintained that the wasps, said to be the principal pollinators of the protogynous *Scrophularia nodosa*, climb downwards. In the tropics, dichogamy is said to affect whole branch systems at the same time (*Nephelium*, Khan 1929; *Persea*, Stout 1926), producing almost a second-order dioecy. Evidently, the effectivity of the phenomenon increases the more general it is.

In plants with diclinous flowers, a static type of second-order herkogamy may be found in contrast to the above-mentioned dynamic or developmental one, viz. when male and female flowers (florets) occupy different positions. A large-scale example is furnished by many conifers, the lower branches of which carry male flowers, the upper female ones. Small-scale examples are furnished by some composites with unisexual florets. Incidentally, unless combined with dichogamy, this type of secondary herkogamy is no guarantee against geitonogamy.

An extreme case is furnished by the inflorescences of *Ficus*, the flowers of which are unisexual and where the two functions of the inflorescence develop at widely different periods resulting in second-order dichogamy (the rather indistinct spatial separation is of little importance here).

The effect of all these second-order mechanisms is to prevent — more or less completely — geitonogamy as well as autogamy. Even if many external mechanisms are far from being 100 per cent effective, and some of those previously described have undoubtedly been misinterpreted, there can be no doubt that a multitude of such mechanisms disfavouring autogamy, in some cases also geitonogamy, do exist.

Dichogamy and the other mechanisms dealt with above counteract self-pollination. In addition, the introduction of stigma and style with angiospermy creates conditions for a variety of types of incompatibility which prevent self-fertilization even when self-pollination has taken place. In a way they replace or supplement the former mechanisms, in some blossoms making them manifestly redundant, e.g. protandry in the perfectly self-incompatible *Butomus* (Pohl 1935). This fact has repeatedly been used as an argument against the current evaluation of the external mechanisms as part of the anti-self-pollination syndrome. The argument is fallacious; phylogenetically, external prevention of self-pollination could precede the development of incompatibility systems, and may remain as a phylogenetic relict long after its usefulness has been superseded. It is difficult to see how the existence of external factors preventing self-pollination should not ultimately lead to physiological self-incompatibility. We shall see later how similar relicts are found also in autogamous flowers.

Self-incompatibility is not invariably absolute; in fact it varies from 100 per cent to a very slight preference for foreign pollen. Incompatibility is a prerequisite for allogamous pollination in many blossoms in which pollen and receptive stigma always must come into contact with each other, e.g. in Fumariaceae, or in many Papilionaceae.

Both incompatibility and so-called pseudo-compatibility systems may be due to various factors. Topographically, incompatibility systems can affect the following functions: (1) pollen—stigma interaction, (2) growth of the pollen tube towards the ovule, and (3) the fusion of nuclei (cf. Linskens 1975).

The nature and rôle of chemical interaction between pollen grain and stigmatic surface are only partly known. Probably various principles are at work. Stigmatic secretions may have both positive and negative effects on the pollen germination process itself. Negative effects manifest themselves both in self-incompatible combinations and in combination with foreign, incompatible pollen (Baker *et al.*, 1973; Martin and Ruberte 1972).

In several plants the pollen—stigma interaction depends on chemical reactions between "substances of recognition" as postulated by Knox *et al.* (1972) for *Populus*, and substances forming chemical barriers which are broken down by enzymes (?) provided by the other partner in the interaction (in *Brassica*, cf. Roggen 1972). Especially the latter mechanism easily explains various previous observations on the effect of wounding the stigma mechanically (*Laburnum*: Jost 1907) or electrically (Roggen and van Dijk 1973), or the enzyme effect of killed compatible pollen, which removes the chemical barrier and thus leaves the path open for other pollen types that would have reacted incompatibly without this aid. The existence of a protective membrane over the stigma has been postulated; the effect of the pollinator would then be to rupture this and thus make pollen tube growth possible. If the membrane is not ruptured during autogamous pollination this would be prevented completely.

However, results of investigations of the rôle of the stigmatic surface give contradictory results. Heslop-Harrison and Heslop-Harrison (1975) have shown that in Caryophyllaceae the proteinaceous pellicule of stigmatic papillae plays a part in activating the enzyme necessary for the penetration of the pollen tube. After removal of the pellicule the pollen grain germinated, but the tube could not penetrate.

On the other hand, Iwanami (1973) obtained increased growth of *Camellia sasanqua* pollen tubes after treatment with cold acetone, and Martin (1969) found that the so-called stigmatic fluid contained very little carbohydrate, much more lipids and phenolics. The latter may act as inhibitors, and adequate enzymes may be necessary to render them inactive. Apparently it is too early to generalize.*

The existence of a chemical barrier would explain various empirical artifices, e.g. the effect of bud pollination (before the incompatibility system has built up; an example in de Lange *et al.* 1974), or the effect of late-anthesis pollination (after break-down of such substances, cf. Ascher and Peloquin 1966). Some unexplained effects of mass pollination by incompatible pollen may also be explained by chemical pollen – stigma interaction. It is more difficult to see if there can be any connection with the alleged difference between autogamy, geitonogamy on the same plant, and clonal geitonogamy.

Massive pollination with incompatible pollen, e.g. self-pollen or pollen from another species, may conceivably have a purely mechanical effect: the incompatible pollen occupying all receptive sites on the stigmatic surface. However, chemical effects are probably more important, incompatible pollen causing a negative recognition effect, which blocks the receptivity of the stigmatic papillae, and lasts for some time after the application of the incompatible pollen, thereby also preventing the germination of pollen that would otherwise have been compatible. In orchids this effect may amount to a permanent poisoning of stigmatic surface or stylar tissue. The self-pollen may therefore not always be the neutral, innocuous stuff mostly assumed (cf. Ockendon and Currah 1977).

Even if the pollen grains have germinated successfully, the growth of the pollen tube can be so slow that fertilization does not take place before the blossom has withered. Not only the speed may be insufficient, but also the direction: incompatible pollen tubes lose their direction and never reach the ovule. These sieving systems are non-existent or comparatively inefficient in primitive, open carpels with a stigmatic crest instead of a regular stigma and style.

On the other hand treatment with chemicals or hormones or high humidity may stimulate germination and growth of the pollen tube and prevent premature abscission of the pistil, thus making possible a pollination that would otherwise not succeed.

In cultivation, many incompatibility effects can be overcome by artifices. Bud pollination can also compensate for slow pollen tube growth; intra-ovarian pollination (Kauta 1960), or even fertilization of ovary cultures (Johri 1971), are among them. Chemical treatment is another.

*For the specially interested we may, in addition to the regular literature, refer also to the Incompatibility Newsletter, published by Euratom Ital, PO Box 48, Wageningen, The Netherlands.
Pollen tubes in stigmatic tissue can be demonstrated by autofluorescence microscopy or by fluorescent or ordinary staining techniques (Campbell and Nelson 1973, Kho and Baer 1970).

Environmental pseudo-compatibilities constitute another group. The inhibitory function may break down for external reasons (e.g. excessive temperatures, day-lengths) or internal ones (e.g. age). In the same way as the artificial pseudo-compatibilities provide information about the nature and location of inhibitory mechanisms, the environmental ones give information about the metabolic states responsible for the incompatibility. Age seems to be a very common factor, a flower starting its existence as strictly self-incompatible becomes gradually less so as the inhibiting principle is being exhausted, so that self-pollination may be effective in the later stages. In many plants external self-pollination mechanisms come into function as well, as in *Commelina coelestis*.

Even if parts of the incompatibility system can be bypassed by artifices, fertilization may still not take place (Lester and Chadby 1965). The final line of resistance against self-fertilization is nuclear incompatibility – and its opposite, genetic or nuclear pseudo-compatibility, responsible for fusion, in a very small number of cases, between nuclei that would otherwise not have fused. For a more comprehensive discussion of the genetical and physiological problems inherent in incompatibility systems we shall have to refer to the general papers by Linskens (1967) and Pandey (1960). Many authors agree that incompatibility is of great age in angiosperms, but that it nevertheless must have arisen polyphyletically, as the three main nuclear systems are found in different families (Lewis 1955).

Pseudo-compatibility may be considered a kind of safety measure: self-incompatibility is the rule, and is given preference by the structure of the flower; but if cross-pollination does not take place, self-fertilization is possible and may produce offspring, even if this offspring often suffers from inbreeding depression. The degree of the latter again varies from only being perceptible by statistical analysis to being so prominent that the individuals are viable under optimum artificial conditions only. It should be noted also that both external and internal self-incompatibility factors may be completely effective in nature, even if not very impressive under experimental conditions. Take, for instance, the common situation in which pollen tubes from self-pollen grow slower than tubes from a compatible individual. If both pollen types are placed at the stigma at the same time, as they are in many allogamous blossoms under natural conditions, the tubes from the other individual will "outrun" the self-pollen tubes and provided there are enough of them their nuclei will carry out all fertilization before the others have arrived. Thus, a condition which would under artificial conditions be without much significance, may in nature be decisive.

A special mechanism for safeguarding cross-pollination is found in some flowers in which the receptivity of the stigma is of short duration. Thus in the ephemeral flowers of *Nicandra*, the stigma withers within one hour after pollination and the style drops off. In *Mimulus* the two lobes of the stigma close on being touched, they open again after a certain period, unless covered by compatible pollen, in which case they remain closed.

Although *heterostyly* is the most widely recognized anti-selfing structure, the functional relation between heterostyly and self-incompatibility is far from clear. Levin (1968) flatly denies that heterostyly in *Lithospermum caroliniense* promotes cross-pollination, and maintains that its only observable effect is reduction of seed-set. Ecologically this appears improbable, and Mulcahy and Caporello (1970) reach the opposite conclusion for *Lythrum salicaria*.

The genetics of heterostyly have been subject of a comprehensive literature which will not be discussed here. The most obvious external manifestation is the existence of two types of flowers – long-styled and short-styled (pin and thrum). The position of the anthers is the opposite of that of the stigma. Additional differences are those of size and frequently ornamentation of pollen grains and size of stigmatic papillae. The more general term *heteromorphy* has therefore been proposed.

In heterostylic plants seed-setting is, to a greater or lesser extent, supposed to depend on the legitimacy of pollination. Legitimate pollination is that between stigma and anthers in the same position, e.g. short style and low anthers (from a long-styled flower). Legitimate pollination perforce is cross-pollination. Some of Darwin's figures may be quoted. If seed-setting after legitimate pollination in *Primula veris* is 100, that after illegitimate is 69 (capsules) or 65 (seeds). In other experiments results from illegitimate pollination vary between 100 and 0. The values are apparently the same irrespective of whether the illegitimate pollination is auto- or xenogamous. However, the result in nature may be different even in those crosses which in experiments give very high yields after illegitimate pollination.

If the morphological expression of heterostyly should have any functional value, its immediate effect should be that pollen of long and short anthers were concentrated on different parts of the body of the pollinator. However, investigations have not yielded distinct differences; on the contrary, pollen loads seem to be fairly mixed and the distribution of pollen on the two types of stigmas random.

Ornduff (1970) has pointed out that the incompatibility reactions concomitant with morphologic heterostyly are the most important part of the system. He has suggested that the main function of the incompatibility system should be to prevent the production of deleterious genotypes while heteromorphy should promote outbreeding between compatible genotypes.

It is fair to state that the function of the morphological expressions of heterostyly is at present not properly understood.

The immediate effect on the pollination process is that of differential growth of pollen tubes. A yellow primrose in one of Darwin's experiments was illegitimately pollinated by pollen from another (yellow) individual and 24 hours later legitimately by pollen from a red-flowered variety. The offspring were red-flowered, indicating a rather extreme retardation of the growth of the illegitimate pollen tubes.

E. B. Smith (1970) measured different speeds of pollen tube growth by pollinating first with illegitimate and after a time interval with legitimate pollen. If the illegitimate offspring is genetically marked and can be recognized, 50 per cent illegitimate seed indicates that the time interval between the two pollinations equals the delay suffered by the illegitimate pollen tube. Obviously, this is a function also of style length.

Levin and Berube (1973) have described two different but closely related *Phlox* species that behave like two heterostylic morphs. Possibly, heterostyly may initiate speciation.

The classical, dimorphic type of heterostyly with two different flower types is known from Primulaceae, Rubiaceae, Boraginaceae, Oleaceae, Plumbaginaceae, Oxalidaceae, Polygonaceae, later also found in Turneraceae, Verbenaceae, Iridaceae, Gentianaceae, etc. (cf. Vuilleumier 1967), Caesalpiniaceae, Capparidaceae, Commelinaceae (Vogel 1955).

There are two aberrant types. The one is the homostylic dimorphism found, for example, in *Armeria* (Iversen 1940). It bears all the characteristics of the dimorphic heterostyly, except that there is no difference in length of style or position of anthers. This indicates that the phenomenon of different lengths of style is a more or less fortuitous complication of the system and that a term like "heteromorphy" would be more appropriate for the general phenomenon.

The other is the trimorphic heterostyly, found in Lythraceae, Linaceae, Oxalidaceae, Pontederiaceae, cf. Baker (1962), which except for its greater morphologic and genetic complication functions in the same way as the dimorphic heterostyly.

The difficulty of finding a functional explanation of the morphological expression of heterostyly cannot be construed to indicate (as has been suggested) that heterostyly is a more or less automatic consequence of incompatibility systems and not a feature of independent standing, *Pontederia* (*Eichhornia*) *crassipes* is trimorphic heterostylous (Barrett 1977), but François (1964) found no distinct self-incompatibility. The general sterility within its various clones must be due to other causes. The investigations of Ornduff and his collaborators have shown that there are all possible combinations between the morphological and physiological expressions of heteromorphy. Ornduff (1972) has described a great variety of stages between complete trimorphic heterostyly and (including) homostyly and autogamy in *Oxalis* sect. Corniculatae. These stages are interpreted as a break-down series, possibly caused by the association of inbreeding and weediness, whereas species of more stable habitats possess a more conservative breeding system. Our ideas about the development and origin of heteromorphy are, on the other hand conjectural, to say the least.

In Plumbaginaceae, Baker (1966) suggested the following developmental sequence: incompatibility—pollen dimorphy—stigma dimorphy—heterostyly. The latter would then be a preventive measure against "nonsense pollination". Finally, heterostyly can be one of the roads leading to "unisexuality", which Baker (1960) found in *Mussaenda*. This was predicted previously·by Darwin and it is important as it explains how dioecy may arise in groups protected against inbreeding by other mechanisms, cf. the corresponding case of *Nymphoides* described by Ornduff (1966).

Galil ˙and Zeroni (1967) have described in *Zizyphus spina-christi* a mechanism which might be considered a time equivalent to the spatial phenomenon of heterostyly. In Israel the flowers are typically protandric and follow the pattern of *Saxifraga aizoides*. However, there are two types, one which is male in the morning and in the female stage in the early afternoon. The other is male at noon and female in the night and morning. Characteristically, the mechanism breaks down towards the end of the flowering period.

The preceding discussion has demonstrated the great variability of self-incompatibility systems in angiosperms. These systems have in common that they are very economical. The complex of alleles causing incompatibility is often so specific as to restrict incompatibility, under natural conditions, to the one individual. This means that not only is there a potential 100 per cent seed-setting against 50 per cent in dioecious plants, but also that the pollen is practically wholly effective if deposited on a stigma of any other individual of the same species. It is readily seen that all of this is a function of angiospermy; in gymnosperms dioecy or dichogamous monoecy are the only possibilities, both of them wasteful and the latter generally not very effective.

In Section 12.2, dealing with autogamy, we shall again return to these questions of self- and cross-pollination.

CHAPTER 6

ABIOTIC POLLINATION

Pollen transfer problems have widely different aspects whether the vector is an animal or an inanimate physical force. In the latter case there is no question of mutual adaptation; there is no relationship between the pollinating agent and the pollination unit; any existing adaptation must be one-sided. With few exceptions abiotic pollination is a wasteful process as the transfer is non-directional. The pollen grains are scattered according to the law of chance, and the percentage of effective pollen grains, i.e. pollen grains reaching the stigma, must equal the area of effective stigmatic surface divided by the total surface area of all plants plus habitat surface within the confines of deposition of the grains, e.g. downwind from the male flower. Even if there is some possibility for re-deposition, which may have effect if the grains are still alive, the percentage of effective grains is almost infinitely small, especially in the case of wind pollination. The transfer in water generally takes place within a smaller volume of the medium, and the dimensions of the problem are therefore not the same.

Wind and water are the abiotic agents of pollination discussed here. So-called gravity pollination will be considered under autogamy.

6.1. WIND POLLINATION, ANEMOPHILY*

Wind pollination is the dominant type of abiotic pollination, comprising perhaps 98 per cent of all known examples and prevailing in several families: Gramineae, Cyperaceae, Juncaceae, and within the orders of Amentiferae and in Urticales. In many families of entomophilous plants there are some few members that have become anemophilous: species of *Thalictrum*, *Ambrosia*, *Fraxinus*, etc. Anemophily is obviously secondary in these genera, and so it is in the monocotyledonous families mentioned above: their immediate phylogenetic ancestors are all entomophilous. As we have already stated, we are of the opinion that within angiosperms entomophily is the more primitive condition, i.e. the immediate ancestors of angiosperms were insect-pollinated, and if angiosperms are monophyletic, the oldest representatives of the class were insect-pollinated too.

Indications that anemophily is derived from entomophily are the possibly relict occurrence of nectaries in blossoms of many anemophiles (*Cannabis*, *Urtica*, etc.; see Stäger 1902), and of specific odours.

Blagoveshchenskaya (1970) suggests that the first angiosperm blossoms may have been undifferentiated, only later on differentiation into distinct anemophilous and entomophilous

---

*Usage differs; the suffixes -phily and -gamy (-philous and -gamous) have both been used in pollination ecology. As gametes are not, as such, involved in pollen transfer, there is some reason to avoid -gamy in pollination ecology and reserve it for fertilization processes. The terms auto-, xeno-, etc., -gamy are still permissible. The terms zoogamy or zoidogamy refer to fertilization by motile gametes, not to pollen transport by animals.

types took place. This is not improbable, but the question is what came first: the differentiation or the establishment of the angiosperm syndrome? The same question can be raised with regard to A. D. J. Meeuse's assertion (1972) that primary anemophily exists in diclinous angiosperms, a hypothesis which we shall not discuss here. At any rate, both anemophilous and entomophilous flowers seem to be present very early in the fossil record of the flower; the catkins described by Crepet *et al.* (1974) from the Middle Eocene at any rate possess morphological features which point towards development from forms with a functional perianth. Specifically, the old problem of primary anemophily in Palmae must be answered in the negative after field studies in the Tropics (cf. Silberbauer-Gottsberger 1973; Essig 1971, 1973).

In comparison with biotic pollination, wind pollination is much less precise, and to achieve its objective it presumes a very high incidence of pollen grains near their source of origin — the dilution factor going by a power of 3 of the distance. In a bifunctional wind-pollinated blossom the receptive surfaces would immediately be overloaded by self-pollen, and only extreme self-incompatibility or dichogamy would prevent self-fertilization. Dichogamy is found in *Thalictrum*, *Potamogeton*, *Plantago*, and *Rumex* (pp.), in all of which anemophily is evidently of recent origin. Ordinary herkogamy is of little help, but an extension into dicliny is more effective. Adding to this the fact that the two functions of the blossom are completely independent and have a tendency of getting into each other's way, as already mentioned; the direction of development via dicliny towards dioecy seems indicated in wind-pollinated plants, as is actually found. A hand-in-hand development of the two syndromes of wind pollination and unisexuality can be followed, e.g. in *Thalictrum* (Kaplan and Mulcahy 1971) or *Acer* (de Jong 1976). Whether anemophily or unisexuality came first, is less interesting. In *Acer*, de Jong maintains that unisexuality came first, and anemophily as a consequence thereof. Unisexuality may be related to other functions of the plant as well, e.g. diaspore dispersal, which is favoured by dimorphy (exozoochory of seed-bearing capitula in *Xanthium*). Morphologically, staminate and pistillate blossoms in anemophiles may be completely dissimilar.

Today anemophily seems to be active together with entomophily in some plants although the two modes of pollination have different relative importance. Thus the anemophilous *Plantago media* is more or less regularly pollinated by insects, even honeybees; also the less conspicuous *P. lanceolata* receives insect visits (Clifford 1962). Nectar is produced in *Calluna* flowers; the flowers are visited and pollinated by insects. On the other hand, great masses of *Calluna* pollen are also spread by wind, and additional wind pollination must be inevitable. The great quantities of pollen in anemophiles attract pollen-collectors, some of which, e.g. certain syrphids (p. 103), may use this material as an essential part of their diet. On the other hand, it has been maintained that some anemophiles deliver pollen that cannot be metabolized by bees. However, it is a fact that pollen loads of collector-bees frequently contain or are exclusively composed of pollen of anemophiles (cf. Sharma 1970). As the architecture of the blossom is adapted for another pollination mode, the pollinating effect of insect visitors to anemophilous blossoms remains doubtful at best. In "unisexual" blossoms it is nil. On the other hand, Pojar (1973) — who does not discuss the pollination effect of bumblebee visits to anemophilous plants — points out that the availability of such pollen may contribute to maintaining a valuable pollinator population which might perhaps not have been able to sustain on the food offered by obligate entomophilous species alone; cf. Chapter 15.

The relative importance of anemophilous and entomophilous pollination varies not only between species, but also between different infra-specific taxa, and also between different habitats, depending on the presence of insects – wind is usually there.

Provided pollen grains are fortuitously distributed, the total output of pollen produced, N, gives a number of "effective" grains, i.e. grains carrying out pollination, n, which is a function of the areas of stigmatic surfaces, a, and the total area of the surroundings, A:

$$\frac{n}{N} \sim \frac{a}{A}.$$

Anemophily, and the syndrome of anemophily can be considered a series of compensating devices. The simplest is an increase of pollen production, N. Although one gets an immediate impression of the heavy pollen production, e.g. by shaking a flowering branch of *Corylus*, it is not easy to give this a quantitative evaluation due to the difficulty of finding a proper term of reference. The most adequate would be the number of pollen grains produced per ovule; but data such as these are very difficult to obtain in plants with unisexual flowers. An example of the inherent difficulties is shown in Pohl's table (Table 1), where the figure for *Corylus* varies, with assumptions, between 3½ and ¼ million. Pohl (1937b) gives the number of pollen grains per ovule for some species; his list is reproduced here with the exclusion of cultivated plants in which the number of ovules may have been subject to artificial selection. Very few comparable data exist so far, and conclusions must be tentative. A conservative statement would be that in anemophiles (and pollen flowers) the number of pollen grains produced per ovule on the whole exceeds that found in nectar blossoms, but that the ranges overlap very much. In anemophiles, the number of anthers per flower is usually conservative, the large numbers being found in entomophilous – or chiropterophilous – blossoms with pollen attraction.

TABLE 1 Thousand pollen grains per ovule
(Pohl, 1937b)

Corylus avellana	anemophilous	2549
Fagus silvatica	anemophilous	637
Aesculus hippocastanum	entomophilous	452
Acer pseudoplatanus	entomophilous	94
Tilia cordata	entomophilous	44
Plantago lanceolata	entomophilous	15
Sanguisorba officinalis	entomophilous	11
Betula verrucosa	anemophilous	7
Polygonum bistorta	entomophilous	6

The figure for *Aesculus* is exceptional among entomophiles.

The effective stigmatic surface, a, enters in the formula for calculating the effectivity of anemophilous pollination, and it is found in many anemophiles that stigmas are greatly enlarged, e.g. in *Plantago*, *Corylus*, Juncaceae, and Gramineae. The feather-like stigmas, e.g. of grasses, represent a special case, as the effectivity of a grid in collecting drifting particles is greater even than that of a leaf of the same gross area (Gregory 1951), and such stigmas therefore represent a great economy in matter. Also the brush-like form of *Typha* inflorescences is aerodynamically favourable. Stigmas of anemophiles must be very effective, frequently sticky, for catching the pollen grains.

The area of other surfaces, A, enters as an inverse factor in the calculation of effectivity of anemophilous pollination, and this area is reduced in various ways, leaving the path to the stigmas unobstructed. The reduction above all affects the perianth, which is in anemophiles always reduced, frequently virtually absent or deciduous. Colour and scent are of no importance in the pollination syndrome, and are also frequently absent. Bracts and perianth are generally green or dark brown to reddish. It has been suggested that dark red may be important for the temperature conditions of the blossom, especially of the bud. Perianths in some anemophiles are still beautifully coloured, e.g. *Plantago media* (brush blossom type), or there may be a distinct scent. These occurrences may be considered evidence for the recent evolution of anemophily from entomophily in these plants.

Further reduction of inert surfaces is achieved by exposure of the flowers above or outside the leaf-mass, e.g. in grasses or *Plantago*, by reduction of leaf surfaces (*Junci genuini* sensu Buchenau) and, in anemophilous trees, by flowering before the leaves are out.

The effectivity of stigmas as pollen-catching surfaces is increased by exposure outside and beyond the surrounding bracts, perianth, etc. It is also important, however, that anthers are exposed, giving air currents free access to carry pollen away. Anemophiles are frequently characterized by long filaments, which bring the anthers outside the surrounding perianth, etc., as in grasses, Cyperaceae, *Thalictrum, Plantago*, etc., and even by explosive anthers, e.g. in *Ricinus*. In some Urticaceae the filaments are under strong tension in the bud stage, being released simultaneously with the opening of flower and anther (see Mosebach 1932). The result is that pollen is thrown actively out in the air into the presumably stronger air current some distance away from the plant. This parallels the explosive phase in the dispersal of many windborne cryptogam spores, especially among the higher fungi.

The other alternative, viz. mechanisms for arresting pollen, is also rather frequent in wind pollinators, even if some instances described may have been wrongly interpreted. In the catkins of, for example, *Betula* or *Corylus*, pollen grains are very effectively arrested between the closely fitting catkin-scales. When the catkin "wags" in the wind, openings form between the scales, and pollen falls out. This is easily seen (and demonstrated) by bringing young catkins indoors and letting them mature there. Practically no pollen comes out until they are set in slight motion. A more primitive type is the stiffly erect male blossom, in which pollen lies arrested between some kind of scale-like appendages until humidity is low and the wind is sufficiently strong to blow (and shake) it out, e.g. in *Pinus* or in *Triglochin*. Stiffly erect, catkin-like blossoms are also found in entomophilous plants (*Salix*, tropical *Castanea* and *Quercus* species) and may here (entomophily being a more primitive condition) represent a lower stage in the development towards the derived, pendulous catkins of the related anemophilous species (*Populus*, temperate *Quercus*) (p. 134).

Anthers generally do not open unless the weather is favourable, i.e. warm and dry. Pollen is rapidly washed out of the air in rain. Anthers dehisce especially when the weather is warming up irrespective of whether this warming-up belongs to the daily changes of temperature or to more large-scale, irregular changes connected with the general weather development. Ponomarev (1966) has demonstrated that the release of pollen in grasses may be almost explosive. The release of pollen of different species at different times during the day limits hybridization. According to Pande *et al.* (1972) (autogamous) barley florets (in India) are open for about an hour, anthers dehisce during the first half of that, and stigmas remain receptive for 6 hours. The turbulences establishing themselves in dry, sunny weather, and especially those establishing themselves over more or less bare ground in the spring of temperate regions, greatly aid in carrying away pollen.

The viability of pollen is influenced by external factors, including ultraviolet radiation. Some of the periods of viability of anemophilous pollen quoted in the literature seem unrealistically short, and it may be dangerous to draw conclusions from these figures, but the shorter it is the smaller is the chance for hybridization (cf. Tikhmenev 1974, Ponomarev and Prokudin 1975). The short periods of viability of the pollen may therefore be a compensatory mechanism, and so may be the staggered short, almost explosive antheses.

One of the factors of major importance in anemophiles is the buoyancy of pollen grains, which increases as the size of the grains decreases. Pollen grains of anemophiles belong to the smaller size classes with a "typical" diameter of 20–30 (–60) μ even if equally small or even smaller pollen grains are found in many entomophiles (total range for angiosperm pollen $ca.$ 10–300 μ). It has been suggested that the smallest pollen grains for physical reasons may have a strong tendency to stick together, which should explain the absence of the smallest pollen classes from among anemophilous plants. The existence of much smaller anemochorous microspores in other plant groups makes this suggestion less probable.

The problem of buoyancy of some relatively large conifer pollen grains has been solved by addition, to the main body of the grain, of one or more air-sacs giving a large volume and surface without appreciably increasing the weight. The result is that the large (50–150 μ) and heavy (30–300 × 10^{-6} mg) pollen grains of $Pinus$ belong to those for which the greatest transport effectivity is realized. Great quantities of pine pollen have been found hundreds of kilometres away from the nearest forests. It also seems that very big anemophilous pollen grains (without air-sacs) have a comparatively low density (Pohl 1937a: 126), thus achieving greater buoyancy. It should be noted, though, that the weight of pollen grains (for figures see Pohl $loc. cit.$, and Dyakowska and Zurzycki 1959) is so much dependent on air moisture as to be a very unrealistic parameter. At any rate, some globular pollen grains become shale-like on drying, thus achieving a much better aerodynamic form.

The typical rate of fall for wind-pollen in calm air is of the magnitude of a few centimetres per second (Rempe 1937: 102).

Contributing to the effectivity of pollen transport in anemophiles and to the buoyancy of the pollen is the fact that grains of anemophiles do not adhere to each other, but are smooth and dry, and are spread separately or in very small groups (Remple $loc. cit.$). In entomophiles, the pollen surface is ornamented and sticky and the grains stick together. Typical anemophilous grains are found in Betulaceae or in Gramineae. The development of pollen grains towards smoothness and dryness (and thinness of the exine) concomitant with the evolution of anemophily has been the subject of special studies in some families, e.g. in Compositae (Wodehouse 1935). The ordinary Compositae pollen grain is very heavily ornamented and sticky; that of the anemophilous genera $Ambrosia$ or $Artemisia$ is smooth and dry. This is a revertence, as pollen grains of primitive entomophilous angiosperms are rather smooth. The heavy ornamentation is a secondary characteristic of highly evolved entomophiles.

The remains of small quantities of oil on the surface of the pollen of many anemophilous plants testifies to their origin from entomophilous ancestors (e.g. $Plantago$ spp. Knoll, 1930b; grasses, Pohl 1930). One may even question whether any angiosperm pollen exines are ever completely free from such oil (Knoll $loc. cit.$; Pohl 1929), though the quantity certainly is very small as in $Corylus$ or $Betula$. In some species in the transition from entomophily to anemophily the quantity of oil may vary individually (e.g. $Alisma plantago$ Daumann 1966). Conifers, as primary anemophiles, may be free from oil.

The non-adherence and even distribution of anemophilous pollen grains are easily demonstrated by leaving some catkins of Betulaceae or grass panicles in a vase in a room with no draught; after some time there is an even deposition of pollen on the table (dark background). A more refined apparatus for studying the degree of coherence during air transport has been used by Knoll (1930b). It consists of a wide glass tube through which pollen grains can fall in calm air, and an apparatus for letting the pollen into the tube without creating any turbulence. "Pollen fall pictures" are obtained from microscope slides exposed in the bottom of the tube and the amount of oil present on the pollen grains can be judged from the traces left on the glass plate on which the pollen is collected (Daumann 1966).

An obvious condition for the effectivity of wind pollination is the presence of wind, as found in open, sparse vegetation or in the top layer of closed, multilayered vegetation types. In dense forest vegetation wind is so slight and infrequent that anemophily is contra-indicated. Under such conditions even smaller pollen grains are too heavy and sink too fast to be spread effectively. Semerikov and Glotov (1971) estimate a maximum pollen dispersal distance for *Quercus petraea* pollen inside a forest to be 80 m, and Koski (1970: 35) calculates the "probable pollination distance" in a Finnish pine forest to be 53 m. Similar data appear in various palynological investigations (cf. Fægri and Iversen 1975).

The correlation between pollination spectrum (the percentages of different pollination agents active in a region) and the more or less windy climate was apparent in the old data of Knuth. In Germany there are 21 per cent anemophiles; in the northern coastal region 27 per cent; on the windy North Sea islands 36 per cent; and on flat sandbanks in the sea 47 per cent. Pollination spectra might be (in some instances have been) worked out similarly for various areas, giving an expression of the availability and effectivity of the various pollination agents in relation to climate and fauna. However, the actual value of such statistics is rather limited, as other factors are of importance in addition to the immediate effects of climate (cf. Kugler 1975).

Another condition for the effectivity of anemophily is some gregariousness of the species concerned. Small populations are always at a disadvantage for pollination, but especially so in anemophiles, which are dependent on a certain massivity of pollen incidence. On the other hand, even isolated specimens of dioecious anemophiles do set some seed (if dioecy is always absolute).

The question of production and transport of anemophilous pollen is of fundamental interest in pollen analysis, and is treated at some length in the relevant textbooks (e.g. Fægri and Iversen, *loc. cit.*). On an ordinary day, turbulence is usually much stronger than the rate of fall of pollen grains, and the gradual attenuation of pollen content in the air is due more to diffusion than to fall-out.

The transport distance of wind-dispersed pollen has been the subject of many investigations, and it can be proved that, even at some dozens of kilometres distance from a forest, the pollen incidence may be sufficient to effect at least some fertilization. Thus Hesselman (1919) in the Bothnian Gulf found, for example, 700 pollen grains/cm² both of *Betula* (a good flier) and of *Picea* (a bad flier, the rate of fall being, respectively, 1.5 and 7–9 cm/sec). Another important example is the wholesale transport, through large-scale turbulence, of closed, dense pollen rains across great distances (see Rempe 1937). Such long-distance pollen dispersals are of very great importance for the pollination and seed-setting of isolated specimens and also for the long-distance transport of genes and effects of local mutations. Record pollen transport distances observed in some anemophiles vary between 150 and 1300 km (both in conifers), but such figures are of very small value. Once it is sufficiently high up in the air, a grain may go practically anywhere, but will probably be dead on arrival. Most grains fall down during the night, from which fact it may be deduced that an average maximum transport distance is some 50 km downwind.

The area of massive pollen deposition is, after all, of greatest interest in pollination ecology. It varies immensely with varying conditions such as wind, height of vegetation, length of period of release, etc., but the order of magnitude will on average vary between a few and some hundred metres — under exceptional circumstances a kilometre.

In contrast to the great output of pollen grains, the number of ovules per flower is generally rather low in anemophiles, frequently reduced to one, so that each flower only produces but one seed, often in striking contrast to conditions in related entomophilous taxa, e.g. the genus *Bocconia* (single- or few- seeded) in Papaveraceae (cf. Pohl 1929). This might be correlated with the character of the pollen; if one of the sticky pollen grains of a typical entomophile lands on a stigma, there will generally be more grains adhering at the same time and a great number of ovules has a chance to be fertilized. But the chance is that the dry, evenly distributed anemophile pollen grains land singly on a stigma. The reduction in the number of ovules must also be seen in connection with dispersal syndromes.

As a small number of seeds per pistil is generally considered a derived character, this would indicate that anemophily is also derived.

Syndrome of anemophily. Flowers unisexual, exposed before leaves come out or exposed outside the leaf-mass; perianth insignificant, small or absent; attractants absent; anthers and stigmas exposed; pollen grain small, smooth, dry, produced in great quantities; pollen-arresting mechanisms frequent, reduction in number of ovules, cf. the summary by Stanley and Kirby (1973).

The concept of wind pollination generally implies that wind, air currents, will transport pollen grains for some distance, at least from one blossom to another. However, by shaking the flower wind can easily cause autogamy in homogamous, hermaphrodite flowers. Such autogamy is of importance also in some commercial crops, e.g. kapok (*Ceiba pentandra*).

Furthermore, wind may play an indirect part in entomophilous pollination by attracting the attention of prospective pollinators to the blossoms in cases like micromelittophily or the *Oncidium* referred to in Section 8.6.

6.2. WATER POLLINATION, HYDROPHILY

Whereas the primitiveness of wind-pollination has been a matter for discussion, or by some has been taken for granted, there has never been any doubt that hydrophily is a derived condition, as the aquatic habitat generally is a derived one for higher plants. Most aquatic plants are still pollinated in air, like their terrestrial relatives. *Nymphaea*; *Alisma*, and *Hottonia* may be quoted as examples of entomophily, *Potamogeton* or *Myriophyllum* species for anemophily, and *Lobelia dortmanna* for autogamy. But some aquatics have taken advantage of the water also for pollination (see also Daumann 1963).

Water pollination may take place either on the surface of the water, ephydrophily, or in the water, hyphydrophily (terms created by Delpino), a further step in the development from entomophily or anemophily. It should be kept in mind that not all pollination taking place in the water is hydrophily. Many small, autogamous terrestrial plants may flower even when submersed, and the autogamous mechanism functions even under those circumstances, generally in an air bubble confined within the flower.

Ephydrophily is unique among abiotic pollinations in that it takes place in a two-dimensional medium. Compared with the three-dimensional media of anemophily or hyphydrophily, this makes possible a great saving of pollen. In general, pollen is released in the water and floats up to the surface where the stigmas are exposed (*Ruppia, Callitriche autumnalis*). The pollen grains float on the surface membrane and spread rapidly, as is easily observed if one wades through a flowering *Ruppia* meadow: small, yellow dots appear on the surface and spread like oil droplets, probably assisted by the oily coating on the pollen surface. According to Mahabale (1968) the pollen of *Neptunia* and *Aeschynomene* spp. forms a foam on the water surface: female flowers emerge into this and are again withdrawn.

In the famous case of *Vallisneria* (Kerner 1898, Section 17.3), the whole male flower is released instead of individual pollen grains; the pollen itself therefore does not touch the water surface. Small depressions in the surface membrane around the emerging female flowers cause the floating male flowers to slide down and the anthers to attach themselves to the stigmas. In accordance with this effective mode of pollination the number of pollen grains per male flower is drastically reduced (72 according to Kerner, *loc. cit.*). According to Ernst-Schwarzenbach (1945) "*Vallisneria* mechanisms" are also found in various Hydro-charitaceae, partly in conjunction with explosive anthers (*Hydrilla*). According to den Hartog (1964), *Lemma trisulca* has a similar type of pollination with the whole plant rising up to the surface; similar pollination mechanisms are found in the genus *Elodea* (Haumann-Merck 1912), the surfacing staminate flowers of which are partly attached, partly flow freely.

Hyphydrophily has been described for a few plants like *Najas, Halophila, Callitriche hamulata*, and *Ceratophyllum*. So far, these are better treated as individual cases, as they do not seem to have very much in common except extreme exine reduction. In *Najas* the slowly sinking pollen grains are caught by the stigmas.

The pollen dispersal unit in *Zostera* is 2500 μ long, and is more like a pollen tube than a grain. It is apparently reactive, twisting itself rapidly around any narrow object in its way, e.g. around the stigma. However, the reaction is purely passive. The morphology of the *Zostera* grain can be considered the extreme of a tendency apparent also in other hyphydrophilous plants: rapid growing out of the pollen tube gives the dispersing pollen grain linearity.

These hydrophilous mechanisms give no guarantee against autogamy. However, the gregarious habit of the plants in question will generally cause allogamy and counteract auto- and geitonogamy unless the whole meadow represents a single clone. Undoubtedly, more hydrophilous pollination mechanisms will eventually be found, but even so we may predict that hydrophily will always remain a curiosity.

Hagerup (1950b) has described rain pollination in a number of flowers in the Faroes. The principle is that, during rain, the flowers fill with water up to a certain level. Pollen grains float on the surface of this water, and eventually reach the stigmas that are exposed at the level of the water. However, Daumann (1970a) doubts if the idea of ombrophily is tenable, as the pollen of some of Hagerup's rain-pollinated species is seriously damaged by water. Primarily, rain pollination, if at all operative, would be a case of autogamy, but allogamy, even if not very effective, would be conceivable with rain-splash. The mechanism would be one of those auxiliary ones, like various other types of autogamy, functioning under adverse conditions where the regular pollination syndrome is ineffective.

It is essential for all forms of hydrophily that the pollen is not killed by immersion in water. Ephydrophily and rain pollination presume water-repellent pollen; in hyphydrophilous plants pollen grains must be able to endure wetting.

CHAPTER 7

BIOTIC POLLINATION. PRINCIPLES

Biotic pollination introduces into the sequence of events a second organism, the pollination agent or the pollen vector, and a certain *relationship* is in some way or other established between the agent and the blossom to be pollinated. The pollinator should visit this particular blossom regularly, and these visits (whatever their cause and outcome) should constitute a regular part of the life activity of the animal. Such a relationship is generally established by means of some kind of *direct* attractant: nectar, pollen, odour, etc. All these will be dealt with later. There may also be an *indirect* attraction, as when insects of prey visit blossoms to profit from the presence of "legitimate" visitors that fall victim to them. The possible part played by such visitors as pollinators has not been adequately studied but one should not overlook the possibility that a more direct insect—blossom relation may have developed out of such a behaviour which, however, presumes that other insects are already regularly visiting blossoms.

Biotic pollination may accidentally take place without any relationship existing between blossom and agent. If a boy climbs in a flowering cherry tree, he may dash the branches and flowers against each other, thus at least causing geitonogamy. Nevertheless, nobody would count *Homo sapiens juv.* as a pollinator of *Prunus cerasus.* Similarly, if a slug creeps across some open blossom, pollen grains may adhere and may possibly be deposited in the next blossom, but unless there is some special reason for slugs regularly to creep across the blossom in question, they cannot be counted as pollinators of that species. Especially, insects abound everywhere, and run in and out of everything; but even if an insect should visit more than one flower of the same species in sequence, and thus cause pollination, it cannot be counted as a pollinator in the restricted sense of the word unless it regularly visits that species for some specific reason which may include an element of deceit. Incidentally, this illustrates the futility of indiscriminate lists of visitors to a blossom. Bohart *et al.* (1970) give a list of all pollinators observed on onion in Utah, classified with regard to efficiency and abundance. Out of 255 species, only 8 were efficient and/or abundant and quantitatively of importance as pollinators; 164 species were both rare and inefficient.

Even with the concept of a definite relationship in mind, it is not always easy to draw the line between pollinators and accidental visitors. Relationships may vary between very loose and general ones to highly specialized in which plant and agent are completely inter-dependent. Similarly, the behaviour of the pollinator may be more or less active. Examples of passive behaviour are furnished, for example, by the insects trapped in an *Arum* blossom; whether they enter that or, for example, a *Darlingtonia* leaf, is ethologically irrelevant. Active behaviour may be very precise, like the activity of *Tegeticula* in the *Yucca* flower, or of bumblebees in *Pedicularis lapponica*, but it may also be rather imprecise, like that of flies in an umbellifer. Even within the group of regular blossom visitors there are many types classified according to their increasing structural and psychological adaptation for blossom visits.

Baker *et al.* (1971) introduce the concepts of major and minor pollinators, the latter being vectors that, although parts of a less-perfect adaptation system, carry out some pollination. Using *Ceiba pentandra* as an example they establish bats as major pollinators, moths, hummingbirds, and bees as minor ones, a number of smaller visitors as parasitic and after anthesis as scavengers, ending up with the cows eating the corollas dropped into the grass the next morning. Especially in the case of long-lived vectors this system can be important to feed them during periods when the "proper" food is not available. In such cases adaptation will be to the major pollinator; the others utilize an existing possibility, which may, however, establish a preadaptation.

Galil (1973a) has formulated the distinction between *topocentric* and *ethodynamic* pollination, the former being distinguished by the fact that because of the construction of the blossom the pollinator moves in such a way as to come − unwittingly − in contact with pollen and stigma. In ethodynamic pollination − known from a few plants only − pollinators actively approach the ripe anthers, load the pollen on to their bodies, then transfer to the pistil of another blossom and again actively deposit the pollen.

The activity spectrum of pollinators is an important factor in biotic pollination. Vectors depend on meteorologic parameters: below a certain temperature, in a high wind, or heavy rain they are inactive. Usually the level of activity rises with temperature until an optimum is reached. Whereas the danger of too high temperatures is remote in temperate climates, pollinators in deserts develop special strategies to avoid the rigours of that climate (Linsley 1962). On the other hand pollinators in a cool forest may follow the sun-specks and avoid blossoms in the shade (Beattie 1971).

Another important factor is periodicity. With few exceptions both blossoms and pollinators live for a short period and in a periodic climate manifestations of both activities are concentrated to short periods. Interaction of vector and blossom depends on synchroneity of periodic phenomena. Both yearly and diurnal periodicity are probably to a great extent environmentally induced, but interacting with endogenous periodicities and the availability of the food that the individual pollinator is interested in collecting. Pollinators who are on the spot when a foodstuff (an attractant) is first presented, can help themselves from a rich source, whereas those who come later will have to contend with left-overs, depending on the continued productivity of the blossom.

The concept of time memory is frequently brought up in connection with pollination processes, especially in relation to bees, which restrict their blossom visits to a certain, often short, period of the day. The word memory may sometimes be deceptive; the timing is probably instinctive, but released by external influences: light, temperature, humidity, etc. Periodicity would be a better term than memory.

Estes and Thorp (1975) discuss a well-documented case of oligolecty. cum periodicity, which can be taken as an example. The blossoms of *Pyrrhopappus carolinianus* open at 0630 and close at 0830−0900 hours. Females of the oligolectic bee *Hemihalictus lustrans* emerge from their burrows at the same time and immediately start collecting pollen from the opening capitula. Anthers which have not yet opened spontaneously are broken up and their pollen removed. Later, after pollen has been deposited on the styles, the remains are collected, but florets visited previously contain very little pollen. After the pollen flights the bees collect nectar, the last flight of the day taking place between 0815 and 0940, followed by retreat and plugging the entrance to the burrow. Males also visit the (nectarless) heads of *Pyrrhopappus* for copulation; they collect nectar in other blossoms.

On days when the opening of *Pyrrhopappus* heads was delayed by adverse weather,

Hemihalictus visits were delayed as well. This may be due to identical reactions to some meteorological parameter, or it may mean that the bee was out all the time, reconnoitering, but not finding any food until the blossoms opened. This presents the question if the end of the pollen foraging period might be an effect of satisfaction, and not of any special timing instinct.

Other insects were observed, but sparingly, in the *Pyrrhopappus* heads. The exact timing and instinctive feeding reaction (breaking open anthers) place *Hemihalictus* at a great advantage in the competition for the food offered by this plant.

This striking, but in no way unique example of symbiosis between plant and pollinator, raises a number of questions. *Pyrrhopappus carolinianus* is a weedy plant, and its present habitat is manmade. Where did the plant grow before man created its present habitat? Did it also then form sufficiently dense populations to support a population of *Hemihalictus*? In case not, has the bee acquired its present timing quite recently, by selection out of a population with less well-defined food preferences and foraging periodicity? Although *H. lustrans* is oligolectic, it is not even today monolectic, nor is *P. carolinianum* monophilic.

Factors regulating the opening and closing of blossoms and presentation of attractants can partly be the simple, external influences, like meteorologic factors, but there are also factors which can only operate in one of the partners: endogenous periodicities of various kinds, especially in the pollinator, the photo-induced red—far red mechanism in plants, where endogenous periodicities may also play a part. However, we shall not go into details here (cf. B. Meeuse 1968; Overland 1960).

The existence and effectivity of an attractant manifest themselves in a succession of visits to a blossom of the same species. This inevitably leads to a discussion of what constitutes a *visit*. Insects pry everywhere, and they may frequently be present in a blossom without this constituting a visit in the sense of pollination ecology. If the presence is to be recognized as a visit, the insect must perform or try to perform certain tasks in accordance with the structure and function of the blossom — suck nectar, collect pollen, etc. A certain degree of preconception is hardly avoidable here. Even so, many cases are doubtful and must be decided by subjective judgement. Is shelter an attractant, and can an insect that is present in a blossom only to seek shelter be considered a regular agent of pollination? The effectivity of the process enters into the picture: an insect does not seek shelter all the time, and may have lost all or most of the adhering pollen since the last time it did.

The nature of the attractant varies a great deal, and so do the psychological reactions of the visitors. Long-range attraction and close-range orientation may have different sensory bases, and, for example, the relative importance of sight and smell may vary widely between different groups of pollinators, diurnal and nocturnal. Studies on pollinators with various sensory organs insensitized frequently give conflicting results. One reason is that the animals, once they are conditioned to blossom visits, use all their senses simultaneously, and may at this stage compensate for the absence of one of them by using the others more intensively, whereas they may be unable to do so during the conditioning stage.

It is customary to separate between *visit* and *approach* in established blossom visitors. In many instances prospective pollinators will approach a blossom, but will turn away at a short distance, apparently because some feature produces an antagonistic effect. A visit under the definition given above is said to commence with the alighting of the insect on or in the blossom. The attraction unit (see p. 21) controls the approach, the pollination unit the visit. Pollination, though dependent on the visit, is not always an automatic consequence, e.g. if the size or behaviour of the visitor is unsuitable. Also the criterion of a visit by hovering pollinators must be given a different wording, e.g. the extension of the proboscis in moths, but then, again, there are hovering blossom-flies which always fly with extended proboscis.

We have already seen how wind pollination may constitute alternatives to entomophily. Similarly, one very rarely finds complete exclusivity within the group of biotic pollination. The ideal interrelationship, one blossom—one pollinator, may be realized in rare cases, like *Yucca* (p. 175), *Ophrys* (p. 74), *Angraecum sesquipedale,* but always remains an exception. Usually, one blossom is visited by many different species, sometimes completely unrelated, even if one group of visitors dominates in each place. One insect species generally visits several plant species. This may be because the flowering season is shorter than the feeding season of the animal, so it is forced to change. In other cases, the same individual may simultaneously utilize different blossoms. Baker (1961a) has pointed out that in some plants, e.g. *Mirabilis froebelii* there seems to be an adaptation to a combined pollination system. In Polemoniaceae, Grant and Grant (1965) have described instances of local difference of emphasis on different pollination adaptations, which may in the long run lead to genetic isolation and then to speciation.

A non-specialized blossom possesses an innate pollination insurance by virtue of its appeal to many different types of pollinators, which can partake of its attractants and substitute for each other as pollinators. On the other hand, it may have to pay for this by the possibility that many of the unadapted visitors will use other sources of food later, and perhaps later neglect the blossoms of that species. This would then mean waste of attractant and loss of genetic material. The undirected visit is of little value to the plant.

The terminology dealing with these relations suffers from ambiguities which have led to some confusion. The blossom—visitor relationships can be seen from two or three different points of view, although the same terms have been used for different types of relationships. In the first instance the relationship can be seen from the point of view of the blossom; whether it is visited by one, a few, or many different visitors, whether it is adapted to being visited by unspecialized or specialized animals. In the second instance the relationship can be seen from the point of view of the animal *species*; whether it visits one, few, or many plant species. And in the third instance the same question may be asked in relation to the *individual* visitors, or even in relation to the behaviour of the individual during a certain period of its life or in relation to a certain activity.

Among the terms used to describe these relationships we have chosen -*phily* to signify the blossom relationship, -*tropy* (-*lecty*) to signify the animal *species* relationship, and *fidelity* to describe the general behaviour of the *individual* visitor—running the risks always inherent in redefining terms. Loew (1884) originally coined -tropy to characterize the nectar collecting activity of bees; -lecty was originally used by Robertson for pollen collecting by the same animals.

The term -lecty has become very popular with entomologists interested in the food habits of insects in general (not only of bees — e.g. Linsley 1958). From a botanical point of view it is important to note that -lecty as now used by entomologists has a much wider meaning than originally and that mono- or oligolecty of a bee in regard to a certain blossom may be nothing but pollen theft, however fixed the relation is.

We shall keep the word *constancy* as a convenient general term describing all these relationships in general. This terminology deviates from our earlier use. However, to restrict the use of a long-abused term like constancy is hardly possible. Successful pollination or a definite pollination adaption is no prerequisite for the development of constancy. Night-flowering *Oenothera* are almost exclusively pollinated by hawk-moths, but many bees are monolectic to their pollen, which they steal (Gregory 1963). The existence of the food source of these bees is therefore dependent on the pollinating activity of a completely different group of animals.

The literature suffers from much confusion because authors do not always realize that a pollinator may be monolectic for pollen but, nevertheless, polytropic in its pollinating activities. The syndrome of monolecty—monotropy in small, solitary bees has been studied by several groups, especially by E. G. Linsley and his collaborators, and we have had the great privilege to discuss the problems with him (in litt.). In this case the constancy (monolecty) is apparently induced with the larval food. Insects carry over into the adult stage substances ingested during larval feeding, including the well-known cases (in butterflies) in which poisonous substances from the larval food plants render imagines inedible as well. It is therefore not unlikely that adult bees may carry with them a taste template to which they respond in collecting food for the larvae. The main part of this food is pollen. On the whole the bond to specific pollen sources seems to be stronger than to nectar sources. This may also be the case in butterflies; according to Ehrlich and Gilbert (1973) the pollen-collecting *Heliconus* butterflies are remarkably sedentary and have very limited home ranges, whereas nectar-collecting butterflies roam more freely (Sharp *et al*. 1974). Sometimes, constancy patterns may be extremely complicated. According to Vogel (1974) female oil bees (*Centris* spp.) are monolectic to *Calceolaria* for oil, whereas they are polylectic (polytropic) to various other species for pollen for the brood and nectar for their own energy consumption. The same pertains to males. Strict oligolecty and oligotropy with regard to more than one type of food that had to be obtained from different blossoms could easily place the animal in an impossible situation.

In Chapter 15 we shall return to all these interconnections: nectar and pollen stealing from one species keeping alive insects that pollinate others, complicated constancy patterns implying several unrelated plant species, add up to a higher interdependence not only between the one pollinator species and the corresponding plants, but covering the biocoenose. The evaluation of adaptations must take into account the whole community within which these adaptations function.

Cruden (1972) has observed oligolectic bees changing to "foreign" pollen in times of dearth. If the larvae can develop on this food the question arises if there will be a different imprint in the imagines of the next generation, or if there is a genetic factor as well. In such cases a selection of types that can survive on an alternative pollen may start a new strain of the pollinating species. A certain genetic element is apparently present: insects react adversely to some types of odour; emerging insects may immediately head for the one specific food plant (in a mixture of species: Levin 1972a).

As a consequence of the different patterns in pollen and nectar collecting it may be useful to have a special term for the pollen-collecting activity of insects, but a more general term is also necessary — many insects collect both pollen and nectar at the same time. Restricting such a useful term as -tropy to bee visits would make unavoidable the creation of a new more general term which would make the older one redundant. Consequently we have chosen to widen the -tropy concept to include all kinds of pollinators.* On the other hand, -tropy has been used (even by Loew) also for the passive relation of the blossom. We have

---

*It would be too cumbersome to say that a bee is polytropic whereas a wasp is poly-something else when we mean exactly the same; some authors have used the form -tropous where we have preferred -tropic.

felt the need to be able to differentiate between these two relationships, and propose a consistent use of -phily, which is already in use for different blossom relations. One of our reasons for using separate terms for plants and visitors is that they do not always develop in parallel. Many orchids are monophilic, but their pollinators are polytropic bees; the species of *Cucurbita* are polyphilic, but the special squash bees pollinating them are monotropic down to species level (Hurd *et al.* 1974), *Nymphaea* is euphilic, but its beetle pollinators are allo- or even dystropic.

The different types of relationship are shown in Table 2. It should be noted, especially, that within the group of euphilic blossoms (and eutropic visitors), a further differentiation is possible, as indicated.

Euphilic/polyphilic. adaptation to regular pollination by representatives of several major taxa of visitors, e.g. *Carnegiea gigantea*: bats, birds, bees (Alcorn *et al.* 1961). Some of these ·may be minor pollinators sensu Baker *et al.* (1971).

Euphilic/oligophilic: adaptations to regular pollination by representatives of one major taxon, e.g. *Aconitum septentrionale/Bombus* spp. (Løken 1950).

Euphilic/monophilic: adaptations to regular pollination by one single species of visitor, e.g. *Ophrys speculum/Campsoscolia ciliata* (Correvon and Pouyanne 1916).

The flower fidelity exhibited by pollinators may have one of three different causes:

Of necessity: there is only one blossom available which can be utilized by the animal in question. This constancy breaks down immediately the situation changes.

Of preference: among the blossoms available, one is selected. This constancy is rarely exclusive, even in one individual, and can change with the coming into flower of a more attractive source of nectar or pollen.

Innate: among the blossoms available the same is invariably selected. This is monotropy.

The main source of confusion (cf. V. Grant 1950b or Free 1966) has been between monotropy, on the one hand, and the two other types of fidelity, on the other. All have been called "flower constancy", but they are entirely unrelated. A monotropic animal is (physiologically, physically, and/or ethologically) unable to utilize any other plant species. Its relation to the blossom is more or less on the lines of a parasite—host relationship. Whether monotropy is always absolute or can be geographically variable remains to be seen; it is easily conceivable that a visitor may in one region be bound to one plant species, in another place where that species is rare it may visit another species. Not enough is known about the habits of solitary bees (the most important group of monotropic visitors, cf. Olberg 1951) to give a basis for the evaluation of geographical replacements in monotropy, even if some scattered information can be found in zoological literature, e.g. the monotropic relations between Cucurbitaceae and specialized bees (cf. Hurd, Linsley, and Whitaker 1971).

On the other hand, monotropy has been found in various primitive pollinators: beetles, gallwasps, and other primitive hymenopters. Misunderstanding of the difference between monotropy and fidelity has lead writers to assume that the first biotic pollination was promiscuous because of the absence of specialized pollinating insects (Davis and Heywood 1963: 437). However, this is an unwarranted assumption. Beetles and flies may have been monotropic — and the corresponding blossoms monophilic — from the beginning, although they were not true flower specialists. This would especially be so if attraction were by deceit, which is a possibility since the relationships to be imitated (food search, etc.) may

TABLE 2. Definitions of terms describing blossom–visitor relationships

Blossom relationship		Visitor–species relationship	
Dealing principally with morphological adaptation of blossom	Dealing principally with character of visits received	Dealing with adaptation for blossom visits	Dealing with character of visit activity
		Unadapted or counter-adapted; visits show no relation to the organization of the blossom, frequently destructive, but may cause pollination: *dystropic*	
No morphological adaptations for guiding visitors; can be utilized by unadapted, short-tongued visitors: *allophilic*	Pollinated by many different taxa of visitors: *polyphilic*	Poorly adapted for utilization of blossoms; the food obtained from blossoms forms part of a mixed diet: *allotropic*	Visiting many different taxa of plants: *polytropic (polylectic)*
Imperfectly adapted to being utilized by animals of intermediate degree of specialization: *hemiphilic*	Pollinated by some related taxa of visitors: *oligophilic*	Intermediate degree of specialization: *hemitropic (hemilectic)*	Visiting some related taxa of plants only: *oligotropic (oligolectic)*
Strongly adapted to being utilized by specialized visitors: *euphilic*	Pollinated by one single or some closely related species only: *monophilic*	Fully adapted blossom visitors, taking their main food from blossoms: *eutropic (eulectic)*	Visiting one single or some closely related plant species only: *monotropic (monolectic)*

have existed previously. In phylogenetically younger insect groups original, primitive polytropy may have developed either into monotropy or fidelity.

Fidelity is an *individual* quality in a polytropic (theoretically also in an oligotropic) animal that as a species is physiologically, physically, and ethologically able to, and does visit any of a number of plant species. The individual may for reasons, which we may describe as psychological, restrict its visits to one single plant species during a longer or shorter period, which may extend from a single flight to the whole of its life-time. One might define fidelity as an individual and (as the case may be) a temporary, monotropy in a polytropic species.

When it is not immediately apparent what is the basis for the restriction of a pollinator to a single species, we have used constancy as a neutral term, which only implies that the visitor restricts itself to a single plant species, but does not indicate whether this is monotropy or fidelity.

Fidelity may be imposed; within the area in question there is only one blossom present in great quantities suitable for the animal in question. This kind of "flower constancy" is purely coincidental and may be found anywhere; it has nothing to do either with constancy or with monotropy. Monotropy is also imposed, but by internal factors. Fidelity, on the other hand, represents preference: the insect has "discovered" how to manipulate a certain blossom in order to get at the attractants, and has "experienced"* that a visit to this blossom generally yields satisfactory quantities of the attractant. This process of conditioning seems especially important for virgin individuals on their maiden flights, during which the subsequent pattern of behaviour becomes fixed (cf. Kugler 1936, 1942). The animal will then visit, by preference, other specimens of the same blossom later, even in a mixed population with other potential sources of an attractant, which, perhaps fortuitiously, does not appeal in the same way. Fidelity presumes a faculty of *memory* and of *recognition* of certain features: blossom characteristics, location, etc., and is therefore restricted to the more highly evolved pollinators, especially bees, but also butterflies. It is an individual property. Inborn colour or odour preferences may contribute to the establishment of constancies (if they do not form the basis of monotropy).

From what has been discussed above, it emerges that pollinators that collect both pollen and nectar may exhibit separate degrees or even types of constancy for the two activities. Usually, there is a stronger binding to the pollen source (oligo- or monolecty) combined with a wider variety of nectar sources. By studying the degree of interspecific or intersubspecific hybridization one may form an idea of the (minimum) of breaks of the constancy rule in the relevant pollination process. This figure seems to be of the magnitude of 1 per cent as shown in various plant groups (*Cicer arietinum*: Niknejad and Khosh-khuni 1972; *Phlox*: Levin and Berube 1972) in cases of sympatric occurrence.

External circumstances also affect constancy. Plants to some degree "compete" for pollinators where flower-produced food is ample, and constancy brings about consecutive visits to and by one species. If food is scarce, insects compete for it and may have to collect also in blossoms which would, under other circumstances, have been neglected. Under such circumstances constancy may have to play a subordinate rôle in relation to the primary

---

*In order not to get involved in terminological discussions we consider to be mainly empty, we refrain from stating anything about the mental faculties of insects. As pointed out by Schremmer (1955), there is no doubt that the reactions of insects change during their attempts to utilize different blossoms. Whether one accepts this as "experience" and "learning" or calls it something else, is irrelevant in this connection. At any rate the effects of the process are very similar to what would under other circumstances have been called learning and memory.

object of securing food, and perhaps even normally monotropic species may be forced to utilize alternative sources of food.

If we consider allotropy as representing the original condition of blossom visiting in primitive insects, blossom relationships develop, among such visitors (if at all), along the lines of monotropy, not of polytropy with constancy. The development of monotropy probably was direct, not via eutropy and polytropy. A monotropic pollinator is restricted to habitats where its fodder plant is sufficiently abundant for its needs. It has thus a much narrower ecological niche than a polytropic pollinator, in which constancy, if present, may develop in relation to different species, and which is therefore not subject to the same restriction of habitat

Two factors seem to be of great importance for the fidelity, i.e. blossom preference; these are relative abundance of the plant in question and the duration of its flowering period. The more dominating the blossom is, the easier a constancy develops (see Arnell's "dominating flowering phenomenon", 1903), and the longer it lasts, the stronger grows the psychological binding of the visitor to the blossom in question and the retention of this binding in its memory. Constancy, like any blossom recognition, apparently develops as a fixation in relation to form, colour, and odour. In experiments, a colour constancy is easily produced, but in nature colour alone is not sufficient to produce a constancy: in a mass occurrence of *Trifolium repens* with subdominant, visually (to us) very similar *Polygonum viviparum* and *Achillea millefolium*, the very abundant bumblebees never made a single approach to the two last-mentioned species (K. F.). Odour seems to be the basis of discrimination in such instances (Manning 1957). On the other hand, there are some very interesting observations of constant pollinators ignoring colour differences between garden varieties of the same species, a subject that should be more closely looked into.

Constancy may develop in relation to two, often very different, blossoms at the same time. If this is still a fidelity of preference, it presumes a rather difficult mental process of recognition. In some cases such double constancy may be an effect of different presentation periods, blossom A being visited during some period of the day when B is not available.

In bumblebees Heinrich (1976) has found a mixed fidelity between a "major" which is usually visited and a "minor" which is kept under supervision by regular, but infrequent visits. In cases of changing yield it is comparatively easy for a bumblebee to switch from a major to a minor, whereas it is more difficult to start utilizing a blossom not previously experienced. A mathematical model for the strategy (by Oster and Heinrich 1976) indicates that in a changing (and adverse) environment a mixed strategy is, according to the model, more rewarding; in a constant environment the pure strategy gives better results.

Contributing to the development of constancy may be the tendency of pollinators (bees and butterflies) to move in the same direction on various flights (Levin *et al.* 1971). Where odour is the attractant the tendency will be to move upwind. A similar effect may be the result of a tendency to remain at the same level above the ground (Levin and Kersten 1973). This effect, which was originally described for *Lythrum*, is of immensely greater significance in the multistoried and more diversified tropical forests (Frankie 1975).

In honeybees, whose flower constancy has been very much discussed, the problem takes on a special character, due to the extreme specialization of work in the hive. Collector bees, "gatherers", stick to what they have been "told" by the scout bees, "searchers", and so, unless they react to new messages from scouts, there will always be a group of collectors whose blossom preference has been fixed by some earlier message from the scouts. New collectors will act upon new messages, whereas the older ones may go on collecting from the old source until this is exhausted, in which case they may change (very distinctly brought out by Table 1 in Percival 1947). In this way honeybees exhibit a group fidelity, dependent on a sense of memory. Pollen analysis of stores gives important information as to blossom preference and constancy in store-collecting insects. All the stores of a bumblebee nest may be completely dominated by one or two pollen types (Fægri 1962). On the other hand, pollen analysis of corbiculae of honeybees in many cases (generally 1–2 per cent, Maurizio 1953) reveals a double or sometimes a multiple constancy, at any rate of pollen collecting. This multiplicity is not enforced by dearth of pollen, but must be one of multiple preference. The species involved may be completely unrelated and dissimilar, and may also include anemophiles.

Constancy develops in pollinators as a response to their own demand for greater effectivity in food gathering. But the phenomenon has very great positive selective value also for the plants which because of it have a greater chance of being visited by a pollinator carrying compatible pollen. For pollination effectivity, some inconstancy is no disadvantage: unless the quantity of pollen is a minimum factor, a mixed load also contains enough compatible grains. Many of the "mixed loads" mentioned in the literature are almost pure, indicating only very occasional visits to other plants or even accidental contamination in the nest or hive. With the possible exception of honeybees it seems that higher pollinators to some extent "experiment" during food-gathering; they visit other blossoms now and then to investigate their properties. Too strict fidelity may be disadvantageous in pollination, e.g. in the example quoted by McGregor et al. (1959) in which one (introduced) honeybee during 3½ hours made 21 visits to 7 flowers on one arm of a self-incompatible, chiropterophilous cactus, thus not causing a single fertilization. Location seems to be an important part of the blossom constancy complex, and self-pollination may frequently be the effect of strict constancy. The importance of this for speciation will be dealt with in Chapter 14. Its importance as a source of errors in researches must always be taken into account.

Inconstancy is apparently the basis for the pollination in certain orchids, e.g. *Orchis*, *Calypso*, the flowers of which seem to be completely devoid of any primary attractant. One must assume a "parasitic mimesis" (Vogel 1975a): these blossoms resemble "real" blossoms so much that bumblebees are deceived and visit them, which would not have happened if constancy was absolute.

For plants, pollinator constancy loses much of its importance in poor vegetation types in which few species flower at the same time: flower-visiting animals are restricted to these few species, whereas in a rich flora they may go from one species to another, wasting pollen picked up previously.

The "experimentation" or "searching" by the insects, mentioned above, may sometimes cause a break-down of blossom constancy. This is easily seen when, for example, *Trifolium pratense* comes into flower: owing to its greater nectar production it suddenly attracts all long-tongued bumblebees that are able to reach the nectar, causing an immediate neglect of other bee blossoms. Blossom constancy is a very utilitarian phenomenon, and only the group behaviour of honeybees may make it seem more complicated than it is.

In highly advanced cases pollination is very precise: the agent visits the blossom in a very definite manner, a special part of its body touches stigma and anther, the pollen of which is deposited as a small dot in a definite place. This is the basis of precision and economy in the process. At the other end of the scale there are primitive insects messing about in primitive blossoms more or less diffusely covered with pollen on legs and body, some part of which will subsequently touch a stigma: the *mess and soil* principle of pollination. Such pollination is found when beetles trample around in a *Magnolia* flower, but also in more evolved pollination mechanisms, e.g. when exploding *Cytisus* throws pollen or carrion insects are liberated from an *Arum* spatha. A completely different aspect of pollination is represented by all-or-nothing cases like many orchids, e.g. *Platanthera chlorantha* placing its pollinia on the eyes or *P. bifolia* on the proboscis of the pollinator, or *Salvia pratensis* hitting just one spot on the back of the visitor.

The structural devices for biotic pollination vary as much as the modes of pollination, and a general characteristic must, therefore, perforce remain rather vague; nevertheless, it is distinct enough to be recognizable. The pollen is generally adapted for adhering to the body of the pollinator; in more highly evolved forms, especially, it is sculptured and coated with a sticky oil, Pollenkitt, which permits an almost acrobatic coherence (Knoll 1930b), and also ensures adherence even to the chitin surface of glabrous insect visitors.* In fall tests, the pollen of typical animal-pollinated blossoms will — if at all — fall in large lumps. The stickiness of the pollen is itself serving as an arresting device; separate arresting devices are generally absent. More rarely the viscid substance is produced by some other organ of the blossom; the pollen itself is then dry and arresting mechanisms occur. Knoll has pointed out that the stickiness of pollen must be correlated to some extent with the size of the pollinator; great lumps of pollen might become too heavy for very small pollinators, or such pollinators might even glue themselves to the anthers and perish. In addition to the ordinary coating of pollen grains, which has the general physical characteristics of a fatty or semi-ethereal oil, more specific substances are sometimes found like the so-called viscin that forms strands between grains (e.g. in Ericaceae or Oenotheraceae), or in the substances keeping the whole contents of a theca together as a pollinium in Orchidaceae and, Asclepiadaceae. Such heavy pollinating units depend on a very effective transport mechanism, and they are, as a consequence of this, combined with some type of viscidium that cements them to the body of the pollen vector. In *Calliandra* (Mimosoideae) there is a similar, but simpler mechanism: the contents of each anther form massulae, each provided with a viscidium that projects from the surface of the anther and cements the massula to blossom visitors.

At the other extreme one finds entomophilous plants that have dry pollen, while the stigma is so viscid that part of the mucus is, on contact, transferred to the body of the pollinator. In such blossoms the mechanism can only function if the insect touches the stigma first, anthers later, and the pollen will adhere to those parts of the insect body that have been in contact with the stigma, and will, presumably, come into contact with the stigma of the next flower. Such mechanisms have been described in orchids, *Monotropa* and *Polygala comosa* (in the latter combined with secondary presentation of pollen). In many species of Rhinanthoideae (insect pollinated) the pollen is dry and falls out of the thecae. However, the anthers are pressed closely together, so that the opposites form closed boxes holding the pollen. When an insect forces its way into the flower, the anthers are forced apart, and the pollen sifts itself over the body of the insect. In some hanging flowers (e.g. *Symphytum officinale*) the whole flower forms a similar apparatus ("Streukegel") with (throat) scales acting as arresting organs as long as the flower is not visited.

---

*Whether the oil coating also protects the pollen grain against being wetted seems less probable. On the other hand, there is an increasing body of evidence that the "recognition substances" and also allergogenic substances may be located in the pollenkitt. In this case, the dry pollenkitt found on the surface of wind-dispersed grains may have a function and not only represent relicts from an earlier syndrome of biotic pollination.

Generally, a biotic pollinator must be in bodily contact with the organs of pollen deposition and reception in order to effect pollination. Cases have been observed, however, (v.d. P.) in which hovering sphingids have caused a whirlwind that brought pollen out from the anthers and on to the insect. Self-pollination must be an automatic consequence of this. Whether any pollen is brought on to the reception part of the next blossom remains to be seen. Cholodny (1944) insists that a similar mechanism occurs with regard to bumblebees in *Salvia glutinosa*.

In *Melampyrum* pollen-collecting bees (*Megachile*) take up a hanging position under the anther, and by vibrating their wings make the pollen grains fall out of the anthers on to the insect (Meidell 1945). Macior (1969) has described an identical mechanism with regard to the workers of three *Bombus* species in *Pedicularis lanceolata*. A further development of this is seen in the way in which *Xylocopa* behaves in *Melastoma* and *Cassia* (van der Pijl 1939). In these blossoms there are both feeding and pollinating anthers. While "milking" the feeding anthers, the bee vibrates body or wings, producing a whole cloud of pollen from the pollinating anthers. Part of it will be deposited nototribically (i.e. on the back of the insect), in spite of the bee sitting on top of the androecium. In *Bauhinia* spp. nototribly is caused by the curve of the filaments of the pollinating anthers, which are already above the back of the bee.

General syndrome of biotic pollination. Blossom in possession of an attractant (real or deceptive) and means for making its existence known, generally by a large and conspicuous (sight or smell) perianth. Pollen grains of variable size, sculptured, sticky, in extreme cases tied together by thin viscin strands or in pollinia. Anthesis and production of attractant synchronized with the activity of the pollinator.

Finally, it should be pointed out that biotic pollination always depends on external circumstances, which may cause the pollinators to be or not to be present or active, and also that blossoms must be in a "proper" state. Thus, Percival (1947) has described how honeybees refuse to collect pollen from flowers that are still wet with dew. Biotic pollination is a two-sided system: there are both the blossom and the visitor. Pollination presents a different aspect to each of the two groups of organisms, but it is evident that unless they fit together, the pollination mechanism does not work. In the study of pollination, therefore, phenomena must also be seen from the pollinator's point of view, which has too frequently been neglected. Also, by a surplus of pollinators, a kind of "pecking order" between species apparently establishes itself (Kikuchi 1964).

The strategy of blossom—insect relationship takes on a different aspect according to whether there is a surplus of visitors, searching for food, or there is a surplus of blossoms producing more pollen and nectar than the animals can cope with. In any given area, this situation can change very rapidly with the coming into bloom of one or a few plant species. In general, plants are adapted to the second situation, advertising themselves as distinctly as possible. On the other hand, those plants are clearly in an advantageous position which flower at a time when there is a surplus of potential pollinators available.

Different groups of pollinators dominate within different climatic regions. Hermann Müller (1881) points out that lepidopters tend to replace bees in the Alps, whereas flies take over in Arctic regions. Hagerup (1943) maintains that ants take over in deserts; in tropical American mountains hummingbirds come in. With specific reference to Polemoniaceae, Grant and Grant (1965) have pointed out that not only do flies take over in cold regions, but in arid zones solitary bees take over from social bumblebees. Of general interest is the demonstration that the same relative distribution of ecological blossom classes recur in geographically separated but climatically similar regions. Where the local population of pollinators belonging to a higher "pecking order" is insufficient in relation to the flora, lower-rank pollinators get a chance. Obviously, this must be of great importance for further speciation.

These problems will be taken up for discussion in Chapter 15, dealing with pollination in the biocoenose context.

CHAPTER 8

BIOTIC POLLINATION. PRIMARY ATTRACTANTS

A blossom—visitor relationship is established by means of an *attractant*. To be effective, an attractant must start in the visitor a reaction chain that creates or satisfies an urge. The three main instinct systems, which can form the basis of such urges, are feeding, sexual, and brood-rearing (including nest-building) instincts. Pollen, nectar, fat oil, water, etc., appeal to the first group of instincts. The second group comprises phenomena of a various nature, and also the known cases of the third group, which are very few, are dissimilar and cannot form the basis of too much generalization.

The food urge is the background for the overwhelming majority of blossom visits; no distinct limits can be established between obtaining food for the insect itself and for the next generation, even if their food demands are frequently qualitatively different. That pollinators obtain food from the plants they visit has always been taken for granted, but the energy relations of the feeding process has been largely neglected until the seventies (papers by Baker, Hainsworth, Heinrich, Raven, Stiles, and others). Heinrich and Raven gave a summary of the problems in 1972 (cf. also Heinrich 1975). If blossom-derived food represents the only energy intake of visiting animals, each visit must give enough energy to pay for the visit itself, for idle time, and for other activities during which the animal spends energy, e.g. nest-building and maintenance of territory. The foraging process in itself must therefore give a relatively large energy surplus.

The energy question contains many parameters: (1) The energy contained in the food and its metabolic availability. (2) The work involved in extracting the food from the blossom. (3) The energy spent to reach the blossom and move from one to the other.

The energy contents of the food naturally depends on its concentration. Proteins and fat, available from pollen grains and sometimes directly from the blossoms, are good energy sources in themselves, but most pollinators cover their energy demands from nectar, i.e. sugar solutions. The value of the sugar solution above all depends on its concentration but also on the character of the sugars, disaccharides giving more energy than the same concentration of monosaccharides (provided both are digestible). Small, ectothermic pollinators have low energy demands and can utilize dilute nectar (obtaining water at the same time), and the amino acids contained in the nectar may be of importance to them as well. Endothermic, heavier pollinators demand greater quantities of a more concentrated nectar and are dependent on mass occurrences of the blossoms they visit (cf. Watt *et al.* 1974).

The work involved in extracting the food depends on the quantity presented at one time and on its availability. It takes more energy to collect nectar out of a deep hole than to lick it up from a shallow bowl. If the efficiency of feeding is defined as the number of calories obtained per time unit and flying expenses, Wolf and Hainsworth (1972) found that for hummingbirds it decreases from 60 to 8 when the corolla tube (glass tube in the experiment) increases from 0 to 70 mm. Very small nectar drops which must be licked up separately exclude large pollinators whose energy consumption in collecting these minute quantities will be too great to give a surplus. Small nectar drops are therefore part of small-animal pollinating syndromes, independently of the availability of the nectar.

Another important expense is travelling, especially flying. Even in a homoiothermic animal like a hummingbird, the oxygen consumption during flying is about four times higher than when the animal is sitting (Wolf and Hainsworth 1971), and for animals that do not maintain a uniformly high temperature this relation is even less favourable. As walking demands much less energy than flying, a (facultatively) walking pollinator can utilize much smaller nectar drops than an obligatory flying one. In this way even large bumblebees can utilize the minute nectar quantities in *Spiraea latifolia* or *Solidago canadensis*: they crawl from one flower to the next in an extensive, dense inflorescence (Heinrich 1973), at the same time saving energy by letting the thoracic temperature drop to the level of the ambient, so that they are incapable of immediate flight (Heinrich 1972). When working the large-flowered *Chamaenerium angustifolium* a bumblebee flies half the time, but when working the small-flowered *Spiraea* only 10 per cent of the time. However, if distances covered are considered, walking is a more expensive way of locomotion than flying (according to Scholze *et al*. 1964).

From the plants' point of view the pollination strategy is wedged between presenting large energy packets to few pollinators or smaller packets to a greater number. A typical case is that of blossoms that have a full nectar load at the beginning of anthesis, or after a night when there have been no visitors.

Even if ambient temperatures are lower so early in the morning that the larger energy packets may compensate for the extra energy expense of working under these circumstances. This explains the occurrence of matinal bees and blossoms adapted to them. Later in the day energy costs are lower, but so also are the benefits, as much of the nectar is already gone. Visits at low ambient temperatures presume large packets of energy at short mutual distances. At higher ambient temperatures the energy loss during travelling is smaller, packets can be smaller, and travelling distances greater. This may be part of the explanation of the "trap-lining" strategy of some pollinators, e.g. the apparently very great flight distances of nectar-feeding euglossine bees (Janzen 1971). Pollinating *Arctostaphylos otayensis* the matinal *Bombus edwardii* maintains a thoracic temperature 35° higher than the ambient one; each flower visit yields nectar corresponding to 1.5 cal for an expenditure of 0.8, in other words a large profit margin. At noon, at higher temperatures, small insects visiting the flowers get only 0.3 cal out of each visit because the energy packets are so much smaller at that time of the day (Heinrich and Raven 1972).

Many small visitors, nectar thieves, are not only deleterious because they deplete blossoms of attractants, but also because their own energy requirements are satisfied by very few blossom visits. Even if they should pollinate they will be ineffective because of the small number of flowers visited, and also geitonogamy will probably be the main result. Again the strategy of the blossom is wedged between giving the pollinator so much energy that it can and will continue its visits and so little that it is obliged to forage for it within a not too restricted range, so that more than one plant (clone) is visited. In relation to the non-adapted honeybees the saguaro cactus (p.51) belongs to the plants giving too much nectar: the (introduced) bee can remain on the same (self-incompatible) individual all the time.

Clustering of blossoms has the advantage of saving energy, which must be weighed against the disadvantage of geitonogamy. Clustering also influences the collecting strategy. Linhart (1973) describes how territorial hummingbirds are attracted to species (of *Heliconia*) with dense clustering of blossoms, both per inflorescence and of plants, whereas non-territorial insects favour plants with few flowers per inflorescence. The latter strategy is energetically weak (can be compensated for by greater nectar output per blossom) but ensures wide pollen dispersal. Sparsely occurring plants with few but nectar-rich blossoms out at the same time are adapted to the trap-lining strategy of pollinators following the same flight path each day. Aggregation of food sources leads to specialization of the predator (in this case the pollinator, cf. Pulliam 1974; cf. also the quantitative studies of the feeding/energy relations in territorial pollinating butterflies by Wolf and collaborators: Wolf 1975 and earlier).

Analyses of energy relations in pollination are still too few to permit geographical conclusions, but Hocking's results (1968) indicate that the energy yields of individual blossoms in the Arctic are not necessarily lower than at temperate latitudes. The lower ambient temperatures are partly compensated for by insects remaining in the warmest parts of the flowers.

However, it is not sufficient that an attractant can satisfy a physiological demand in the visitor; the urge must also be created in the visitor to avail itself of the attractant in question. Thus, we may distinguish between *primary* attractants that satisfy demands like those for food, etc., discussed above, and *secondary* attractants that start a reaction chain by direct or indirect action of the sensory apparatus of the visitor.*

Primary attractants are in themselves useless unless accompanied by a secondary one; some organization must be present to advertise the presence of the former. The two attractants may have the same origin, e.g. pollen as food (primary) and pollen odour as a secondary attractant. But they may also have a different origin, e.g. concealed nectar and a strikingly coloured perianth, the latter working on the instincts of a pollinator because of its correlation with the primary attractant of nectar. Odour, which is usually a typical secondary attractant, is a primary one in so-called "perfume blossoms".

Both types of attractant will be considered together in the following. With some exceptions (deceit, perfume blossoms) at least one of each group is generally present in a blossom, but attractants within the same group are not mutually exclusive. The ultimate result of the combination of attractants in a given pollination unit, is to start a reaction chain that leads to pollination, i.e. a positive result as seen from the point of view of the blossom. It may also lead to a positive result for the animal, really providing food or a brood-place, but instincts may also deceive the pollinator, as in the cases of pseudo-copulation, pseudonectaries, or carrion blossoms. Such blossoms start instinctive reaction chains for the positive conclusion of which there is no possibility, but which may in extreme cases prove fatal for the individual pollinator (*Nymphaea*, *Pinellia*) or its brood (*Stapelia*, *Aristolochia*).

Deceit in pollination ecology is primarily due to lack of correspondence between primary and secondary attractants. Pollinators are attracted by "false promises", i.e. structures which actively influence reaction chains so that certain actions are taken which are functional under other circumstances, but not commensurate with the possibilities presented by the blossom. In a few blossoms even that does not seem to exist. With due reservation for the

---

*A reviewer has criticized this terminology, alleging that colour and/or odour are observed first, and should therefore be considered primary, whereas nectar, etc., come in at a later stage in the pollination process. We do not feel that this is an important point, but should like to point out that, logically, the primary attractant is the basis for the existence and function of the secondary one. In spite of scattered (parasitic) examples of the opposite it is difficult to conceive of blossoms being regularly visited if they do not present the promise of an award, i.e. what we have called the primary attractant.

(not very great) probability that future investigations will come up with the missing attractant, we may point to the 200-year-old problem of finding an attractant in the *Orchis* flower. A contributing factor may be that in orchids even a single pollination visit is sufficient to produce thousands of seeds, which are widely disseminated. As pointed out by Heinrich and Raven (1972), cross-pollination can be effected in such plants even if they are widely separated. For such blossoms it is not only unnecessary to maintain an energy balance in their pollinators, but the bonds between pollinator and blossom need not be (but are nevertheless frequently) very strong.

Thus, many flowers are actually "cheating" in their production and display of attractants and make visitors behave as if the real thing, be it some kind of food or sexual partner, were present. These instinctive reactions, being without possibility of fulfilment, cause the visitor to carry out the pollination. As in all other kinds of parasitic behaviour, this cannot be carried to extremes, otherwise it becomes self-defeating. In particular, this kind of deceit presumes that the imitation does not outnumber the model.

A special type of deceit is found in some pollen flowers which are either strongly protandric, like *Exacum*, where the anthers are still quite fresh-looking, but empty by the time the stigmas ripen, or in a monoecious plant like *Begonia*, in which the stigmas are remarkably like the anthers. Female flowers are much fewer than male ones in *Begonia* (cf. Vogel 1975a).

In a biocoenosis context theft of attractant: nectar, pollen, i.e. its removal without pollinating, can be considered a parasitic behaviour on the part of the thieves. Similarly, attraction by deceit is a parasitic behaviour on the part of the blossom towards the visitor.

Even if primary and secondary attractants are, in their typical state, easily kept apart, they are so interdependent and intergrade so much that they are better treated under one heading.

Gregarious flowering seems to be important in establishing relations; the effect of the attractant is accentuated by its being produced simultaneously in great quantities. Such gregarious flowering is known in *Coffea*, in orchids, and in *Passiflora*. It is not always strictly seasonal, but may depend on other external factors, or on an innate rhythm, like the 7-year period in *Strobilanthus* or the various rhythms found in other plants, e.g. the (anemophilous) bamboos.

8.1. PRIMARY ATTRACTANTS. I: POLLEN

Despite some opinion to the contrary (Downes 1971: 248) there is reason to consider pollen the original attractant. In the form of microspores, pollen existed before the flower came into existence, and, as we have seen, pollen of cycads is utilized as food by beetles. Microspores and pollen of more primitive extinct groups were presumably utilized in the same manner by contemporary insects and other small animals during earlier geological epochs. However, pollination drops and possibly also extrafloral nectar and other sources of sugar may have been available before the advent of the angiosperm flower. There is no reason to presume that such sources of sugar were neglected, if available.

As an attractant, pollen has the disadvantage that it does not work in female blossoms, preventing the development of unisexuality or causing difficulties if flowers are already — primarily or secondarily — unisexual (cf. the discussion of gymnosperms). The conclusion of this would be that dioecy could hardly have occurred in the first angiosperms if they were entomophilous, unless pollen attraction was combined with oviposition in the ovary.

Whereas this is a well-known combination in present-day blossoms it may be surmised that it is a secondary, modern development. The same reasoning pertains to pollen attraction in protogynous flowers with no nectar — a problem that has not been adequately solved.

Pollen attraction is apparently more selective than nectar attraction. As a consequence, the pollination process may be more economical. Instances have been reported when the pollen (odour?) of certain plants has proved unacceptable to local pollinators.

When functioning as an attractant, pollen is generally well exposed and available also to visitors with a primitive organization. This in contrast to nectar, which is frequently well hidden and available to very specialized visitors only. The behaviour pattern in pollen collecting is therefore often simpler than in nectar collecting, even when the same animal carries out both activities.

For the biochemistry and physiology and for the morphology of pollen we refer to relevant texts (Stanley and Linskens 1974; Fægri and Iversen 1975). Roughly speaking, the pollen grain consists of three concentric layers — exine, intine, and protoplast. Of these, the intine is a pectine and cellulose membrane, probably indigestible. The high-molecule substances making up the exine are extremely resistant against everything, and are also indigestible. Utilization of the protoplast thus presumes either destruction of the outer shells, or digestion by diffusion through existing openings. The latter seems to be the case, for example, in bumblebee larvae, in the recta of which exines are found entire but empty (Fægri 1962). It is therefore often very difficult to state for certain whether pollen is intentionally ingested by any particular insect, or whether it is a more or less accidental admixture of the food, for example taken up with the nectar passed through the digestive tractus.

Nectar hoarding of honeybees and bumblebees and the production of honey are so well known as to make us forget that in brood-rearing insects pollen collecting for the brood is generally more important than nectar collecting, especially in solitary bees (but even when dealing with these insects one should not neglect nectar collecting).

Pollen is a rich source of food, especially of proteins. Analyses give 16—30 per cent protein, 1—7 per cent starch, 0—15 per cent sugars, 3—10 per cent fat, and 1—9 per cent ashes. Pollen is eaten directly (chewed) by beetles and primitive lepidopters and in a more indirect manner (digestion by diffusion) by other insects. Insects with brood management — bees, bumblebees — use great quantities of pollen for their larvae. Pollen exines are found in the digestive tracts of flower-visiting mammals, e.g. bats, and there is reason to believe that it constitutes a source of protein food for these animals as well.

On the other hand, it is doubtful if pollen can deliver enough energy to sustain a pollen-collecting bumblebee during its collecting flight. Heinrich (1973) assumes that it must forage on the more easily assimilable sugar of (stored) nectar before starting the flight. This throws new light on the nectar foraging of pollen-collecting honeybees who have been assumed to take the nectar only to moisten their pollen loads.

According to Free (1955), bumblebees collect most nectar and little pollen early in the morning. During the day, pollen collecting intensity increases, nectar falls off. This may be seen as a replenishing of the energy store before the real work of the day begins. According to Brian (1957) *Bombus agrorum*, which collects both nectar and pollen, does not carry nectar out of the nest. *Bombus lucorum*, which specializes in pollen collecting, does so, or it interrupts pollen-collecting flights by visits to nectariferous blossoms. Bees returning from pollen-collecting trips need a rest to "refuel" before flying out again.

The alternative is to collect some nectar during the pollen-collecting flights. According to

Raw (1974) *Osmia rufa* queens collected some nectar from other blossoms before collecting pollen from *Quercus*, but those who visited *Ranunculus* did not. However, the *Ranunculus* flowers did not yield enough: also these bees had to forage for nectar elsewhere. This throws light on the simultaneous presentation of nectar and pollen in some blossoms: the nectar may simply serve to provide the pollen-collectors with the necessary energy during their work. This again leads to the question of what were the conditions in the primitive assumed pollen-presenting flowers. However, the energy demands of slow-moving, crawling beetles is relatively low and could presumably be satisfied with pollen alone.

In the developmental metabolism of honeybees, pollen also plays a vitamin rôle inasmuch as it contains substance(s) that stimulate the secretion from the hypopharyngeal gland of nurse bees (Doull 1973). Similar effects may occur also in other, less well-investigated blossom visitors.

Like nectar, pollen is presented at certain periods only, also in insect-pollinated plants. Where nectar is the chief attractant, pollen presentation must be synchronized with nectar presentation. Percival (1955) has shown that also in "dry" pollen flowers anthers dehisce at certain periods only, and has distinguished between different types, from "early morning" to "night" crops. The anthers within a single blossom may dehisce simultaneously (*Rosa pimpinellifolia*) or within a few hours, but they may also expose themselves gradually during a longer period, e.g. more than a week in *Anemone* (*s.l.*) spp.

There has been some doubt as to the general existence of a specific pollen odour, or odours, which at close range might lead the visitor to the source of pollen. However, von Frisch (1923, 1924) demonstrated that honeybees differentiate between the odour of the pollen and of the flower in general, and can be trained on them separately. von Aufsess (1960) has later elaborated on this point. Pollen odour alone is insufficient in plants with protogynous or diclinous flowers, and the odour may also change between a male and a female phase in dichogamous flowers (cf. review in Free 1970: 21).

Where pollen is the attractant, it is frequently produced in great quantities (Table 1), the order of magnitude corresponding to that of anemophiles, e.g. (insect-pollinated) *Rosa* and *Papaver* species. Such pollen flowers are easily recognized by their numerous stamens. However, many stamens are also found in flowers that possess nectar in addition, e.g. *Ranunculus* or *Helleborus*, whereas wind pollinators often have few stamens per flower. In the bat-pollinated blossoms of *Parkia* and *Adansonia* both pollen and nectar are produced in enormous quantities. Without entering upon the question of possible relicts we may state that such plants seems to possess a dual system of primary attractants. Similar cases are known from other plants as well.

Producing great quantities of pollen, anemophiles are potential sources of this attractant, and are utilized (Louveaux 1960), even if the typical characteristics of anemophiles, the long, flexible stamens, the dryness of the grains, etc., make their pollen less suitable for insect collection. Bees are reported to collect pollen from grasses, and even beetles seem rather adept at utilizing pollen of anemophilous plants (Porsch 1956). Some of the pollen grains of anemophiles found in stores in bee-nests or beehives may have come there accidentally, but Maurizio's investigations (1953) have proved that such pollen is also collected systematically. The pollinating effect even of habitual visits to anemophiles remains doubtful. In a plant like European *Quercus* it must be nil.

Pollen attraction is in principle dystropic: the "expected" outcomes of the process is the destruction of the pollen. The reason why this can function as a pollination process is the great quantity of pollen and the inability of the visitor to clean itself of all of it. This a difference in principle between pollen attraction and other pollination syndromes in which treatment of the (smaller) quantities of pollen is either neutral (most cases) or directly aimed at pollination (Galil's ethodynamic pollination).

The power of discrimination with regard to pollen is variable. Oligolectic/monolectic bees apparently discriminate between two types: the one acceptable and the others. In honeybees there are, on the one hand, indications that they are completely unable to recognize the nutritive value of the pollen collected (that of Amentiferae and conifers is said to be nutritionally poor), and may collect material of no value altogether: flour, *Puccinia* spores, flakes of paint or black coal dust (Wahl 1966; Percival and Miliner in Mohr 1962). The possibility that these substances are — partly — mistaken for propolis, not for pollen, should not be left out entirely. On the other hand, there are pollen types which are definitely rejected, and honeybees may make frantic efforts to free themselves, e.g. of *Gossypium* pollen. Cazier and Einsley (1974) describe how nectar-collecting bees and wasps actively avoid contact with the pollen grains of *Kallstroemia grandiflora*, the flowers of which they visit from the underside. This syndrome is remarkable because there are other bees which do not avoid the pollen or even collect it. Cases of pollen toxic to honeybees have been described (cf. Stanley and Linskens 1974: 104). If such toxicity should prove general, one might interpret it as a mechanism for protecting the pollen from being robbed in blossoms with nectar attraction.

In Australia the dominant, primitive Colletidae seem to be connected with the dominant Myrtaceae, but imported European honeybees often ignore these endemic plants and rely on uncommon pollen sources, often anemophiles (Michener 1965; Blake and Roff 1953–6). The great number of pollen grains produced in entomophilous pollen blossoms may result in some accidental wind pollination.

In the same way as there are exclusive pollen blossoms and pollen collectors, there are also exclusive nectar blossoms and visitors which are interested in the nectar of the blossoms only, e.g. higher butterflies. In honeybees specialization of work is carried through to such an extent that one individual will collect pollen or nectar exclusively at a given time. On the other hand, bumblebees may collect both nectar and pollen during the same visit, often going through a complicated and acrobatic sequence of movements, twisting their bodies round the anthers to achieve both objectives.

In the life of the plant, pollen may thus have a double function, both the original, as microspores and this secondary one, as an attractant. These functions are mutually exclusive, and only the great excess of pollen in relation to ovules permits both of them to be carried out. All utilization of pollen except for pollination purposes is in itself pollen theft. The only difference between pollinators and those animals characterized as pollen thieves, is that the latter do not pollinate as they remain outside the flower proper, or are too small or too dexterous to come into contact with the stigma as well. A group of special significance in this respect is the primitive oligotropic (oligolectic) bees, the pollen-collecting activity of which in more highly organized blossoms frequently does not cause pollination. Various features in the flower have been described as devices to stave off thieves, but this is a field that is sorely in need of a modern, critical investigation.

Insects which carry pollen away to feed their young possess various collecting devices. Some bees collect pollen in the crop and regurgitate it again. In most bees pollen is carried externally between stiff hairs which may be found either on the abdomen or on the extremities. The collection techniques vary somewhat, some abdomen collectors also collect with the stiff abdominal hairs that constitute the transport apparatus. More frequently, the hairy animals collect pollen with the whole of their surface and afterwards, usually during the travel to the next blossom, transfer it to the transport apparatus. The extremities possess special cleaning brushes for combing out pollen from the pubescence of the rest of the body. The greatest specialization is met with in genera like *Apis* and *Bombus* in which the outside of the tibia of the third pair of legs is developed into a *corbicula*, a special apparatus for the temporary storage of and transport of pollen. The weight of a pollen load carried in the corbicula of a honeybee is of the magnitude of 5–10 mg (Maurizio 1953), comprising some 50,000–1,000,000 individual grains. These grains are lost for pollination, but even the most highly evolved types of bees are unable to clean all pollen out of their pubescence; they are even contaminated by pollen from plants that flowered before the emergence of the individual insect in question (Fredskild 1955). This pollen must have attached itself to the insect in the hive or nest. It is probably always dead; viability of insect-transported pollen does not seem to last as long as 12 hours (Kraai 1962).

The development of pollination into a more and more precise mechanism influences the way in which pollen is deposited on insects – both on insects that visit blossoms in utilizing pollen, and those that are dependent on other attractants, and on which the deposition of pollen is incidental. In primitive blossoms and primitive insects, deposition is diffuse – no region is preferred but the underside will for reasons of gravity receive more. In others, even highly evolved blossoms, pollen deposition is sternotribic, but with increasing adaptation to higher bees there is also an increasing tendency for pollen deposition to become nototribic. Nototribic pollen deposition is one of the devices which in highly evolved blossoms ensure that pollen is placed on a part of the insect body from which it is not easily groomed away, like zygomorphy forcing the pollinator to occupy a certain position, resupination bringing the anthers into the upper part of the blossom, vibration or even explosion (Loranthaceae, *Cytisus scoparius*). But one also sees that nototriby is again lost if pollination is taken over by more primitive pollinators (in Mentheae, *Nigritella*).

The double function of pollen is in some blossoms reflected in a differentiation (heterandry or heteranthery) within the androecium, between feeding anthers producing pollen for consumption, and fertilization anthers producing "proper" pollen, e.g. in the flower of *Lagerstroemia indica* or, more primitively, in *Verbascum thapsus*. In addition to food and pollinating anthers, a third category, which has been described in some species of *Cassia*, presumably serves the landing function only.

In *Tripogandra grandiflora* and *Cassia* spp. the pollen produced by the feeding anthers is degenerate (R. E. Lex 1961), and in *Commelina coelestis* the "feeding anthers" do not (any more) seem to produce any pollen. Only observations in the natural habitat can decide whether pollinators utilize the milky fluid in these anthers or whether these staminodes have also lost the feeding function, and function only as advertising organs. The difference between the vividly coloured feeding anthers and the dull-coloured pollinating ones is very striking in some of the genera mentioned here. The end result in this line of development is represented by dioeceous plants, the female flowers of which possess anthers that not only do not produce viable pollen, but are empty: another example of attraction by deceit. The attraction of pollen feeders to nectarless protogynous flowers in the female stage – presumably by the sight of still unripe anthers – can also be classified as a mild, temporary deceit.

The tendency of feeding anthers to cluster in one part of the flowers induces zygomorphy also in families with predominantly radiate flowers, e.g. *Exacum*.

In contrast to the ordinary "dry" pollen blossom with its great number of anthers and quantities of easily available pollen, those with feeding anthers often produce comparatively little pollen. Moreover, in many cases they do not offer it freely. The anthers may be pore opening (p. 17), or possess other devices which demand manipulations to squeeze out the pollen. This is often connected with vibrations of the whole insect body, not only the wings. In orchids pollen has lost its attraction function, which would be incompatible with the demand for precision. This may explain the occurrence of pollen surrogates in the attraction syndrome (p. 71).

There is no reason to doubt that pollen attraction may in some cases be secondary, developed from a previous nectar attraction syndrome. However, this would be exceptional (cf. Hiepko 1966). One might perhaps generalize the following evolution:

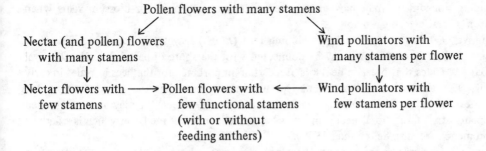

8.2. PRIMARY ATTRACTANTS. II: NECTAR

While pollen is an organ with a long history in the evolution of plants, nectar is apparently something new, and is in the main restricted to angiosperms. The principle of attracting insects by the exudation of a sweet liquid is found in other plants too: fungi, gymnosperms; but these cases have nothing except function in common with the nectar of higher plants.* There is no real homology between the pollination drop of gymnosperms and the nectar of angiosperms, even if both have a function in the transfer of microspores. All these instances of "nectar" production in lower plants have the same ecological function as the floral nectar in the higher ones, viz. that of attracting animals. They can therefore not be brought into the discussion of the origin and function of the extra-nuptial nectar secretion (see below), if we presume that the latter has a different ecological function.

Pollen (microspores) plays a very important rôle in the life-cycle of both the more primitive plants and of angiosperms. Except for some few apomictic taxa, pollen is an indispensable stage in the life-cycle. Even if it is not equally important, the pollination drop in gymnosperms has a similar, obvious place in the life-cycle of those plants. In contrast, nectar as now known in angiosperms seems completely superfluous, a waste of good assimilation material, and its presence does not seem to make sense unless seen in the context of pollination. Also, in contrast to pollen, the production of nectar is independent of the fructification organs of the blossom, even if it is habitually localized at or near them.

However, nectar production hardly arose in connection with pollination; there is ample evidence that some nectar-like substance was produced by plants before pollination came

---

*That they may form the basis of honey (Maurizio 1942) as do also frequently the liquid excreta of aphids (honey-dew), may be mentioned as a curiosity and as an illustration of the importance of availability of the sugar and the non-importance of the taxonomy of the producer. Ordetz (1952) describes an *Ephedra* honey, the basis of which should be the pollination drop.

into existence. As a matter of fact, even today very much nectar, thin sugar solution, is secreted extra-nuptially, i.e. independently of the floral region proper.

If the present main function of nectar in the blossom is considered original and primary, the existence of extra-nuptial nectar production is logically very difficult to accept, as its effect should be to distract visitors from the blossoms. To "explain" its occurrence, rather complicated hypotheses have been put forward, none of which have been very convincing, especially if one considers the very refined structures sometimes observed, e.g. in *Pithecellobium* (Elias 1972).

It is logically more consistent to presume that nectar production of some sort existed also in the ancestors of angiosperms independently of and previous to pollination. Downes (1974) has maintained that the existence and utilization of exposed sugar (solutions), including honey-dew, may have preadapted older insect groups for blossom visits when blossom nectar became available.

Active nectaries occur also in developing fern (*Pteris*) leaves (F. Darwin 1877), where their existence cannot in any way be connected with the spore dispersal function. Percival (1965: 81) states that bracken nectar is of a certain importance for the bees in Wales in early spring. The existence of such nectar production suggests that the many extra-nuptial nectaries found in higher plants, on leaves and bracts especially, may have originated independently of the floral nectar production, indeed, that it is the latter which is a derived phenomenon (cf. also Schremmer 1969).

Loss of assimilates by young, developing organs seems to be a rather common phenomenon. Flowering shoots and young fruits of Cactaceae can be seen covered with bead-like sugar drops in all areoles. Diffuse exudation of nectar from young leaves is well known by orchid growers. However, in nature sugar secreted in this way is rarely seen, as it is immediately removed by insects if not washed away by rain.

Various hypotheses have been put forward to explain the physiological significance of the voluntary loss of assimilates represented by extra-floral nectar production. They are not much more convincing than the ecological hypotheses referred to above, but at least it gives a consistent pattern to consider nectaries as something that originally developed independently of the flower, and floral nectar production as the utilization and further development of an already existing organization.

The easily available sugar of extra-nuptial nectaries is much sought after by sugar-consuming animals. Especially in the Tropics extra-nuptial nectar (Zimmermann 1932) forms an important part of the diet of many insects, e.g. primitive hymenopters. Many typical blossom visitors also visit extra-nuptial nectaries (Knoll 1930a), and negative observations (e.g. Springensgut 1935) may be due to the fact that plants have been studied outside their natural region, as well as to an erroneous primary conception of the function of these nectaries. There is a whole, independent complex of utilization of extra-nuptial nectar, which has, on the whole, been considered to be without primary ecological significance for the plant, but to be more like an innocuous parasitic behaviour. Even apart from the question of ant-guards (cf. Bentley 1976), we are not convinced that this is right.

Ford and Forde (1976) have described a pollination syndrome based upon extra-nuptial nectaries. Nectar-eating birds visit active nectaries at the base of phyllodes occurring near the inflorescences of *Acacia pycnantha* (and only there). During these visits the birds brush against anthers and stigmas and presumably carry out pollination. Nectar production is synchronized with anthesis. The larger a pollinator is the greater can be the distance between the source of nectar, on one hand, and the position of anthers and stigma, on the other, for pollination still to be possible. The utilization of extra-nuptial nectar in a pollination syndrome consequently presupposes that the pollinator is relatively large, larger at any rate than pollinators utilizing floral nectar. This, again, means that these large pollinators have a high energy consumption, which in turn presupposes a high productivity of any extra-nuptial nectaries involved in pollination syndromes.

However, even if nectar and nectaries are thus possibly phylogenetically older than pollination, and even if floral nectar production and its attendant ecological function only represent the utilization of an already existing mechanism, there is no doubt that nectar production acquired a new aspect as the need for a floral attractant of this type arose, and the frequency, variability, and productivity of floral nectaries far outweigh those of extra-floral ones. The occurrence of extra-floral, but nuptial nectaries (i.e. nectaries occurring outside the flowers, but playing a rôle in the visit of animals to the blossoms) is also more easily understood on this assumption. In the cyathia of Euphorbiaceae they are especially well developed, and form the only attractant produced — as such — by those blossoms. In *Poinsettia* they are so big and productive that they are even utilized by birds.

Some aberrant cases can be mentioned. In Malpighiaceae there are often semi-extrafloral nectaries on the outside of the calyx, to be reached between the long claws of the petals. Similar nectaries also occur on foliage leaves. In some Pedaliaceae, tropical Papilionaceae, and Tiliaceae there are nectaries consisting of entire but reduced flowers without any other function (van der Pijl 1951, 1954). Like all extra-floral nectaries, these are avidly sought by ants.

Nectaries (Bonnier 1879; Dvorak 1968) are usually formed by densely packed cells. The ways in which nectar is secernated vary. Some nectaries have a smooth surface with a thin-walled epidermis through the walls of which nectar exudes diffusely. In others, the walls are more or less papilla-like, but the mechanism is otherwise similar. In tissues with thicker epidermal cell walls there may be an attenuated part through which nectar exudes, or the cellulose walls may even degenerate or break up. Similarly, secretion may take place through the cuticula, or the cuticula may separate from the cellulose walls, or it may break up. Such differences may be found between closely related species, and may be of taxonomic importance (Daumann 1935). In other plants, nectar exudes through stomatalike, more or less immobilized apertures. Some extra-nuptial nectaries are completely aberrant, like those of *Opuntia monacantha*, formed by the tip of a spine. The term nectary therefore does not cover a morphological concept, but an ecological one, a certain function.

Likewise, the physiological origin of the nectar is not uniform. Ordinarily, nectar seems to derive from the phloem, and to be a modification of the solution passing through sieve-tubes (Agthe 1951; Lüttge 1960; Frey-Wyssling and Häusermann 1960). Similarly, nectaries are frequently located at or near the end of phloem strands.

Some nectary-like organs in pollen blossoms (*Vitis*) are in reality odour-producing. *Vitis* is functionally dioecious, the pollen of the so-called hermaphroditic flowers being sterile and functioning as attractant only (Brantjes, pers. comm.).

In other, aberrant instances, nectar is apparently produced by degenerative decomposition of certain cells or cell complexes. In Araceae, where the stigmatic papillae secrete a nectar-like substance (Knoll 1922; Daumann 1930b), the cell contents only degenerate. In the extra-nuptial nectaries of *Opuntia monacantha* already referred to (Daumann 1930a), the whole cell complex of the tip of the spine degenerates.

The question may be raised if substances like these, or the stigmatic exudates described by Baker *et al* (1973) should be called nectar even if they, wholly or partly, serve the same purpose.

A corollary of the hypothesis that nectaries existed before and independently of any pollination syndrome is that the occurrence of more or less rudimentary nectaries in pollen blossoms or anemophilous blossoms does not immediately prove that such blossoms are derived from blossoms with nectar as an attractant, as has been taken for granted by earlier authors.

In most cases nectar remains on the nectary until utilized, but many examples of secondary nectar receptacles are known, generally in the shape of long spurs into which the nectar runs (*Linaria*, *Viola*). In some instances, what have been described as nectaries may be such secondary repositories. Specialized pollinators seem to prefer narrow nectaries, probably because there is less competition.

Nectaries have a very characteristic surface, generally dark yellowish green and shining. In some blossoms there are organs resembling nectaries in outward appearance, but not producing any nectar. These pseudo-nectaries, which are well exposed (real nectaries are frequently concealed) have been controversial subjects since they were first described. It has later been possible to interpret the ones in *Ophrys* blossoms as parts of a sexual syndrome, but even after very careful experiments one is still left with an unsatisfactory feeling that those remarkable pseudo-nectaries of *Parnassia* have not been finally dealt with; some of the discrepancies are possibly due to experiments being carried out with conditioned insects.* However, there seems to be experimental evidence (Daumann 1960) for the assumption that pseudo-nectaries attract at least dipters, and that they are an important part of the attraction apparatus of some blossoms (Kugler 1955a).

Nectar secretion is to a large extent dependent on the physiological state of the plant (Huber 1956). But even in healthy well-nourished plants nectar production shows a more or less pronounced autonomous rhythm, corresponding to the periodicity of the pollination process. Thus, many night-pollinated blossoms are dry during the day; in other plants nectar may be amply present in the morning, but there is no further production during the day, and the blossom is dry in the afternoon. In others nectar production starts in the afternoon. Nectar is generally produced more freely in dry (but not too dry) warm weather.

In some plants, nectar production goes on after anthesis (Sernander 1906; Daumann 1932), sometimes possibly as part of a seed-dispersal syndrome. In *Myrmecodia* post-floral nectar forms part of the food for the ants inhabiting the plant. As the primitive extra-nuptial nectaries dealt with above seem to be first and foremost connected with the young and developing organs, post-floral nectar secretion, being connected with a senescent organ, would seem to represent a completely different phenomenon.

Generally, the sugar concentration of nectars varies between 25 and 75 per cent (Percival 1961; Gottsberger *et al.* (1973). The relative proportions of glucose, fructose, and saccharose (other sugars are subordinate) vary, but these qualitative variations do not seem to be correlated with different classes of pollinators. On the other hand, concentrations and quantities of nectar are important in the energy budget of pollinators. A great deal has been written about this. However, concentration is not a very meaningful parameter in pollination ecology. The concentration of fresh nectar in a recently opened blossom may tell something about the physiological conditions of nectar production, but this concentration in many cases changes fundamentally under meteorological influence. In very dry weather nectar may crystallize on the nectary, if the latter is exposed. On the other hand, nectar may also be washed away by rain unless protected. The main protective measures, the same as for

*But what is an unconditioned insect in nature?

protecting pollen, are: flowers closing in bad weather, closed or drooping flowers, narrow tubes into which water does not penetrate, hair coating keeping water away, etc.

Low sugar concentrations mean less food (energy) in the nectar, but, on the other hand, pollinating insects need water, too. The increased viscosity concomitant with high nectar concentrations may be detrimental in some pollinator relationships. Baker's table of sugar concentrations in virgin nectar from Jamaican plants (1975, data from Percival) show remarkably constant mean values in various pollination groups: 20–25 per cent, and very great variations inside the groups (5–50%).

The chemoreceptors of pollinating insects are sensitive to sugar, and generally the threshold value is lower than 10 per cent. According to Free (1965) a more concentrated nectar has selective value if there are too few pollinators about. In the further development of nectar into honey, sugar concentration rises, saccharose is inverted, and small quantities of other substances, e.g. formic acid, are added. Pollen grains occur in honey (Zander and other textbooks), some of them being picked up inadvertently during nectar gathering, others coming, as contaminations, from the pollen storage.

Nectar is not a pure carbohydrate solution. Baker and Baker (1973, 1975) have demonstrated variable quantities of amino acids and lipids in most nectars investigated (whereas the occurrence of protein was negligible), apart from what may have been brought into the nectar by pollen grains falling into it. Not all groups of pollinators are supposed to use this as a major source of nitrogen; they may obtain that from other sources available to them (bees, flies, birds). Apparently butterflies may be the group to which the nectar nitrogen is potentially most important. The nitrogen requirements of butterflies is not particularly well known, but Baker and Baker (1973) suggest that, calculated on a nitrogen basis, 12–25 blossom visits would in some cases suffice to cover the daily requirements. If sapromyophylous blossoms offer nectar, this is remarkably rich in amino acids. Whether the quantity and composition of nectar amino acids are of importance as differentiating characters in pollination syndromes, seems uncertain.

Some curious instances have been described of nectar (and pollen) being poisonous to pollinators. Some of these cases are due to misinterpretation, like the case of large-scale killing of visitors by insects of prey lurking in the flowers or even near extra-nuptial nectaries (Knoll 1930a), or of nectar being poisonous to "illegitimate" pollinators, e.g. to bees visiting bird blossoms. Other cases are less easy to understand (Maurizio 1950a). The occurrence of certain rare sugars may be the cause of some of these reactions (Geisler and Steche 1962). Two of the suspects are mannose and lactose, enzymes for the metabolism of which are apparently absent (redundant?) in higher hymenopters (Sols et al. 1960). Demianowicz (1964: 278) describes a case in which honeybees under experimental conditions refused to collect nectar from Nicotiana rustica unless they were at the same time fed with sugar syrup. Once the syrup had been exhausted the bees stopped visiting Nicotiana as well, in spite of ample nectar and favourable external conditions. The question if a honey is poisonous to man is completely redundant for pollinators. Toxins seem to be responsible for that effect (Carey et al. 1959; Maurizio 1975; cf. also Pryce-Jones 1943). According to Leach (1972) the toxin of Rhododendron nectar is acetylandromedol.

The attraction of nectar is so great and it is so eagerly sought that occasional nectarless blossoms are visited if they occur together with nectariferous ones. In this way the pollination of nectarless strains by nectarseeking visitors can be secured by interplanting between nectariferous ones (Bohn and Davis 1964).

Of all attractants, nectar has the greatest general appeal to all groups of animals –

vertebrate and invertebrate — and any available source of nectar is likely to be utilized by any animal that can get at it. So there is always the danger of nectar theft, especially by insects so small that they can get at the nectar without touching anthers or stigma, depriving the flower of its attractant without doing any useful work. Ants, especially, are frequently thought of as nectar thieves, and their smooth hard bodies are certainly not too well suited for the transfer of pollen. Besides bigger insects operating the flower mechanism in a precise manner, there may be smaller visitors, operating more haphazardly, but nevertheless carrying out mess and soil pollination due to their great number and their prolonged activity in the flower (cf. the activities of *Taeniothrips ericae* in *Calluna*).

A special form of nectar theft is performed by insects, the probosces of which are so thin that they reach the nectar without coming into contact with anthers and stigma, and are so long that the rest of the animal stays outside the blossom, e.g. when a typical bee flower is visited by a butterfly (Schremmer 1953).

If an insect is sufficiently small, it will keep clear of anthers and stigma and its activities will consequently assume the character of theft in any kind of blossom with nectar attraction. On the other hand, larger visitors will cause some mess and soil pollination in primitive, open blossoms anyway because they can hardly help making contact with anthers and stigma; theft, therefore, will rarely occur in such blossoms. The irony of the situation is that theft is conceptually possible only in complicated blossoms, the structure of which is interpreted as countermeasures against it. The visits establish themselves as theft by circumventing just these structures.

The distinction between simple theft and housebreaking exists in pollination ecology, too; thieves that cannot creep into the flower and steal nectar that way, may bite or prick a hole through the perianth and get at it from the outside. This is what some bumblebees with powerful mandibles do, leaving their very characteristic oblong holes in the corolla tube. Once an opening has been made that way, other insects, which would otherwise have used the ordinary entrance, take the same route (Schremmer 1955), even lepidopters and birds (Porsch 1924). The latter may themselves use their beaks for making holes. This throws the pollination system out of balance, and such blossoms may be assumed a total loss. That this is not always so has been shown by Meidell (1945) for *Melampyrum*: insects obtain nectar through the holes bitten in the corolla tube, but to get at the pollen the same insects must still perform in the "proper" manner — with pollination as a result. A similar case has been described by Macior (1966): queens of *Bombus affinis* perforate *Aquilegia* spurs and steal nectar through the perforations, but they collect pollen the "correct" way. Whereas queens do both, workers carry out only one or other of these activities. Again, Hurd and Linsley (1963) describe how a solitary bee (queen) bites a hole in the buds of *Proboscidea arenaria* and enters the unopened flower to collect pollen. Nectar is collected the "proper" way from flowers in full anthesis. Pollination ensues. Males collect nectar only, but there may also be a rendezvous attraction (p. 75).

One reviewer pointed out that nectar thieving is not necessarily deleterious (which we have not indicated): the smaller quantity of nectar left in the blossom (of *Trifolium pratense*) may force the legitimate pollinator to work harder to get its share, and consequently pollination is better ensured. This may be so, but if a blossom has been punctured, subsequent visitors will tend to enter by that way and no pollination takes place. Also, this observation was made in a cultivation community where the ecological equilibria are already disturbed. We are not convinced that the same effects will appear in an undisturbed biocoenosis.

Some insects are habitual nectar thieves, and have apparently lost the instinct for a "proper" blossom visit, e.g. the notorious *Bombus mastrucatus*, which is known to bite a hole even in corollas into which it might easily have crept and reached the nectar the ordinary way. In the tropics *Xylocopa* species are habitual nectar thieves. cf. also the nectar stealing in *Kallstroemia grandiflora*.

Once discovered in a blossom, nectar is in itself a powerful attractant, and may to some extent replace other, visual attractants. Blossoms yielding much nectar frequently have less conspicuous corollas than their relatives with less nectar (cf. the large, wide-open flowers of *Rubi fruticosi* and the small, half-concealed, brush-like ones of *R. idaeus*). Such comparisons are only meaningful if they are made between taxa occurring together and competing for the same pollinator(s).

8.3. PRIMARY ATTRACTANTS. III: OIL

Vogel in 1969 (cf. 1974) announced the discovery of a previously unknown or misunderstood primary attractant, viz. fat oil. In various species of Scrophulariaceae, Iridaceae, Orchidaceae, Malpighiaceae, and Krameriaceae there are oil-secreting organs, elaiophors, which may either consist of trichomes or of special epithels. The elaiophors are formed by different parts of the blossoms in the various taxa. Some genera, like *Calceolaria*, seem to specialize in this type of attractant (free fatty acid in *Krameria*, Simpson 1977).

The oil is collected by (solitary) bees of the family Anthophoridae (subfamily Anthophorinae). The females collect oil on their extremities by means of a complicated system of absorbent brushes and sharp edges for squeezing the oil out of the oil-pads again. The system varies between genera.

The oil is brought back to the nests and mixed with pollen. A lump of this mixture (earlier mistaken for a nectar—pollen mixture) is placed in each brood-cell and an egg placed on top of it. Apparently the females do not take oil themselves, but some males may feed on oil (Vogel 1974: 513). Both sexes mainly rely on nectar for their own food. The plants from which nectar and pollen are collected are different from those delivering the oil.

As pollen contains appreciable amounts of fat, bee larvae must possess fat-digesting enzymes. The transition to this very lipid-rich diet therefore does not imply any major change in their metabolic system. Nor is it remarkable that insects utilize the energy-rich lipids instead of the poorer carbohydrate solutions, especially for larvae, to whom the more difficult and slower assimilation of fats does not play a decisive role. While pollen had a definite function in the plant before it became an attractant, and there are also indications that nectar existed earlier, the development of fat as an attractant is extremely difficult to understand, fat being a scarce commodity in the plant world. A very close co-evolution between plant and animal must have taken place. This syndrome has hardly anything to do with the diffuse film of oil providing slipperiness elsewhere or with terpenoids, even if the latter are also connected with a pollination syndrome (p. 70). We also wonder if the oil on the outside of pollen grains comes into this picture, even if it sometimes occurs in great quantities and has been supposed to have a nutritive value.

So far, oil-collecting bees have mainly been observed in South America; in Malpighiaceae the Old World representatives have nectar, whereas the New World ones have elaiophors. Some older observations on aberrant blossom visits of various insects can perhaps be reinterpreted as misunderstood observations of oil collecting. It remains to be seen how widespread this syndrome is outside South America (cf. Vogel 1976).

8.4. PRIMARY ATTRACTANTS. IV: OTHER SUBSTANCES

Animals will collect any substance they can use wherever they find it. Thus, other substances than nectar, if present in blossoms, may attract pollinators. Some such substances are unspecific and may be collected in various places, e.g. various types of nest material of which wax is said to be collected from the labellum of the orchid genus *Ornithidium* (Porsch 1905). Many of the observations belonging here are isolated and may be misunderstood, but there are at least two syndromes that deserve some discussion.

Perfume. Whereas most odours act as signals and belong to the secondary attractants, Vogel (1966) has described a completely different function which might be called a perfume syndrome, to distinguish it from the regular function of odours: euglossid male bees collect odoriferous substances in special organs on their hind legs. The substances, in the form of oil droplets, are swept up by and into the pubescence of the front legs and by means of a special apparatus on the middle legs transferred to the collector organ – the enormously enlarged tibia of hind legs, the interior of which is filled by a cotton-plug-like structure (branched hairs) and which communicates with the exterior by means of special canals. The odoriferous substances are usually scooped up from various flowers (polytropic), and the males seem to become more or less drugged during the process. The transference from front to hind legs can only take place in the air, so the insect performs short flights in between, often returning to the same flower several times. Such behaviour was first observed (by Dodson and Frymire, 1961a) in *Catasetum*, but has later been found also in a number of other orchid genera, Araceae (*Spathiphyllum, Anthurium*), and in *Gloxinia speciosa*.

After having filled its perfume containers, the male insect takes up an isolated position and apparently again emits the odour, most probably as part of a sexual syndrome, but the details are not known.

The total impact of this observation has not yet been assessed. Previously there were observations of male bees visiting various orchids, scratching calli in the flowers, and becoming more or less "drugged" during the process. This "narcotic" effect has so far been interpreted in relation to the male itself, and it was presumed that it derived some kind of satisfaction from it. After Vogel's observation of perfume collecting it will be necessary to investigate in more detail all these exclusive male visits to see if there are two different principles involved – attraction of females and self-satisfaction (cf. p. 76).

This syndrome is remarkable for mobilizing a group of prospective pollinators which are usually not activated, viz. hymenopter males. Their feeding instincts are much lower than that of the females (they would rank with butterflies), but instead other instincts are utilized. It is part of this syndrome that just in these taxa the males live very long, half a year or so, which is very rare in hymenoptera. Williams and Dodson (1972) have put the fragrance dependence of male bees in connection with the very extended flights performed by some of them, the so-called trap-lining. If the objective of such flights is to obtain other substances than those yielding energy food, their extension is more easily understandable.

Food bodies and tissues. In the brood-place syndrome food-bodies serve the nutrition of the pollinators' brood. More rarely, such foods serve the attraction of the pollinator itself even if that function has often been postulated. Many such data are vague and apparently offered because no other explanation could be found, e.g. the case of the enigmatic *Orchis* attractant (cf. Daumann 1971).

Many other tissues described as food-bodies remain equally enigmatic. The function of some of them seems satisfactorily substantiated, e.g. those on the carpels of *Victoria amazonica* (Knoch, 1899) or on the staminodes of *Calycanthus occidentalis* (Grant 1950c).

Similar feeding organs are known from other primitive families, but scattered observations of food-bodies and food-tissues have been made also in more highly evolved taxa, even orchids. Most food-bodies at least are apparently "meant for" chewing and can therefore most easily be utilized by beetles, and by primitive representatives of other classes. If Delevoryas' interpretation of the *Cycadeoidea* blossom is correct (1968), one might envisage a food-body pollination even in this group of plants, cf. Crepet's observation of traces of boring organisms in the flowers. In modern blossoms they have an exact counterpart in *Prosopanche* (Bruch 1932). The autogamy presumed by Crepet (1972) is not necessary: dichogamy would ensure xenogamy, like in *Ficus*. If the pollination syndrome has been like that, it represents a remarkable case of blossom—pollinator co-evolution at a very early phylogenetic stage; although there is no reason to believe that *Cycadeoidea* was an ancestor of modern angiosperms, the syndrome throws light also on the origin of angiosperm pollination and flowers.

Development of food-tissues is dependent on the general status of the plant, more or less undernourished greenhouse specimens failing to develop tissues present in healthy plants in the wild. This accounts for some conflicting information in literature. Also, at any rate in orchids, many "calli" are not eaten, although they do contain much "food". However, these substances seem to be metabolic material for odour production (van der Pijl and Dodson 1966: 24).

In the trap flowers of some Araceae, the insects are said to eat food-tissues during their confinement. The food thus provided may serve to keep the insects alive and happy, but it hardly qualifies as an attractant — except that it might be a counter-deterrent. A special case is that of *Amorphophallus variabilis* (van der Pijl 1937b), which does not seem to have a physical trap, but where the pollinators (beetles) stay in the blossom, tempted by the occurrence in the bottom of the spathe of an edible substance of great attraction. They stay there for days until the substance starts decaying and loses it attraction. The anthers dehisce some time before the animals leave, so that the beetles are thoroughly powdered before leaving.

Some other, more or less aberrant, sources of food presented by blossoms may be mentioned here. The liquid exuded by the stigma hairs in *Arum nigrum* is devoured by insects trapped in the blossom (Knoll, 1926). The composition of this liquid is unknown.

From *Eria* and some related orchids Beck von Managetta (1914) described a pollen-like cell powder occurring on the labellum. He concluded that this is a pollen imitation and functions as a food for pollen-eating visitors; orchid pollen is usually not utilized. Whereas this has not been tested by observations in nature Dodson and Frymire (1961b) have confirmed a similar observation in *Maxillaria* by examination of the intestines of euglossid bee females.

Food-bodies have been observed also in vertebrate-pollinated blossoms.

8.5. PRIMARY ATTRACTANTS. V: PROTECTION AND BROOD-PLACE

The concept of "Obdachblume" ("shelter blossom") is frequently met with in the literature. For the time being this explanation is too obviously resorted to when no other can be found. There is no doubt that bumblebees frequently stay out at night, and seek shelter in blossoms (males do this regularly), but the effectivity of this as a pollination process seems doubtful even if certain preference may be observed (*Peponapis* in Cucurbitaceae). Vogel has informed us that he found insects hiding in *Serapias* flowers, and at the same time van der Pijl has never been able to find insects in flowers of this genus during day-time. Horovitz (1976) maintains that in *Anemona coronaria* "the important pollen vectors are those insects which use the flowers for shelter and move indiscriminately from male-stage flowers ... to female-stage flowers ... a minority of bees such as male ... *Eucera* sp. ... and range otherwise from pollen eating beetles to predaceous anthocorid bugs." For a discussion of temperature conditions inside blossoms and their possible significance, cf. p. 86.

A substantiated case of shelter blossom from temperate regions is that of *Calluna vulgaris*. *Calluna* flowers certainly show adaptation to visits by larger insects, and are avidly visited by bees, but these visits may be redundant because of the activity of *Taeniothrips ericae* (Hagerup 1950a; Haslerud 1974), which lives most of its life inside the blossoms. Animals that do not actively tend their brood, place the eggs where the food is, also if that happens to be in a blossom. While *Calluna* represents a combination of shelter and brood-place, *Trollius europaeus* exemplifies chiefly the latter. Here pollination is still of the mess and soil type. A more refined procedure, a continuation along the same developmental route, is represented by *Yucca* and the *Yucca* moth as unravelled by Riley and later observers (Baker 1961b; Powell and Mackie 1966). It is sometimes maintained that the adaptations met with in pollination ecology are not really mutual; plants adapt to (static) animals as they adapt to wind or water as pollinating agents. There is no better refutation of this than *Tegiticula* ("*Pronuba*"), the ethology and morphology of which is completely nonsensical except as part of the common pollination syndrome ("interspecific symbiosis", Alpatov).

Other examples of perfect mutual adaptation between blossom and brood-place-seeking insects are the famous cases of *Ficus* spp. distinguished by very complicated sequences of generations both in plant and pollinator. Another much simpler relation has been described by Bruch (1932) from *Prosopanche burmeisteri* (Hydnoraceae). The very thick, fleshy blossoms of this parasitic angiosperm are evidently the normal breeding-ground for the larvae of the pollinating beetle, which restricts their feeding to the external parts of the blossom with no damage to its further development. The interesting part of this story is the close similarity between the *Prosopanche* and *Cycadoidea* blossoms and the probability that the latter was pollinated in a similar way, as stated independently by Crepet (1972).

In no other part of pollination ecology is the balance as precarious as in the brood-place syndrome, the balance between mutual benefit and destructive behaviour on the part of pollinator or blossom. The former may eat the whole seed-set and the latter may induce "false oviposition".

The role of certain moths (*Dianthoecia*, *Hadena*) as parasitizing pollinators of certain Caryophyllaceae, and of *Lycaena* spp. in other families, has been known for a long time (Kerner 1898). The effect is mainly destructive — the grubs eat the developing ovules. However, the destruction is not always total: some capsules may be left (no oviposition, eggs fail to develop, etc.) and develop seeds. Also in this case pollination is necessary for the insect; without it there would be no food for the developing grubs inside the gynoecium. According to Stirton (1976) another member of the Hadeniae is a seed parasite and at the same time the only observed pollinator of some South African Liliaceae.

In the interrelationship between *Hadena bicruris* and *Melandrium album* described by Brantjes (1976), the following factors are pertinent:

(1) the female must feed before each oviposition;

(2) both females and males feed on both staminate and pistillate flowers;

(3) oviposition (in nature) takes place on pistillate flowers only.

If each blossom visit gives a volume of nectar (or other food) v and a moth needs V before ovipositing, it will have to visit V/v blossoms. Provided the male does the same, there will be twice that number of visits of which, however, only one half again goes to a female flower. If each oviposition period yields e eggs, and each larva eats c pistils out of C per flower, we get

$$\frac{V}{v} \gtreqless e\frac{C}{c}.$$

If there is equality, the moth's requirements are covered, but there is no gain for the plant. If the left-hand side is smaller, there will not be enough food, and some larvae will starve, and only if it is greater will there be a net pollination gain for the plant. In the case of *Yucca*, v, V, e, and C are all equal to one, and the "pollination surplus" derives from the fact that c is smaller than unity (the larva does not eat the whole pistil). In *Trollius* $(C/c) > 1$ and again there is a gain. In *Melandrium album* the outcome for the plant seems very uncertain, and probably other pollinators are responsible for the necessary seed-set. In that case this would bring up the problem of the pollination strategy of the community as a totality as contrasted to the individual strategies. This problem, which has been very much to the fore in recent discussions, will be dealt with elsewhere (cf. p. 156). In this context the *Hadena–Melandrium* relationship might be considered a parasitic behaviour, and as such it comes under the usual constraint that it cannot exist unsupported: there must exist alternatives, otherwise the process would run down. Whether such a relationship can develop towards a co-adaptation, such as *Yucca*, is conjectural and will not be discussed.

It is difficult not to see pollination syndromes like these as stages in an ecological evolution, which may have taken place in various insect groups and has taken place in many plant groups, leading from a more or less accidental, more or less dystrophic relationship to the perfect cyclic symbiosis where blossom and pollinator follow each other and are mutually interdependent.

Gottsberger (1970) found pollinating (curculionid) beetles breeding in Annonaceae blossoms following various patterns, from the larvae pupating inside the receptacle to others living on in the deciduous, very thick tepals. However, as blossoms are not the ordinary substratum for these larvae, the relationship is not a proper symbiosis although it serves the plants' purposes excellently.

Silberbauer-Gottsberger (1973) and Essig (1971, 1973) describe pollination through feeding and breeding of dystrophic beetles in the still closed spatha of various palm genera. The same was reported by Heusser (1912) for the oil-palm.

This type of breeding-ground attraction is independent of the organization of the angiosperm flower, as witnessed by the identical phenomenon in cycad cones (Wester 1910). Species observed by Gottsberger as breeding in *Zamia* cones were found as imagines inside Annonaceae flowers. The specificity of attraction characteristic of ordinary pollination syndromes is apparently not developed.

A somewhat more distant relationship exists between pollinating microdipters (*Forcipomya* sp.) and the cocoa plant. The insect breeds in decaying pods, and it has been suggested that too ruthless cleaning in the plantations may impair seed production in the next generation (Dessart 1961). However, *Theobroma* is not native in Africa, from where this was described, and relationships may be different in America. On the other hand,

Winder and Silva (1972a, b) report similar conditions from Bahia with litter and bromeliads as alternatives.

A special type of attraction, related to the shelter/brood-place syndrome is the use of blossoms by predators lurking in them. Horowitz *et al.* (1975: 38) describe the potential pollination of *Anemone coronaria* in Israel by a predaceous heteropter. This syndrome is found also in the Arctic (Kevan 1972).

In deceit attraction the process may (in rare cases) go as far as to induce oviposition in the blossom. Usually the eggs have no chance of developing, but again in some blossoms the eggs do develop into a new generation of pollinators. In this way the supply of pollinators is secured. This occurs in *Artocarpus heterophyllus* inside an otherwise anemophilous genus (van der Pijl 1953), a kind of prelude to *Ficus*. Other examples are *Alocasia pubera* (van der Pijl 1933), *Rafflesia*, and *Philodendron*, which is pollinated by beetles (Hubbard 1895). Most of these syndromes can be considered developments from a simple sapromyophilous one. The special case of the mycetophilous gnats is described p. 105.

In this connection we refer again to Rattray's (1913) observation of beetles eating pollen in the male cones of *Encephalartos* and depositing eggs in the female ones. Apparently, the two types of cones act on different instincts in the visitor; a unique case, so far as is known. However, one may ask oneself what would attract egg-laying insects to the male flowers of other dioecious plants. To man, the male cones of *Encephalartos* are odoriferous and the female cones inodorous.

8.6. PRIMARY ATTRACTANTS. VI: SEXUAL ATTRACTION

The discovery by Pouyanne (first published by Correvon and Pouyanne in 1916) that *Ophrys speculum* attracts a male hymenopter by sexual attraction,* imitating the female, remained for many years unnoticed, although it was in itself sensational and in addition solved the age-old problem of the attractant of the *Ophrys* blossom. However, the challenge has been taken up both by botanists and zoologists (Wolff, 1950, 1951; Kullenberg 1973a and earlier; and various papers in Kullenberg and Stenhagen 1973) who have been able to corroborate the original observations and deductions. The *Ophrys* flower acts upon the sensory organs of sexually unsatisfied male hymenopters in such a way as to cause the same instinctive reactions and actions as those leading to a copulation. This latter act takes place in such a position and in such a way that the insect touches the rostellum with its head or abdomen, and carries off the pollinia. Its sexual urge evidently not satisfied, the insect continues to the next flower and repeats the performance. One generally talks about the blossom imitating the female insect, and this imitation having the psychological effect described. Strictly, this is not true, as many of the acts take place before the females emerge, the male can therefore not have any mental picture of the females to be imitated. On the other hand, the outward similarity between the blossom and the insect is remarkable. Size and texture are easily explained as necessary for the pseudocopulation process, and the "furriness" of the labellum as well — acting as a tactile stimulus. But what about the imitation of eyes, (cf. below) and of antennae in *O. insectifera*? Do they provide any non-visual stimulus? And even more so the blue spot on the labellum of many of these

---

*The term "pseudosexuality" is redundant. There can be no doubt that the chain of instinctive reactions started in the insect is based upon true sexuality. "Pseudo-copulation" makes sense.

blossoms, imitating, to a human eye, the reflection in the superposed wings of a hymenopter at rest? The ultraviolet reflection should not be forgotten!

The "pseudonectaries" of *O. insectifera* constitute an excellent example of the syndrome effect; in themselves they are inexplicable, as testified by the wealth of literature trying to explain their function. In the syndrome of insect similarity, the "pseudonectaries" come into their own without difficulty, as the eyes of the pseudofemale. In *Calochilus* and other genera they are even better developed. (Fordham 1946).

While there is a striking similarity in form and structure between the flower and the insect female to be imitated, there is apparent identity with regard to odour. The same substances have been identified in emanations from both blossom and insect. That such substances are also active is demonstrated by the ability of insects to find hidden flowers. Stoutamire (1974) relates how male wasps followed and entered his car when he had flowers of this type lying on the floor behind the seat.

So far, the sexual attraction syndrome seems to be restricted to the orchid family, where it has now been found in several genera in addition to *Ophrys* (cf. van der Pijl and Dodson 1966; 135–140). The "victims" include wasps, bees, and flies, and even a sexually excited beetle has been reported.

Another case of sexually conditioned pollination has recently been found by Dodson and Frymire (1961b) in South America, where there are bees (*Centris* spp.) the males of which maintain a territory from which they chase other bees. When the flowers of *Oncidium* spp. vibrate in the wind, they are mistaken for bees, chased (swiftly stricken), and pollinated during the attack.

A counterpart of the "pseudoaggression" of *Centris* is found in some other orchids (*Brassia* and *Calochilus* spp.). Female scolicid wasps (*Campsomeris*) mistake the labellum of these flowers for insects preyed upon, and sting into it, causing pollination (van der Pijl and Dodson 1966: 38,142; and later observations by Dodson). One might call this a pseudoparasitic attraction, belonging, of course, to the great group of deceptive devices.

It is important in the evaluation of these syndromes that sexual attraction activates for pollination a group of animals, viz. males, which because of their weaker feeding instincts are less likely to be attracted by regular attractants.

Within the general concept of sexual attraction one may tentatively establish a class of *rendezvous attraction,* referring to the simultaneous presence, in a blossom, of both sexes of a pollinator, with ensuing mating. Insects are well known for chosing specific places for mating, and the frequent use of blossoms for this purpose cannot, in most cases, be construed as part of a pollination syndrome. Rendezvous pollination can be established if (a) the pollination of the blossom by these insects is dependent on the presence of the other sex there not only due to the accidental meeting because of the existence of some other attractant, and (b) if the copulation ensures or improves the effectivity of the pollination. Some cases have been described which may belong here. Female bees are oligolectically attracted to *Proboscidea* blossoms, and also visit them for nectar. Males are polytropic for their nectar feeding, but await the females in the *Proboscidea* blossoms where copulation takes place (Hurd and Linsley 1963). Gottsberger (1970) has described a similar behaviour of beetles in various flowers of Annonaceae, and Zawortnik (1972) has described the flowers of *Mentzelia tricuspis* as a rendezvous for *Megandrena mentzeliae.*

A more complicated case is that of the thynnid wasps described by Stoutamire (1974). The males pick up the flightless, terrestrial females from the ground and mate with them either in the air or on some blossom. Thereafter the female is released to her terrestric life again. The labellum of various Australian orchids (*Caladenia*, *Drakaea*) imitate these females, and in their attempts to abduct it, males collide with the column and receive the pollinia on the thorax.

There is also a territorial rendezvous. Unfortunately, the word territory is ambiguous. Its usual sense is the one above: for a "private" area, strictly defended, usually by the male. In social hymenopters the word also signifies a non-exclusive area where a large number of males collect, usually at a certain time of the day. This regular concentration of males is assumed to increase the sexual attraction. Inasmuch as blossoms may enter as markers in such territoriality, this may be of importance for pollination. Vogel (1972) has described a territorial flight pattern in *Eucera tuberculata* during which *Orchis papilionacea* (which has no known attractant) was pollinated.

Trichosteta fascicularis, a big beetle, is frequently found hiding in the bottom of the permanently half-open inflorescences of certain *Protea* species (e.g. *P. barbigera*), which do not seem to attract pollinators any ordinary way. The possibility should be investigated if here is another case of rendezvous attraction. *Trichosteta* is frequently seen in copula in Proteaceae inflorescences (*Protea*, *Leucospermum*, K.F.).

The relation between perfume collection and rendezvous attraction may become extremely complicated. Dodson *et al.* (1969, cf. 1975) maintain that euglossine females are not attracted by the scent, but "somehow" by the odorous display of the males, which again attracts more males. Female bees are supposed to be attracted by the loud buzz of the displaying males more than by the attractants, some of which are deleterious if taken in.

CHAPTER 9

BIOTIC POLLINATION. SECONDARY ATTRACTANTS

9.1. SECONDARY ATTRACTANTS. I: ODOUR

Like nectar, odour in its rôle in the blossom is a new phenomenon in the history of the higher plants, and does not belong to the flower in its capacity as a set of specialized sporangiophores. There is no reason to believe that either the dispersal or the sexual organs of ancient vascular cryptogams were odoriferous. On the other hand, odours do function as secondary attractants also in some cryptogams: Phallaceae, Splachnaceae, etc. These odours are imitative (see below); where absolute odours occur, e.g. in Characeae or Jungermanniaceae, their ecological rôle, if any, seems to be that of a repellant.

Most probably, odour plays a major rôle in releasing instinctive reactions in animals, including man, and especially so in insects. It is therefore likely that odour is a major attractant. In keeping with this a great deal of chemical energy is spent in odour production as shown by the disappearance of, for example, starch during the odoriferous phase of anthesis (cf. Meeuse and Buggeln 1969). Incidentally, the heat generated will cause the odour to rise, and experiments have shown that an ascending stream of odour attracted more insects than a descending one. The heat promotes evaporation.

A priori we cannot presume anything about the chemical senses of insects, but on the whole they do not seem to differ greatly from those of man except that some compounds that affect man's senses strongly and adversely seem to be without significance for insects. Conversely, insect senses are probably much keener than man's with regard to other compounds, especially those that convey a message about food or the presence of sexual partners.

There are reasons for believing that odour is an older secondary attractant than are the visual stimuli. At any rate in plants as primitive as Cycads, odours are prominent, and are very strong in many beetle-pollinated blossoms — beetle pollination being considered primitive. Tropical nights are filled, almost beyond belief, with the fragrance of beetle-pollinated blossoms: *Victoria*, *Amorphophallus*, *Myristica*, Cyclanthaceae, Magnoliaceae, Annonaceae, and others. Sometimes blossoms that emit such strong odours are of very primitive construction (Leppik's amorphic type) or very small, like those of *Osmanthus fragrans*, which pervade the atmosphere with a strong smell of peach. One may often observe that olfactory and visual attraction act as substitutes for each other.

The way in which a visitor approaches a blossom tells a great deal about which secondary attractant is the more important: if the attraction is visual, the visitor will fly in a more or less straight line towards the blossom, independent of wind direction. If it is olfactory, the approach is less regular, the insect flying against the sensory gradient which means (a) that it will approach from downwind, and (b) that it will reach the zone of maximum concentration, pass through it, turn back, and pass through it again, all the time approaching the object in a zigzag route. Dung and carrion beetles have their own peculiar way of behaviour following olfactoric attraction (dropping suddenly down from a flight).

The question as to whether a given insect is attracted by sight or by odour can be determined by a simple glass-tube test; the blossom is enclosed in an open glass tube from the ends of which the odour emerges. Insects generally fly towards the flower in the tube indicating that they react on visual attraction. The experiment generally shows that allowing for important exceptions it may be stated as a general rule that in day insects visual attraction is primary where long-distance attraction is concerned. Odour, probably together with surface texture, is a very important factor for short-distance recognition (i.e. within a distance of the order of magnitude of 1 cm or less). Experiments with scented and unscented blossom models, together with the originals, show that whereas approach is in such cases governed by sight, visits are determined by odour (Manning 1957). It is therefore not unexpected that, especially in very large or "difficult" blossoms, odour production is localized to special parts or even separate organs, "osmophores" (Vogel 1963), which may imitate nectaries.

The rhythm of odour production, especially in night-flowering plants, indicates that in spite of the very keen eyesight of nocturnal pollinators, scent is a major means of locating blossoms, cf. the drab, indistinctive colours of some night blossoms. And there are some species even among day-flowering plants in which odour must be the major means of advertisement: *Reseda odorata*, *Vitis*, etc. (Knoll 1928).

Systems of classification of odour have been proposed, but are highly unsatisfactory as here one is dealing with subjective sensory reactions not directly and unequivocally related to chemical or physical conditions — as are colour or sound; however, see Amoore *et al.* (1962). Most of the terms in the proposed classifications remain rather loose and descriptive like aminoid or terpenoid; and instead of entering upon them, we shall divide odour into two functional classes, viz. absolute and imitative odour.

By absolute odour we understand an odour that — to man — has no direct, immediate counterpart outside the sphere of blossoms. In rare instances similarities occur between absolute odours, or between such odours and others derived from elsewhere, but such similarities like that between the odour of violets, *Trentepohlia iolithus*, and the *Iris florentina* rhizome, or between *Saussurea alpina* and vanilla, are probably fortuitous. At most, the similarity, if also a chemical one, indicates that the odoriferous substance is produced by similar metabolic processes. If the similarity of odours is not due to identical chemical composition the possibility must be left open that the similarity is not apparent to animals with different olfactoric senses, e.g. insects.

The reaction to absolute odours may be instinctive or acquired, but sooner or later it is grafted on to a reaction chain connected with feeding or sexual activity. The absolute odour functions in the context of the insect's relationship with the blossom and will not start any reaction chain outside this interrelationship. On the other hand, a "wrong" odour, absolute or not, may serve to warn the insect that it is in a wrong place, probably mislead by morphologic similarities.

Imitative odours have their counterparts elsewhere, and they establish the same reaction chain in the insect as is usually triggered off by the similar odour emanating from a different source than the blossom. These reactions, which are functional in the original context, are usually meaningless or deleterious in the insect-blossom interrelationship, e.g. the pseudocopulation in *Ophrys* or the visits of dung and carrion insects for feeding or sometimes even oviposition in blossoms, the scent of which imitates the smell of decaying protein or fruit.

In both cases the operative element is the effect on the insect—pollinator, not the chemical composition of the odour. In experiments it is obviously possible to imitate an absolute odour, i.e. to present a pollinator to substances which would have released reactions like flower visits, independently of the presence of any flower. This can be done by isolated fragrances from flowers. In such experiments they will function as imitative odours as they release reaction chains aimed at a non-existing source of emanation.

Imitative odours are main operators in the attraction by deceit syndromes.

A complicated case of imitative odour is represented by some blossoms like *Veratrum album* var. *lobelianum* (Daumann 1967; other examples p. 104) the smell of which to man resembles that of sapromyophilous blossoms, but which offer nectar at the same time. Whether this could be ranked as deceit or not, is mainly a semantic question.

Carrion odours often act very specifically, which means that different insect groups may react to different parts of the odour complex in question. In some plants the principle of attraction at any rate seems to form a very close parallel to carrion attraction, but there is no odour perceptible to man (e.g. some Cypripedilinae).

Another type of deceptive odour attraction apparently exists in *Arum conophalloides* (Knoll 1926). Its pollination follows the general *Arum* pattern, but the insects attracted are exclusively small blood-sucking midges — and the females only (males do not suck blood). Evidently the odour emanating from these blossoms imitates that of the skin of mammals on which the insects generally feed.

The great difference between absolute and imitative odour is that in the development of the former there are only the two partners — plant and visitor. In the latter case there is a third element, too: the pre-existing scent to be imitated. The phylogenetic paths leading to one or the other are therefore different.

In their action on visitors, odours may be classified into those with direct, respectively indirect, effect. A direct odour immediately activates an already existing path of reaction, e.g. the reactions of carrion insects to the smell of putrescent protein, whether this comes from a blossom or from real proteins. On the other hand, the smell of an apple blossom in itself hardly starts any reaction in a bee — only after the insect has "experienced" that this scent means a source of nectar does it react to it. The typical "blossom" odour is therefore indirect.

With due allowance for the present state of knowledge about the senses and reactions of insects, the two examples mentioned above are rather definite, but there are many borderline cases less easy to classify.

Kullenberg (1956a) tentatively advanced the hypothesis that flower fragrances resemble the smell of visiting insects themselves. A causal—evolutionary evaluation of this hypothesis might easily lead to a nonsense discussion of which came first. Does the blossom imitate the smell of insects or have the latter acquired the smell of blossoms?. The answer is simple enough in sexual attraction syndromes; in others, if they exist, we must await further information (cf. below). The (to man's senses) bat-like smell of many bat-pollinated blossoms may come in as part of this hypothetical "Kullenberg syndrome". Is this an imitative odour? Bats are social animals, and the smell is apparently a means of keeping the individuals together (v.d.P.).

Imitative odours are not necessarily aminoid. Many beetle-pollinated blossoms possess a strong fruit-like odour. Does this imitate the smell of fruits and thus activate the innate instincts of fruit-eating beetles or beetles descending from fruit-eating ancestors? In the latter case: how many generations will pass before this contact backwards is lost and the smell becomes absolute?

Conceptually, things are much simpler when it comes to pheromones. To insects, odours are much more important as signals than to man: insects live in a world of odours, and their actions are triggered by odours more than by any other stimulus. Especially odours connected with the breeding system of insects may be very specific and have very specific effects. Odours are important components in the sexual attraction system, and by entering into this system flower fragrances may trigger sexual reactions in susceptible individuals. For *Ophrys* the nature of the active substances has been studied by Kullenberg (1973a) and his collaborators (Priesner 1973). γ-Cadinene is apparently one of the key substances. Electroantennograms show strong reactions on this substance and also strong and differentiated reactions on various substances isolated from the labellae of *Ophrys* blossoms. The absolute identity of these substances with the actual pheromones (suggested by Kullenberg in 1956b) is not definitely established.

The extremely strong attraction effects of pheromones can be imitated by aromatic and terpenoid substances which can be used as baits for attracting male euglossine bees from afar. Those of them that carry orchid pollinia (identifiable to genus, sometimes to species) can give important information about orchid pollination even in the absence of the actual flower (Dressler 1970).

The pheromone observations fit into a more complex syndrome, viz. the perfuming of courtship territories as described by Kullenberg (1973): male bumblebees mark a territory with scent from their cephalic glands by brushing their mouth-parts against leaves, twigs, etc. In the course of this process they also perfume themselves: "bumblebee males . . . function as veritable perfume-brushes" (Kullenberg *et al* 1973: 25). The difference between this marking of a territory and the perfume collecting described by Vogel (cf. p. 70) is mainly that in the latter case the males do not produce the perfume themselves (or only to a lesser extent) and that they are not known to deposit it afterwards. At any rate, such perfuming may constitute important parts of rendezvous attraction syndromes.

Odours are not only means of attraction. They are also very important in differentiation. Hills (1972) maintains that differential fragrances constitute the main barriers against interspecific hybridization in *Catasetum*. The fragrances in this case depend on mixtures of 39 different aromatic substances and a change of only one of them may constitute a differentiating barrier. Such changes may easily arise by mutation; the resulting breeding isolation may be the starting point for speciation, even before morphological differences have developed (Dodson *et al*. 1969). According to Loper and Waller (1970), the differences in odour quality between races of alfalfa causes preferential visits by bees. Whether this is innate or a result of the communications system of honeybees is unknown (cf. Waller *et al*. 1974). On the other hand, Kullenberg (1973) maintains that there is little specific difference between *Ophrys* fragrances and their effect on various sexually attracted pollinators.

Williams and Dodson (1972) have shown that male euglossine bees are attracted by one chemically pure substance among the ones isolated from orchid fragrances, whereas the addition of a second fragrance may have a negative synergistic effect. This emphasizes the dual role of fragrances as attractants and as isolating mechanisms at the same time. As the rôle played by fragrances is in this case correlated with the breeding syndrome of the bees, some care is indicated in generalizing with regard to more regular odour attraction. The question to which – if any – extent odours also act as repellents (outside the sphere of pheromones), is controversial. In the literature there are many references to honeybees and bumblebees being repelled by aminoid and skatoloid odours from fly blossoms, by sphingid odours (p. 116), and by the fatty acid smell from bat flowers. There is better evidence that higher hymenopters "label their empties" by leaving odour traces in blossoms visited (van der Pijl 1954). The marks are of short duration (cf. Simpson 1966; Nunez 1967). Avoidance of visited flowers is easily observed. Apparently, hawk-moths do not possess the same mechanism and can be seen visiting the same (empty) bossoms at short intervals (Gregory 1963: 407).

9.2. SECONDARY ATTRACTANTS. II: VISUAL ATTRACTION

Visual attraction by means of enlarged "semaphylls" is probably not a primary part of the biotic pollination syndrome. The first angiovulate blossoms may well have been aphananthous with odour as their only secondary attractant — if any. This does not make them anemophilous. It has been argued that inconspicuous primitive Winteraceae blossoms should be anemophilous, but this has been refuted by Gottsberger (1974) for *Drimys brasiliensis*: the blossoms have a sweet odour and are pollinated by pollen-eating beetles. The enveloping phyllomes, which are much older than the angiosperms (occur in cycads, even in *Polytrichum*), may also in angiosperms have arisen as a response to the demand for protection or arresting visitors. The attraction function may therefore be secondary.

The two means of visual attraction are colour and shape — inclusive of size. In its function as an advertising organ, the larger the blossom is (within reasonable limits), and the more it contrasts with the surroundings, the more effective it generally will be. The details of this attraction process have been the subject of very interesting studies on the senses and reactions of insects; studies to which we shall return later.

Generally, the perianth constitutes the advertising organ of the blossom. If it is differentiated, the inner members, the corolla, usually assume that function. Unlike blossoms of anemophilous plants, entomophilous flowers with visual advertising effects vary immensely in size. Those of *Rafflesia arnoldii* are cup-shaped, almost 1 m across; those of *Galium hercynicum* or *Conopodium majus* about 2 mm. The large flowers are sufficiently sensational to give the desired effect alone, the small ones form attraction units by combining into integrated inflorescences, pseudanthia.

Primitive pseudanthia are very loosely organized. Further development takes place by contraction into a smaller space — *Trifolium*, *Taraxacum* — and later by differentiation between the members of the inflorescence. In *Orlaya grandiflora* the central flowers of the umbel are radiate, the peripheral ones asymmetric inasmuch as their centrifugal petals are much larger than any others, thus surrounding the umbel with a row of large petals. In many composites, the daisy type, the differentiation goes further as the outer florets are morphologically completely different from the central ones, much larger and differently coloured. The ultimate step is found when the peripheral florets have lost their sexual function altogether and function as attraction organs only, as in many composites or in the looser inflorescence of *Viburnum* spp. With such a differentiation the inflorescence has assumed the same organization as the individual flower; the attraction function is peripheral and the "sexual" functions central. In pollination ecology it is impossible to keep apart such inflorescences and single flowers of a similar size.

In racemose inflorescences a similar differentiation may also be found, e.g. in *Muscari comosum* (Knoll 1921); the upper flowers are violet, whereas the lower ones are yellow-brownish, nectariferous, sterile, barren, and fertile, resulting in a clear differentiation between approach to the upper, sterile flowers, and visits to the lower, fertile ones. Morphologically, the sterile flowers are the central ones, in contrast to the examples above.

In *Primula vialii* the dark-purple buds are important advertisers in contrast with the paler bluish flowers (both colours may belong to the same discrimination group).

From such inflorescences in which there is a functional differentiation between the individual flowers it is an insignificant logical step to those in which extra-floral parts like bracts function as advertisement organs: *Houttynia, Cornus, Poinsettia*, or the Araceae. Still, there is no significant difference between this and the large, single flower.

There is a parallelism in development between flowers and inflorescences. Thus the integration is rather low in the most primitive flowers, too, cf. the independence of floral members in lower Polycarpicae. Like the floral members in a more developed flower, the individual flowers lose their independence in the most integrated inflorescences. The primitive, flat amorphic inflorescence in Compositae functionally corresponds to the flat sporophyll system of early angiosperms. From this point a redifferentiation takes place under the same influences that steered the development of the euanthium. It is therefore natural that many of the same visual types occur both in the evolution of the flower and of the inflorescence, cf. the similarity between the flower of *Anemone* and the inflorescence of *Cosmos*. The visual differentiation and integration in Compositae is an expression of functions: protection (involucrum), attraction (ray florets), pollen reception and nectar production (old, peripheral disc florets), and pollen presentation (young, central disc florets). In some bird-pollinated South American Mutisiae (*Barnadesia* spp.) there is a further differentiation inasmuch as nectar production is restricted to a central group of specialized disc florets.

Just as there must be a lower size limit, beyond which the blossom loses in attraction value for the pollinator in question, there is also an upper limit. In Knoll's experiments (1922) with *Macroglossa* (Sphingidae) the optimum size of (undifferentiated) attraction units seemed to be a square of *ca.* 15 mm side length. Above 30 mm reactions were less active, and interest was concentrated to the edges of the unit, not to its central parts. Whereas dystropic insects, which run about everywhere, have an equal chance of finding any source of food that is openly displayed, it is evident that eutropic visitors have difficulties in locating sources of food that are displayed within a large, undifferentiated area, but that they will easily find them under the influence of leading structures.

Kugler (1943) has shown that the distance at which a blossom attracts visually the interest of a bumblebee is directly proportional to its diameter, provided it is isolated. But neighbouring features seem to have a cumulative effect, with the consequence that the combined effect of several flowers is greater than that of the individual ones taken together. This rule of cumulative effects is also valid for finer details within the blossom.

Size is not the only form element entering into the picture. Studies have shown that *segmentation* and *depth* of blossoms are of great importance. When blossoms are presented simultaneously, higher pollinators especially, such as bumblebees, are more attracted by more finely segmented and deeper blossoms. By placing a glass plate over the blossom or a blossom model, one can show that the visual impression of depth is decisive. The reason for this preference may be an instinctive or acquired "knowledge" that deep-lying sources of nectar (which cannot be utilized by non-specialized pollinators) are usually richer. We shall return to the preference for more segmented blossoms.

The *colours* of blossoms have been the subject of intense studies from the point of view of plant physiology, and the chemical composition of, and to some extent the metabolic paths leading to, various colouring substances are known (see textbooks of plant physiology, and also Benl 1938). As a rule, colours through the range pink—red—violet—blue are due to anthocyanins, yellow–red–purple to carotinoids or flavones. White is due to multiple reflections in the intercellular spaces between uncoloured cells, black to similar reflections between layers of complementary colours. Such reflections also contribute to the visibility of blossoms and to the saturation of colours; cf. the appearance of a blossom from the intercellulars of which air has been driven out (e.g. by alcohol). Green is not counted as an independent colour in pollination ecology. Apart from possible ultraviolet effects, green blossoms are invisible to pollinators and are for their attraction dependent on odour: sweet in *Reseda odorata*, foul in *Veratrum album* var. *lobelianum*. Chlorophyll is frequently present before anthesis, but generally breaks down. Very rarely, green flower-colours are due to other substances than chlorophyll. Substances responsible for reflections in the ultraviolet seem to belong to distinct chemical groups (Dement and Raven 1974).

Some colours are not light-resistant and fade during anthesis. In other blossoms colour develops or changes (apart from fading) during anthesis, e.g. many Boraginaceae in which the colour changes (with increasing pH) from red to blue. In *Myosotis discolor* the main colour is even produced during anthesis, so that the flower starts yellowish, and later runs through red to blue. In spite of many attempts nobody has yet succeeded in giving a general ecological interpretation of the significance, if any, of such changes. In *Aesculus hippocastanum* the nectar guide changes from yellow to red at the end of anthesis when no more nectar is produced. Kugler (1936) has shown that red-marked flowers are not regularly visited, but again it is difficult to give an ecological interpretation of this and similar observations (see also Süssenguth 1936; Vogel 1950). On the other hand, Gottsberger (1971) has shown that a light fading, difficult to observe for the human eye, indicates to hummingbirds the onset of nectar production in *Malvaviscus*. Vogel (1954, cf. Ingram 1967) reports a remarkable change of colour in *Gladiolus grandis*: the flowers (of one subspecies) are brownish during daytime, blue at night.

Post-pollination or post-fertilization developments usually start with wilting of the blossom whereby it loses its attractiveness. This development may take place connected with or independent of the pollination or fertilization process itself. When fertilization has taken place, the advertising function of the blossom is redundant and, inasmuch as it represents a loss of energy, should disappear. Furthermore, if there is also a dearth of pollinators, repeated visits to an already pollinated fertilized flower would, from the flower's point, represent wasted pollinator effort and time. It is therefore not unexpected to find extremely rapidly developing post-pollination phenomena in orchids in which pollination events are perforce rare (van der Pijl 1972). It must be left aside if the post-pollination phenomena are caused by a chemical influence from the pollen itself or from the developing ovary (cf. Harrison and Arditti 1976 and references). The *Aesculus* syndrome referred to above would fit into this picture even if there is in this case presumably no dearth of pollinators. The difference caused by the fact that spent blossoms remain attractive from a distance may be understood in the light of collaboration within the inflorescence: the red-dotted flowers would then correspond to, for example, the sterile flowers in *Muscari comosum*, which attract pollinators from afar and lead them on to the operative part of the inflorescence after the primary approach. Wilting would spoil the attractiveness of the inflorescence as a whole, no change would be detrimental to pollinator economy/efficiency. Other colour changes should be investigated if they function in a similar way as that of *Aesculus*, e.g. the fading of the nectar guides in *Arnebia echioides* or the change of the colour of the guide from yellow to red in *Lupinus pilosus*. As the latter is a pollen flower, this would represent a reverse syndrome from the one suggested for *Exacum* (p. 62) or *Saintpaulia*.

Surface texture is of importance, both mechanically and for light reflection. A rough, hairy surface gives better foothold for visitors, especially when they have to land on vertical or near-vertical surfaces. By contrast, extremely smooth surfaces with some slippery coating occur in unscalable places or may cause visitors to lose their foothold and fall into traps. The same principle is found both in the leaf traps of insect-catching Sarraceniaceae and Nepenthaceae and in the blossom traps of some Araceae.

As the bumblebee proboscis is a very sensitive organ, the animal will shun surfaces too rough for it. For instance, in *Pedicularis* flowers, leading structures are formed by echinate surfaces which create, as it were, barbed-wire entanglements to the pollinators. Probably the hairy parts of many corollas are avoided for a like reason. Besides those dubious cases, where hairs have been interpreted as measures against nectar thieves, there are well-authenticated ones in which hairs bar certain routes of entrance or exit, thus keeping visitors in temporary confinement or forcing them to proceed in some specific manner (*Pinguicula alpina*).

Different surfaces have different sheens, resembling fat, glass, silk, velvet, etc., according to whether the epidermal cells are isodiametric or elongated, smooth or bulging, dry or covered. A whole classification of such sheens has been established, but their deeper significance is mostly obscure, except when they enter into a specific syndrome (e.g. *Ophrys*). Kugler (1942) has shown that shiny surfaces attract inexperienced visitors more effectively than mat ones.

Whereas we may judge by analogy about the visual perception of vertebrate pollinators, that of invertebrate ones is more problematic, and it is impossible here to deal with the enormous and constantly growing zoological literature about these problems (cf. Burkhardt 1964). Responses to colours may also be conditioned, and may be different for various functions, e.g. feeding, mating, ovipositing. Inborn colour preferences in pollinators may affect speciation in plants. Experiments with flower models of different colours* have shown that, on the whole, the visual range of the spectrum in the most important pollinators does not differ greatly from that of man, although there is a slight shift towards shorter wavelengths. The power of discrimination is apparently limited; bees recognize four colour groups: yellow, blue-green, blue, and ultraviolet, but no red (Kühn and Pohl 1924). Most red blossoms contain a blue component, so they will appear blue or ultraviolet to a bee. As far as known, other insects more or less resemble bees in this respect. In some moths the power of discriminating between colours at low light intensity is greatly superior to that of man; they do not lose their sense of colour in weak light. Besides the spectral composition, insects recognize the saturation of colours. Some nectar guides are simply variations of saturation of the colour (Knoll 1922).†

*Such experiments are not easy to conduct because visual definition by the human eye is insufficient. Colours must be defined spectrally, also with regard to ultraviolet. Ultraviolet photographs of flowers, e.g. Daumer (1958), frequently show patterns invisible to man and different from those seen by the human eye. The red of the poppy flower is invisible to the bee, but it can (and apparently does) see it by means of the strong ultraviolet reflection apparent also in photographs. The ultraviolet sensitive television camera has made possible a much better study of the ultraviolet patterns in blossoms, and one may also see them working (Eisner 1969).

†A very common type of experiment in pollination ecology consists of conditioning insects to blossom models of definite characteristics − colour or form − by giving sugar solution in such models and not in others differing in one or more characteristics. The power of discrimination can be tested in this way, and also the retention of learning, the "memory". It has been proved easier to condition insects to some colours than to others; there may be a colour preference. Such possible preferences have also been registered in virgin animals, where they cannot have been acquired, but must be inborn (Knoll 1922). This preference has nothing to do with the power of discrimination.

It has also been proved by experiments that higher insects react to contrasts, either between blossom and surroundings, or within the blossom. A preference for the more segmented pollination unit may be caused by the more vigorous background contrast. Vogel (1954) maintains that nocturnal insects react more strongly to contrasts between blossom and background (strong dissection, etc.), whereas diurnal pollinators react to contrasts within the blossom. According to Dufay (1961) noctuids (moths) prefer the shorter wavelengths, whereas according to Knoll (1922) *Macroglossa* is not interested in ultraviolet. A very important type of contrast within the blossom is the visual nectar ("honey") guide – a figure in contrasting colour giving emphasis to the operative part of the blossom. The contrasts may be in brightness or saturation, or they may be spectral; in evaluating such contrasts one should keep in mind the restricted spectral colour discrimination of insects, but, on the other hand, not forget that (to us invisible) ultraviolet nectar guides occur (Daumer 1958). Nectar guides were originally recognized by Sprengel, and their significance has been proved experimentally both by Knoll (1922) and by later experiments (Kugler 1930, 1938, 1962, 1966; Manning 1956). The nectar guides may be concentric; more or less circular, like the white throat of *Myosotis* flowers, the orange blotch in the gullet blossom of *Linaria vulgaris*, or the zones of a *Chrysanthemum carinatum* inflorescence. More frequently a nectar guide system consists of radiating lines pointing toward the centre of the flower. This line system may be rather simple, as in many *Geranium* flowers, or form a very complicated pattern on a background that contrasts against the rest of the attraction unit, as in *Iris sibirica*. In many *Veronica* flowers there is a combination of radial and concentric nectar guides. In some flowers, e.g. *Digitalis purpurea*, the nectar guide consists of a pattern of blotches.

Knoll (1922) exposed blossom models with and without nectar guides, with concentric and excentric lines, under a glass plate. Insects trying to get at the presumed nectar touched the glass plate with their probosces, leaving "fingerprint" traces that could be developed. The traces always showed definite relations to the nectar guide pattern. Nectar guides are chiefly found in the more highly evolved pollination units, and in units appealing to higher pollinators. Especially noteworthy is their absence from most carrion flowers; the relevant insects are presumably unable to utilize the information contained in a nectar guide.

As similar guides may be found in pollen blossoms, the term "nectar" guide is slightly misleading.

Because of difficulties inherent in the analysis, little is known about the localization of scent in blossoms, but odour guides (olfactory nectar guides) are known, both from direct observations (spadix in *Arum*) and from observations of the behaviour of pollinators (T. Lex 1954). Odour guides may be coincident with visual nectar guides, but they may also be invisible to man, e.g. in white flowers of *Convolvulus arvensis*. By training bees separately to the odour of odour guides and that of the rest of the flower, von Aufsess (1960) demonstrated the difference of the two. According to this author odour guides may be even more frequent than visible guides. Not all localized odour production comes under the heading of odour guides; exceptions are found, for example, by the odour emanating from the median perianth tip of *Arachnis flosaeris* (v. d. P.) or by blossoms in which odour is the only attractant. Vogel (1963) has later been able to localize odour production in various blossoms.

Many groups of insects have an aversion against darkness and go towards a positive light gradient. A visitor creeping into a bell-shaped blossom will have to move away from the light. This apparently does not matter in white or very light-coloured blossoms. Besides, many of these are hanging, so that the bottom of the bell-shape turns towards the light, and will seem lighter even in a self-coloured blossom, as in campanulas. But in a dark-coloured upright blossom, e.g. *Gentiana acaulis*, this does not apply, and it is interesting to see that the lower part of the corolla is whitish, letting the light through. Similar window panes,* more or less discrete, are found in many other blossoms in which the pollinating insect is obliged to immerse itself or at least part of its body, not only its proboscis. A very elegant example is that of *Cypripedium calceolus* (Troll 1951); in the subgenus *Paphiopedilum* the labellum is straight, and the exit is directly visible from the bottom of the labellum. No windows are therefore necessary and none are found. The phototropic reactions of insects are apparently much dependent on their general psychic state; they are stronger if the insects are in a state of shock, which may be presumed to exist in trap blossoms (see *Arisarum vulgare*, p. 173).

The light effect of windows is sometimes strengthened by deflexion of the light from above in such a manner that the windows are also lighter than the background immediately behind them, e.g. in *Arisaema laminatum* (van der Pijl 1953). It is also frequently accentuated by dark-coloured rims surrounding the windows. In *Ceropegia*, basal windows serve to lead insects towards the gynoecium (Vogel 1961).

9.3. SECONDARY ATTRACTANTS. III: TEMPERATURE ATTRACTION

Not only light conditions inside blossoms depend on the structure of the blossom, but temperature conditions also frequently vary from those outside, sometimes probably as a secondary effect due to strong physiological activity (odour production, B. Meeuse 1966, 1975) in more or less closed blossoms (*Cycas*, Nymphaeaceae, Araceae, Palmae). Whether the thermogenic respiration (cf. B. Meeuse 1975) in such blossoms has other ecological effects in addition to that of production and volatilization of odour substances, is unknown. High temperatures inside blossoms may also be due to absorption of energy from the sun with resulting differences of about 10° (Büdel 1959). The ecological effects of this are largely unknown, and presumably differ in cool and hot climates. The general idea of a "warm shelter" as an attractant is hardly tenable, being unnecessary in the Tropics, and realized during day-time in fair weather only (when no shelter is necessary) in temperate regions. However, in the Arctic, Hocking and Sharplin (1968) have observed mosquitoes basking in open flowers at a temperature up to 6°C higher than that of the surroundings. The insects in question are dark-coloured, and take up a position near the focal point of the heliotropic flowers. The ecological significance of this has been distinctly demonstrated by Kevan (1975): temperature is here a secondary attractant, in some respects almost a primary one. As insects are reluctant to leave their warm abode, this attractant is not particularly effective from a pollination economy point of view.

9.4. SECONDARY ATTRACTANTS. IV: MOTION AS AN ATTRACTANT

Except for the small movements of anthers or stigma on touch (the ecological significance of which is frequently doubtful), or the more spectacular explosive mechanisms, blossoms have always been considered as fixed, which many certainly are not in nature. It is well known that even many vertebrates hardly react visually on immobile objects, and the insect

---

*Unfortunately, the term window is used in two different senses in pollination ecology. Ref. window openings, see p. 104.

eye seems even less adapted for discovering such objects, being more sensitive to changes in the illumination of the ocelli in time than to spatial differences. We may assume that a big surface of saturated colour, like a *Taraxacum* or a *Tulipa* blossom, is sufficiently sensational to be discovered also by insects without exhibiting any motion. But there is hardly any doubt that when it comes to smaller blossoms of less distinctive colour, relative movement plays a great part. The whole blossom may be in motion, like Vogel's (1954) micromelittophilous blossom; or parts may be in motion, like the filiform appendages of many myophilous blossoms. As a third possibility, owing to the specific peculiarities of the insect eye, open inflorescences like those of umbellifers, or *Galium*, will appear to flicker to an insect flying by (Wolf 1933). The rôle of such apparent movements in pollination ecology is still largely unknown; it must be left to future investigations to study the details of this principle, which is certainly not restricted to myophilous blossoms (Vogt 1966).

CHAPTER 10

STRUCTURAL BLOSSOM CLASSES

The observation that pollination units from different taxa and of different morphological value are pollinated in the same way naturally led to attempts to classify blossoms according to their mode of pollination and their functional structure. Many such systems of structural "flower classes" have been proposed; most of them, even some proposed by prominent scientists, suffer from extremely bad logical flaws. Consequently, there is no reason to enter upon them in detail; the more recent systems by Werth (1956a) and Leppik (1957) are at least logically fairly consistent. In Section 5.2 we described the structure and function of the blossom in general terms and stressed the development towards greater precision. The system proposed here intends to classify these tendencies and give concrete indications of the main levels reached.

The background of the difficulty is that there is not one basic principle upon which a classification can be based, but at least three major ones, partly independent of each other, and therefore impossible to combine into one linear system.

Pollination units may be classified according to the *attractant*: no attractant, pollen, or nectar, alone or combined, brood-place, etc., as in the preceding chapter. Subdivisions, if necessary, can be based upon the mode of presentation. This would give a perfectly valid classification, but one that is primarily a classification of attractants.

A second system could be classification of the *mode of pollination*: wind, water, insects, etc., with subdivisions according to details in the process. This classification could be perfectly consistent, too, and would come nearer to the central problems than the preceding one; but it would mainly be a classification of pollinators and their behaviour. In reality these two classifications underly the discussion in the preceding and following chapters.

The third system could be a classification according to the combined *functional structure* with regard to the individual processes of pollination: deposition and reception of pollen, long-distance attraction, presentation of attractants, etc. Even if it is thus dependent on a variety of parameters, we find it more suitable than other viewpoints for the definition of the blossom classes. But the important thing is to avoid mixing these three approaches, as this only results in such muddled systems as those found in older literature. The systems of Werth and Leppik (*loc. cit*), both classify according to functional structure, and the former is so closely related to the system proposed here that we shall not enter upon details of difference in the practical delimitation of groups. But we wish to make a reservation with regard to the theoretical basis of Werth's classification; our classes are purely morpho-logical—typological with no phylogenetic or orthogenetic bias. Of course, the evolution of the types may be discussed − if wished − but their definition should be static.

Leppik's system divides blossoms into type classes presumed to correspond to evolutionary levels. Of these classes, the amorphic has no definite shape, number, or symmetry. In the haplomorphic the shape is open, bowl-like, but definite number and symmetry are still undeveloped. The actinomorphic type is flat, with a great and indefinite number of symmetry axes, which are in the pleomorphic reduced to a few that are recognizable by more highly evolved pollinators as "form numerals".* The two highest groups, the stereomorphic (with depth effect) and zygomorphic, are more or less self-explanatory.

Except for the last two, all Leppik's type classes fall within our group of dish-bowl-shaped blossoms, and may form a good complement to our classification in that group, which is rather comprehensive. On the other hand, we cannot accept Leppik's system as a whole; his two last groups are in their turn too comprehensive. Formal morphology still plays too great a part in his scheme in relation to functional aspects (traps, closed flowers), taxonomic phylogeny (Sippenphylogenie, Zimmermann) comes too much into the picture (Compositae), and important types are missing (brush blossoms, flag blossoms). We therefore propose the following classification of blossoms (flowers and inflorescences alike), reiterating that our classification is purely typological with no phylogenetic background, and refers to pollination condition in general. Another point to be kept in mind is that there is nothing like a pollinator-in-general; this concept involves a great, but unavoidable generalization.

I. Blossom *open* during anthesis — with due allowance for blossoms like *Plantago*, in which stigmas project out of a closed flower.
II. Blossoms *closed* during anthesis (cleistopetalous) — by definition cleistogamic flowers are excluded.
III. Blossoms forming *traps*, i.e. temporarily closed for exit or difficult to get out of. They are more or less open for entry.

Groups II and III are small, and will be dealt with rather summarily, whereas group I needs a further subdivision into types which may intergrade. Some blossoms — of orchids especially — are so irregular as to defy any attempt at classification.

I.1. Blossoms inconspicuous, no optical attraction.
I.2. Blossoms conspicuous, advertising.
    A: Dish- to bowl-shaped blossoms.
    B: Bell- or funnel-shaped blossoms.
    C: Head- or brush-shaped blossoms.
    D: Gullet-shaped blossoms.
    E: Flag-shaped blossoms.
    F: Tube-shaped blossoms.

1. Reduced pollination units without optical attraction are above all found in plants with abiotic pollination — Betulaceae, Urticaceae, *Zostera*, etc. There is a small subgroup of insect-pollinated blossoms that belong here because, apart from the possible effect of ultraviolet radiation, their only attractant seems to be olfactory, e.g. *Reseda odorata*, *Vitis*, etc., the flowers of which are greenish-brown and do not stand conspicuously out from the background. There is also a group of night-pollinated blossoms with small corollas and dull

---

*It should be noted that according to Kugler's experiments (1938) bumblebees prefer an actinomorphic form (*Cichorium*) to a pleomorphic one (*Geranium*), presumably because of the stronger segmentation of the former.

colour, e.g. *Hesperis tristis*. Recognizing our ignorance with regard to the senses of night insects, we may possibly include these plants here as well. Further, some carrion blossoms at any rate approach this group. Many autogamous, but still chasmogamous blossoms belong here, too, obviously with transitions to closed blossoms.

2. The classification of conspicuous pollination units is based upon the position of sexual organs in relation to the centre of the blossom. In A they are more or less diffuse to concentric, in B distinctly centric, in C diffuse, and D and E excentric. To F we shall return below. The position of visitors changes correspondingly. In A and C it is indifferent, in B excentric, in D and E (sub)centric.

A. Dish- to bowl-shaped blossoms may be exemplified by *Caltha*, *Rosa*, which are bowl-shaped. The completely flat, dish-like single flower is not too frequent (cf. *Rubi fruticosi*, *Tulipa silvestris*), but the type is very common in compound units: heads of composites (*Chrysanthemum*), or umbels (*Conopodium*, *Viburnum*). The reproductive organs (flowers) constitute the – more or less extended – centre of the blossom, and the insect works on or from the top of them. In others these organs may be elevated (*Magnolia*, *Passiflora*), and insects work among or beneath them. With reservation for some instances of zygomorphy, cf. *Aconitum*, to which we shall return under point F, this group on the whole consists of radiate units. Generally, the abdominal side of the pollinator touches anthers and stigma, and pollination is of the mess and soil type.

B. The difference between bell and funnel lies in the contour: convex in the former (*Campanula*), straight or concave in the latter (*Calystegia*, *Zantedeschia*).

In addition to the morphological elements constituting the bell or funnel itself, there is generally a narrower or broader, flat rim (bell, *Gentiana acaulis*; funnel, many Rubiaceae, e.g. *Bikkia comptonii*). The rim has obvious advertising functions, e.g. in *Narcissus pseudonarcissus*, in which the beaker is formed by the paracorolla; the rim is at the base of the blossom. The rim also has the important function of serving as an alighting platform. In a rimless bell or funnel type there is no opportunity to land outside, and the insect must crawl into the blossom at once, which is therefore large in relation to the insect. If there is a rim, insects can land on that and immerse only their heads; such blossoms form transitions to the tube type (*q.v.*).

The bottom of the funnel type is narrower than that of the bell, but the difference is not very important. In both types the pollinator is obliged to immerse itself in the pollination unit; it cannot remain on the top of it, as in the former group. The sexual organs are generally elevated above the centre of the unit, on long filaments and styles. Any part of the pollinator may come into contact with the sexual organs; an insect alighting on the central column in a *Campanula* flower will pollinate with its abdomen, and one climbing down the walls may do so with its back. An extreme example like that of the *Ficus* syconium also belongs to this type.

A very important subdivision of the bell-beaker type depends on the position of the anthers and stigma. In the bell-type proper, they are central, and an insect landing on the corolla and crawling down to reach the nectar will get pollen on its back. The alternative is formed by blossoms in which the anthers and stigma are more or less attached to the corolla, and pollen may be deposited sternotribically if the blossom does not approach the gullet type and is plagiotropic.

C. The brush type defines itself. *Thalictrum* (some species are anemophilous!), *Salix*, *Phyteuma*, *Eupatorium cannabinum*, and many Proteaceae may be quoted as examples. The external surface of the pollination unit is, exclusively or partly, formed by the "sexual" organs, while the perianth is reduced or split into filiform segments interspersed between the sexual organs. More rarely, the brush is dominated by perianth segments or by bracts as in *Castilleja*, where bracts dominate the attraction unit. Pollinators that alight on brush units naturally carry pollen on their abdominal side, fluttering pollinators — as usual — on their heads. Brush blossoms are generally pseudanthia, but may be single-flowered, e.g. *Rubus idaeus*, or within Myrtaceae.

The reduction of tepals in brush blossoms can facilitate the transition to anemophily, but is not in itself a consequence of such a transition. The transition found, for example, in Fagaceae: *Castanea–Lithocarpus–Quercus* is one example, another can be found in the genus *Thalictrum*: *T. dipterocarpum* (approximately bowl-shaped)–*T. aquilegifolium* (brush)–*T. alpinum* (anemophilous).

D. In the gullet type, sexual organs are restricted to the functionally upper side (cf. below) of the pollination unit, and pollen is deposited on the back (upper side of the head) of the pollinator (nototribic). The typical gullet blossom is that of temperate Labiatae (*Salvia pratensis*) and Scrophulariaceae (*Mimulus*), which is strongly zygomorphic. Other representatives are most basitonic orchids. In all these examples the unit is a whole flower. In the *Iris* flower there are three typical gullet units. In *Acanthus* parts of the calyx play a major rôle in formation of a gullet, and in *Mimetes hartogii* (Vogel, 1954) the gullet is formed by an inflorescence. In *Centrosema* the flag has inverted and turned into a gullet. In relation to the typical gullet, the blossom of the (southern) Ocymoideae is "upside down" and functions as a flag.

Again, there is an important subdivision of the type. Whereas the typical gullet has a landing-place, a lower lip, there are some gullet blossoms adapted for hovering visitors, in which the lower lip is absent or at any rate bent back and non-functional, cf. the case histories from *Salvia* (p. 196).

E. The flag type shows the same division of the pollination unit into an upper and lower part, but in this case the sexual organs are found in the lower part, and the insects carry the pollen on their abdominal side (sternotribic). The typical flag blossom is formed in the majority of Papilionaceae. Outside this family it is found, for example, in *Corydalis* and *Pelargonium*. However, these highly evolved, more or less closed blossoms represent the ultimate development of a type that is well represented also in a more primitive stage. At first glance, the blossoms of *Dictamnus* belong to types A or B, but the zygomorphic disposition of anthers and stigma characterize them as primitive representatives of transitional types towards the flag blossom. Like gullets, flag blossoms are generally single flowers. *Pedilanthus* is an example of a strongly modified flag-type which is an inflorescence (Dressler 1957).

In the two last groups (D and E) the terms "upper" and "lower" refer to the relative position of the pollinator. If both pollinator and blossom reverse their position at the same time, the type remains unchanged as sterno- or nototribic. *Centrosema* is an inversed type because only the flower is turned upside down, not the pollinator. The fact that some of the pollen and generally also the stigma of *Cytisus scoparius* end up on the back of the pollinating insect, is a queer exception; but this is still a good flag type, the pollen coming from down behind, not from overhead. Similarly, *Fumaria* or *Dicentra* may be reckoned as a kind of duplicate flag-type.

F. Logically, the tube-type is not comparable with the preceding ones, because the criterion is not the disposition of sexual organs but the accessibility of nectar and the consequent exclusion of all visitors with mouth-parts shorter than the effective tube length.

Tubes may be central, as a corolla tube, subcentric, as a spur (*Linaria* or *Gymnadenia*), or excentric, as in *Delphinium* and *Aconitum*. The sub- and excentric types may be convergent with (*Linaria*) or divergent from (*Aconitum*, *Viola*) the main axis of the blossom. Tubes and spurs are usually single, double in *Aconitum*, and *Aquilegia* has multiple, subcentric convergent spurs. Many composite heads are also multitubular.

The tube appears as a type in its own right, with no limb in *Castilleja*, *Kentrosiphon* (with laternal entrance; Vogel 1954), *Symphytum officinale*, etc.

Usually the tube or spur is formed by the corolla, but the spur in *Impatients* and the tube in Silenoideae are formed by the calyx. In some Mimosoideae fused filaments form a tube. In *Pelargonium* the tube appears to be an axial structure. As the main function of the tube is to hide the nectar, nectaries are frequently formed in tubes, or even form tubes within other blossom types.

Like the two preceding types, tube blossoms are generally single flowers, but may be inflorescences (*Mimetes hirta*).

From an evolutionary point of view the tube type is a continuation of the typological line dish—bowl—bell—funnel, the unit becoming relatively narrower all the time. However, a qualitative difference is that a pollinator cannot enter a tube; it has to stay on the outside and suck up the nectar by means of a long proboscis. It must carry the pollen on the front part of the body, generally on the head.

The typological development within different lines towards more complicated blossom shapes generally goes together with a development from choripetaly to sympetaly; simple types are choripetalous, the more refined ones sympetalous. But exceptions abound, choripetalous tubes in *Malvaviscus* or Silenoideae (synsepalous) or gullets in *Acanthus* mainly formed by free members (cf. Schremmer 1960). The flag type is generally choripetalous, but may be sympetalous as in *Trifolium*. It should also be remembered that the fantastic flowers of orchids are basically choritepalous.

By way of comparison, gullet and flag types provide for landing of visitor and deposition/reception of pollen, whereas the tube type in itself provides for the latter function only. Landing, if any, will have to take place outside the tube, even the blossom. In gullets the landing and pollen deposition functions are separated. They are combined in the lower part in flags, the upper part of the blossoms being left with only an advertising function. Table 3 summarizes the subtypes of group I.

Thus, while there may exist *transitional* types between the preceding ones, the types themselves cannot exist in combination. The tube, on the other hand, can also exist in combination with other types as a spur or as a corolla tube. *Narcissus poeticus*, *Primula*, and *Myosotis* have flat corollas, type A, combined with tubes, a very frequent combination called, wrongly, "salverform".* *Oenothera* species have a funnel, type B, on top of a tube; many funnels continue downwards in this way.

In the tube blossom the existence or non-existence of a rim is of even greater importance than in the bell-funnel type because the absence of a rim prevents a visitor from alighting. And as the tube is too narrow for visitors to crawl into it, all non-hovering visitors are thus excluded unless they can find some adjacent perch. We may therefore subdivide this type into two subtypes, tubes in a strict sense, and what we might call the *trumpet* type, which is a tube

---

*A salver is a *flat* dish, and a salver-shaped blossom should therefore belong to type A. The German term *Stielteller* is adequate, but untranslatable. "Hypocraterimorphous" is hardly usable in practical work.

TABLE 3. Main blossom types and their way of functioning

	Dish	Bell–funnel	Brush	Flag	Gullet	Trumpet	Tube
Visual attraction	Diffuse, if any	Corolla or substitute	Diffuse	Standard	Both lips	Margin	Generally supplemented by other parts
Alighting	Diffuse	More exact	Diffuse	Carina	Lower lip	Margin	None
Guiding	None (or traps)	Some	None	Symmetry, marks on standard	Symmetry, build of lower lip	Towards a central opening	Automatic
Displaying of attractant	Diffuse, open	Halfhidden, ± centralized	Diffuse, open	Well hidden, entrance to be forced	Well hidden	Hidden	Deeply hidden
Pollen deposition and reception	Diffuse, inside	± central, inside	Diffuse, outside	Sternotribe by carina	Nototribe, in upper lip	Central, inside	Varied
Primarily adapted to insect behaviour	Primitive (beetles)	Crawling in (bees)	Alighting visitors with longer mouth-parts (bees, butterflies, birds)	Alighting, forcing their way in (higher bees)	Alighting, forcing their way in (higher bees)	Alighting, not crawling in (butterflies)	Hovering or perching on adjacent structures (moths, birds)

with an extended upper rim, on which visitors can alight. All the examples mentioned above are examples of such trumpet subtypes with a more or less pronounced salverform rim. In *Epidendrum o'brienianum* (which has a tube formed by gynostemium + labellum) the other tepals form a salver at the base of the tube, upon which visitors alight (v. d. P.).

But the tubes may also be combined with other blossom types into more complicated shapes. In many brush units the nectar is hidden at the bottom of tubular florets as in *Eupatorium*. A combination of gullet and tube (spur) is found in the long-spurred orchids and in *Linaria vulgaris*, a corolla tube in *Lonicera periclymenum*. Flag and tube are combined in *Trifolium*.

The two other main groups, traps and closed blossoms, form parallel series to group I. The closed blossom may be a simple, radiate, bowl-like flower that has closed, as in *Trollius europaeus*, or it may be derived from a complicated, zygomorphic asymmetric gullet type, like *Pedicularis sceptrum-carolinum*. Similarly, a trap may be a bowl (*Nymphaea*) or a complicated tubular structure (bell type: *Arum*), sometimes incredibly complicated as in *Aristolochia grandiflora* (Cammerloher 1923). Traps may be more or less active (orchids) or passive (Araceae), but transitions occur (Nymphaeaceae). Obviously, some types, e.g. brush blossoms, cannot develop into these two other main groups.

Some closed buds are opened forcibly by the pollinator, and stay open afterwards (Kennedy 1973). Others remain closed, also after having been entered, and are in most cases complicated structures that demand both strength and intelligence from the visitor, and therefore chiefly operated by bumblebees; but exceptions occur, e.g. *Trollius europaeus* or those tropical Annonaceae into which beetles are irresistibly attracted by the odour. This also raises the very difficult question about how pollinators originally find out that there is an attractant inside a closed blossom. In the beetles one may presume a direct olfactoric reaction, but that can hardly apply to the way bumblebees, for example, are attracted to the intricate, closed flowers of *P. sceptrum-carolinum*.

It should be noted that blossoms may be closed in relation to one pollinator, open in relation to another, e.g. *Linaria vulgaris*, which must be opened by bumblebees, whereas moths are able to get at the nectar without opening the blossom (and without pollinating! Knoll, 1922).

Traps seem to occur in two connections. In some blossoms trapped animals have been attracted by deceit and have every reason for a hasty departure if they were not detained (*Arum maculatum*, so-called *Kesselfallenblumen*, cf. Vogel 1965a). In others the trapped animals are in the blossom for "good" reasons, but are, we may assume, too stupid to be properly co-operative (*Calycanthus*). Most traps operate on the principle that the detaining structures cease to function after some time, leaving the imprisoned visitors free to depart. In a few blossoms the visitors perish in the trap after having pollinated (Vogel 1965).

Other blossoms we shall call *semi-traps*; structures force the visitor to take certain actions in order to be released from the blossom, but do not enforce any time interval; examples are the bristles in the bottom of the *Pinguicula alpina* flower, or the exit route from *Cypripedium calceolus* or *Stanhopea* (Dodson and Frymire 1961a). In many Annonaceae, *Calycanthus*, and *Amorphophallus variabilis*, visitors remain voluntarily for nutritional, sexual, or "psychological" reasons until the blossom has changed or the urge is satisfied.

The border between such semi-traps and ordinary guiding structures in open blossoms in indefinite. Other semi-traps work on the principle that the visitor loses its foothold, either because parts of the surface are slippery (*Gleitfallen*) or because the organ to which it clings turns over (*Kippfallen*) like in many genuine traps. However, in contrast to the genuine traps there is no organized time lag between entry and exit; the visitor leaves the blossom immediately, forcing its way out if necessary.

In the preceding system we have tried to classify pollination units on the basis of the salient points of pollination in general. The system does not purport to divide blossoms in levels expressing anything about the sensory levels of insects. Thus, while some of these groups are stereometrically defined (A, C, flat; B, F, deep), D and E may run from flat types (*Dictamnus*) to very complicated, more or less closed ones (Papilionaceae). It is obvious that depth and functional zygomorphism of pollination units develop independently of each other, and that a linear grouping cannot be established. On the other hand, some of these types, in order to be worked successfully, presume a higher "flower intelligence" in the visitor than to others. The extreme flag and gullet blossoms are inconceivable without the sensory and bodily characteristics of bees, and the frequent occurrence of traps in fly- and beetle-pollinated blossoms reflects lack of "flower intelligence" in these animals. The stereometric sense also differs in different groups of pollinators.

The general form of the blossom exerts a very strong compulsion on a visiting animal, especially on the more highly evolved pollinators. A bumblebee that visited the upper side of the stigmas of an *Iris pseudacorus* (licking some unidentified substance), landed each time "correctly" for pollinating, i.e. on the outer tepal, after which it had to creep up onto the stigma. Obviously, it was under such a strong compulsion from the general form and nectar guides, of the blossom, that it was psychologically unable to effect a landing in the best place for its immediate purpose, though it was physically quite possible (K. F.).

It has been maintained — and experiments seem to indicate this — that many details of form are irrelevant, either because they are beyond the comprehension of visitors or because these ignore them. This does not, however, imply that such structures must always have been "useless"; it is conceivable that at any rate some of them may have had a special function in another syndrome. And there is no doubt that under experimental conditions insects visit rather crude blossom models and mutilated blossoms apparently without discrimination as compared with the perfect specimens. It is questionable, however, whether this would apply under natural conditions with competition entering. There are also some other obscure points here, most of them referring again to the problem of conditioning and recognition by means of more or less complete syndromes, cf. Knoll's experiments (1922) in which *Macroglossa* first visited (*Linaria vulgaris*) flowers with a nectar guide, and only afterwards those that had been operated on.

As a last point we should like to stress the parallelism in development of blossom types. The same types appear in many, frequently unrelated, families, often based upon different morphological elements: in one case a flower, in a second an inflorescence, in a third petals, in a fourth the filaments, etc., carry out the same function. Examples are found in gullet blossoms which in Labiatae are formed by the corolla, in *Anapalina* by the perianth, in *Acanthus* by corolla plus sepal, in *Mimetes* by a whole inflorescence, and in *Iris* by a tepal and a stigma.

There is no linear development leading from one type to the other, and one does not generally represent a higher phylogenetic stage than the other. The series from type A to type E represents a series of increasing complication, and many phylogenetic developments have gone in that direction, but reversions of direction did also frequently occur.

CHAPTER 11

ANIMALS AS POLLINATORS

It is customary to refer to butterfly flowers, bee flowers, etc. This usage reflects the fact that large animal groups are morphologically and consequently ethologically similar, and tend to find their nourishment in the same way. If they get it, or part of it, from blossoms, they tend therefore to utilize blossoms of similar types. The terms are also based on the, more or less subconscious, idea of harmony between visitor and blossom. Lack of this harmony will prevent pollination and may prove fatal to the visitor. At the best, nothing comes out of such a visit; too small visitors may do some damage reaching the nectar or stealing the pollen without effecting pollination; at the worst too strong or rapacious visitors may destroy the blossom, too weak visitors may stick to the stigma or be unable to extricate themselves from some narrow crevice or tube, etc.

This principle of harmony is admirably expressed by Vogel's (1954) concept of *Stil* (style, pattern), marvellously well demonstrated by his plates. Recognition of these harmonious types and patterns permits pollination ecologists to draw detailed inferences about the pollination of plants observed outside their natural habitats, perhaps even in a herbarium, but such conclusions can only be hypotheses requiring verification by observation of actual conditions (Fig. 3). Thus, there is no doubt that the pattern of *Lonicera periclymenum* is phalaenophilic which does not, however, prevent pollen-eating syrphids landing on an anther or style from being at least locally equally important as pollinators (K. F.), and even very long-tongued bumblebees may with difficulty work the blossom. Needless to say, the same pattern may develop on the basis of different morphological elements.

A kind of reverse effect of this principle of harmony is the correspondence between the pollinators of a certain plant or plant group, notwithstanding their mutual taxonomic position (Kikuchi 1962).

Observed interdependence between blossoms and pollinators always depends on this harmony. But the argument cannot be reversed; the absence of a possible pollinator may be due to other factors. Kugler (1955a) has shown that the apparent restriction to fly pollination for some plants that belong to shaded, damp places is due to the fact that flies abound in such biotopes, where, for example, hymenopters are rare; obviously, harmony or no harmony, an insect can only pollinate within its own biotope.

But it should not be forgotten that all such type designations are extremely generalized. Flies may be anything from a non-specialized omnivorous creature like the house-fly to some of the most specialized pollinator types known, with a proboscis comparable in length to those of butterflies. Bees may be anything from primitive, solitary types, to the highly organized honeybee, etc. In the following we shall mainly deal with the "typical" representatives of the group, and refer to the others at appropriate places.

With this restriction to the "typical" representatives we have summarized the blossom preference of the main groups of animal pollinators in Table 4. Dystropic flies should be

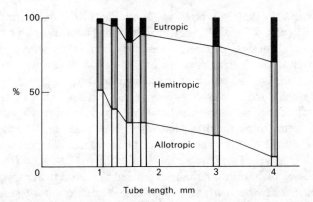

FIG. 3. Harmony between blossom and pollinators: the diagram shows how, with increasing length of the corolla tube in North American *Aster* species, the number of allotropic visitors goes down, because they cannot reach the nectar. At the same time, the number of eutropic increases, indicating their preference for the long-tubed flowers. The example is important because the blossoms are otherwise very similar to each other (after Graenicher 1909; cf. also Brian 1957).

TABLE 4. "Harmonic" relations between pollinators and blossoms

Blossom class	Pollinator	Colour preference
Dish bowl	Beetles	Brown
Bell beaker	Wasps	Drab
	Flies	White
Brush	Bats	Yellow
Gullet	Bees	Blue
	Moths	
Flag	Butterflies	Red
Tube	Birds	Green

entered in the table in the same place as "Beetles". On the other hand, carrion beetles conform with the entry "Flies". "Bees" include both bumblebees and blossom-flies, and "Birds" also diurnal moths. Further, food attraction is the only type of attraction considered, and closed and trap blossoms are left aside. The colour preferences reflect statistical evidence and have no reference to ethological reactions.

It should be emphasized that even with these reservations, Table 4 gives the main types only. There is, of course, nothing remarkable in butterflies visiting a yellow blossom, but their preference seems to be for other colours. "Green" only includes cases in which that colour appears as a bright, vivid contrast in the blossom, excluding the inconspicuous units of the *Reseda odorata* type. Furthermore the "red' is the vivid, bright colour.

In Table 4 classes of blossoms have been entered in a systematic, typological sequence. The other definitions have been arranged to give the shortest connecting lines. It will be seen that pollinators fall into two groups: the first half comprises unspecialized groups (individual taxa may be specialized), the second half the specialized. Similarly, the colours in the upper half of the column are all dull, in contrast to the vivid colours in the lower half of the column. It might also be added that the further down we come in the diagram the more exclusive is nectar as an attractant.

It cannot be too strongly stressed that the generalizations inherent in the blossom-type concept must not become a mental straitjacket and preclude the appreciation of visits by other groups of visitors, the adaptations of and to whom is less apparent. Such visitors may pollinate or not; in the latter case they may deplete the blossom of attractants and play a negative role in the pollination game. Also, the same blossoms may be visited in a different manner according to which attractant is operative for this special visit, e.g. pollen or nectar in many Rhinanthoideae. Not only may different castes of the same visitor species behave differently (e.g. bumblebee queens and workers), but even the same individual may participate in different syndromes in the same blossom.

In the following, we shall first deal with insects, which constitute the great majority of known pollinators, and later with the group of vertebrate pollinators.

11.1. INSECTS (INVERTEBRATES) AS POLLINATORS

The life of an insect is more or less distinctly divided into two phases: a larval phase during which the insect grows and develops its organs, and an imago phase during which no more growth takes place, and which is the time of mating.

The food requirements of insects differ greatly during the two phases; the larvae need a balanced diet to grow and develop; imagines need energy food to keep their activities going. Some imagines are equipped with so much reserve food upon emerging that they do not feed at all, e.g. the *Yucca* moth; others eat a great deal. Their food preferences and requirements vary greatly, but the important thing is that theoretically imagines ought to be able to subsist on carbohydrates only as these should provide the necessary energy. The main exception is the protein necessary for egg-laying, but even that may have been carried over from the larval stage.

In honeybees some development of organs takes place in the adult state under the influence of protein food, which also increases the longevity of the animals (Maurizio 1950b).

So far as is known, only imagines are of interest as pollinators.

The "blossom intelligence" of pollinators, i.e. their ability to perceive, discriminate between, and remember the characteristics of blossoms, varies enormously, reaching its peak in the higher bumblebees and the honeybees. The differences existing between different types of butterflies seem to be differences of structure rather than of "intelligence". The differences between primitive insects and their highly evolved relatives parallel the differences between the main taxonomic units, but, for example, a primitive bee may be both structurally and ecologically further removed from its most highly evolved relatives than from a primitive wasp or even a fly.

Leppik (1957) has advocated the view that the different stages in blossom differentiation (amorphic, etc.) correspond to different levels of pollinator intelligence, and that a pollinator cannot successfully work a blossom of a higher "intelligence level" than its proper one — but it may easily work one of a lower level. Similar ideas have been proposed, for example, by Loew in respect of his allotropic—eutropic series. Such rules represent interesting and useful generalizations, but they are in no way infallible. Insects developed before flowering plants came into existence.

Plants form the food of a majority of insects, which may eat almost any part. The soft, often succulent and sweet blossom tissues are a favourite food of many species, as garden owners know only too well. Such more or less destructive visits may effect pollination, and many insect groups that are mainly looked upon as predators may in reality be of some importance in pollination, e.g. hemipters (especially Capsidae). According to Whigham (1974) non-flying staphylinid beetles are regular (exclusive?) pollinators of *Uvularia perfoliata*. Porsch (1958) has collected some of the available evidence about the blossom visits of representatives of groups that are usually not reckoned as pollinators. Most of them, in accordance with Leppik's principle, are found in rather primitive blossom types, but even such a primitive and unadapted insect as *Chloroperla torrentium* has been observed as a pollinator of *Listera ovata*, which, in any blossom classification, must rank in a high level.

11.1.1. Beetles and Beetle-pollinated Blossoms. Cantharophily

The rôle of beetles in pollination has largely been overlooked, and beetle flowers are still too little known, even after Diels' demonstration (1916) of cantharophily and its importance (Grinfeld and Issi 1958; Meeuse 1959). One reason is the fact that the general syndrome of beetle pollination is rather uncharacteristic; owing to their lack of specialization, beetle flowers are frequently overlooked, and the visits of beetles in blossoms considered accidental — which they sometimes, but far from always, are. Another important reason is that typical beetle pollination is rare in the European, extra-tropical flora, and prevailing concepts in pollination ecology still depend largely on the European tradition. Beetle pollination is as characteristic of the tropical zone as bee pollination is of (semi-arid) temperate regions.

Beetles (Coleoptera) constitute one of the oldest groups of insects. They were already numerous at the time we assume that higher plants first came into existence, during the Upper Jurassic or Lower Cretaceous, whereas the higher hymenopters and lepidopters, so important in pollination today, had not developed. So, if insects had any function in pollination at that time, we may safely assume that beetles stood at the cradle of the flower. And we have already seen that beetles had acquired the pollen-eating habit at lower phylogenetic stages (cycads, probably also microspore-eating in cryptogams) and potentially also in earlier geological epochs. Today, while some beetles are more or less accidental visitors of blossoms, others are habitual visitors, and have acquired adaptations for blossom visits.

There is every reason to believe that the orthognatic position (perpendicular to body axis) of the mouth-parts seen in most beetles is the original one. In animals with the ethology of beetles, this position limits the length of the mouth-parts, and the orthognatic beetles are therefore, if interested at all, only able to lick up nectar from very open flat blossoms, but they are very well adapted for chewing pollen. According to Matthes and Schika (1966), the mouth-parts of some beetles are modified for pollen chewing. Such beetles are dystropic, or allotropic in some cases. In some groups the position of the mouth-parts changes to prognathic (parallel to the body axis); this gives a much better position for probing into the crevices in which nectar is hidden ("flower-beetles are acuminate in front"; Olberg 1951). The extreme is reached by the genus *Nemognathus*, in which the length of maxillae reaches or even exceeds that of the body of the insect. Such beetles are evidently eutropic as the morphology of their mouth-parts restricts them to deep-lying nectar and prevents them from utilizing most other sources of food. Even if very little is known about their psychology, they must certainly have a definite "idea" of the depth of a blossom.

However, most adaptations found among beetles are rather crude. They have gnawing mouth-parts, and their instincts are evidently very primitive; they are unable to land precisely in a blossom, hence they can only handle the simplest blossoms, chiefly of the dish-and-bowl type with pollen (food bodies), or very open, accessible nectar as an attractant. Pollination is by mess and soil, while the big, awkward, and, as insects go, heavy animals move about. Any blossom constancy will be imposed. An exception should be made for carrion beetles, which will be considered with sapromyophily. The smooth bodies of most beetles are badly adapted for pollen transport (Kendall and Solomon 1973), but in typical blossom-visiting species not only are the mouth-parts metamorphosed towards a shape more suitable for pollen collecting, but also the bodies develop hairs or scales to which pollen grains can adhere (Fuchs 1974).

It has already been suggested that protection against the rude handling of ovules by these awkward animals may have been one of the major selective advantages of angiospermy and of epigyny in ovaries. Gottsberger (1974) proposes that the original flower-visiting beetles were small and gentle, even calling them anthophilous (*loc. cit.* 467). We doubt that such beetles, if they occurred, were the ones to spur the development of the flower. As pointed out in Section 8.5, the flower-breeding beetles represent a completely different phenomenon, and cannot be accepted as part of the evolution of ordinary pollination syndromes.

Tropical beetle flowers certainly represent an original phase in the development of pollination (van der Pijl 1960, 1961, 1969) in which both pollinator and blossom have been able to retain their primitive character because of the protective homeostasis of the biotope, which also offers food for larval (and imaginal?) stages. It is remarkable how frequently these primitive beetle—blossom relationships are based upon deceit: the insects are irresistibly attracted by the strong odours (fruity in Annonaceae, fermenting in *Calycanthus*, or spermatic) notwithstanding whether or not the blossom also offers some compensatory, primary attractant (pollen, food bodies).

Magnolia is one of the primitive genera for which beetle pollination is an almost automatic assumption. This has been corroborated (cf. Heiser 1962). Higher pollinators, like bees, are ill at ease in these large blossoms which have no guide structures (like a small child in a big bed).

Another genus in which beetle pollination was suspected is the generally primitive *Degeneria*. This has now been corroborated by Takhtajan (1973: 24 and pers. comm.). However, in this blossom there is a further development towards saprocantharophily.

In the European flora the frequently neglected beetles are found on flat, open inflorescences with exposed nectar (and pollen), e.g. umbelliferae or *Sorbus aucuparia*, the latter with a strong aminoid (trimethylamine) odour component.

In tropical and subtropical forests, various Fagaceae with a strong spermatic odour are beetle-pollinated. Grant (in litt.) found that in California *Lithocarpus densiflorus* is pollinated by beetles feeding on the smelly, sticky pollen. They perch on the female flowers at the base of the catkin.

These four instances exemplify the two types of beetle pollination distinguished previously by Delpino (1868–75): the big, single flower, frequently amorphic or haplomorphic sensu Leppik, and the inflorescence. In some cases the latter is obviously a more advanced type, which is also found in more advanced families: umbellifers, *Cornus*, *Viburnum*.

We agree with Gottsberger (1974) that too frequently the discussion of the phylogeny of the flower has been sidetracked by the showy, terminal *Magnolia* flower (which is a highly specialized beetle blossom; Thien 1974), and that the earliest flowers were probably more similar to the smaller ones in inflorescences like those of Winteraceae and similar families. Their main attractant was probably odour, not visual. Especially if the line of development is projected back to a pre-angiosperm stage the attractant – if any – can hardly have been visual.

This raises the question if beetle pollination may in some cases be secondary, i.e. represented in taxa, the ancestors of which were pollinated by more highly developed insects. There is hardly any doubt that this is the case in *Leucadendron discolor*, the (musty and drab) inflorescences of which have returned to the magnolia stage so completely as to cast doubt even on the morphological character of the *Magnolia* flower (cf. Fægri 1965). Similarly, Grant and Grant (1965) found revertence to beetle-pollinated, flat inflorescences in Polemoniaceae (*Ipomopsis congesta*). Probably the same holds true for most of Delpino's second type.

There is a gradual transition to the aminoid odours attracting carrion and dung-beetles, which are here treated together with sapromyophily.

The famous *Amorphophallus titanum* is pollinated by a large silphid beetle (*Diamesus*) regularly bound to carrion in the same biotope. *Victoria amazonica* attracts, imprisons, and feeds (with food bodies) its large dynastid pollinator, *Cyclocephala hardyi* (cf. Valle and Cirino 1972), which is also found as a gnawing pollinator in the flowers of some Lecythidaceae (Prance 1976).

Traps force beetles to stay longer in blossoms (cf. case history of *Calycanthus*). The chances of successful visits by these unco-operative insects are thereby increased, cf. the aquatic aroid *Cryptocoryne griffithii*, in the 30-cm long underwater tube of which beetles "overnight in an underwater cabaret" (Corner 1964). Protogyny is an essential element in the function of this type of trap blossoms.

Nectar-feeding beetles are obviously a late development, and there is no special blossom type corresponding to the highly developed flower beetles. In spite of the odour of decaying protein in some cycad inflorescences we would consider carrion attraction as secondary in angiosperms.

In addition to food attraction, beetle visits to blossoms may also be caused by rendezvous attraction (Porsch 1950: 290), and there are indications that regular sexual attraction may play a rôle (Kullenberg 1973b).

Syndrome of cantharophily. Pollination units with few visual attractions, no special or definite shape, no depth effect, no nectar guide, generally large, flat, cylindric or shallow, bowl-shaped — sometimes closed — easy of access (beetles being poor fliers). Colours dull, frequently greenish or off-white. Odour strong, fruity, or aminoid. Attractants open, easily accessible, pollen, food-bodies, or nectar. Sexual organs exposed. This syndrome should not obscure the fact that beetles are habitual pollinators also of more evolved blossoms, even orchids (Cady and Rotherham 1970: 38).

11.1.2. Flies and Fly-pollinated Blossoms. Myophily

Greater variation of methods and habits of pollination is found among Diptera than among any other insect group. Where habits and mental capacity are concerned, many primitive flies can be ranked with beetles. From these dys- or allotropic types, there is a gradual transition to more highly evolved blossom visitors, e.g. the wasp-like syrphids which still have a short, typical fly-like proboscis, and which collect and chew pollen; further to the bumblebee-like Bombyliidae, with probosces approaching 10 mm in length, enabling them to utilize rather deeply hidden nectar. The final step is represented by the highly evolved South African tabanids and nemestrinids, with a proboscis length approaching or even exceeding the 50-mm mark (Vogel 1954). Both with regard to physical capacity and ethological possibilities, these specialized flies are comparable to the corresponding groups within bumblebees or hawk-moths, even including the ability to hover when visiting a blossom.

Unspecialized flies are restricted to more primitive blossoms. The operative factors are the shortness of the proboscis and the small size of the animal. They are chiefly found therefore in blossoms with open nectar or very short tubes, like umbellifers, *Hedera*, *Ilex*, *Veronica*, *Circaea*, etc. Although these flies are more or less omnivorous, they generally visit blossoms for nectar exclusively, but conditions vary. Among specialized "bee-flies" Knoll (1921) has shown that whereas *Bombylius medius* also feeds on pollen, its close relative *B. fuliginosus* in blossoms feeds on nectar exclusively. It remains to be seen if such rules always hold true, or whether they are modified by external circumstances.

As flies do not nurse their brood, they only take food for their own consumption, which means that they chiefly need carbohydrates for their energy metabolism. Hemitropic forms, like syrphids, may obtain their protein from pollen; dys- or allotropic forms generally get their protein from other sources which do not demand as much chewing.

Flies are not as busy collecting food as those insects that nurse their brood and must collect for their young too. As they generally utilize many different sources of food, their pollinating activity is irregular and unreliable. The absence of local opportunities for breeding may also reduce the extent of fly pollination, as in the case of *Theobroma*. Absence of opportunities for breeding results in irregular and unreliable availability of pollinators; only rarely is this compensated for by the plant itself offering facilities for breeding. In an orchid genus like *Pterostylis*, where each species has its own midge pollinator, the survival of species may have depended on the development of autogamy or even of cleistogamy. On the other hand, less specialized blossoms suffer from random fly visits, which going between different species, rarely bring compatible pollen to any one in sufficient quantity.

On the other hand, flies may be important under certain climatic conditions because they are present at all times of the year, unlike the strictly periodic and more demanding bumblebees and bees. Some plants flowering at odd times of the year, e.g. *Hedera helix*, may in northern countries be completely dependent on flies for their pollination, and this may also apply to plants in adverse climatic conditions (the Faeroes, Hagerup 1951; the Arctic, Kevan 1972). The proximity of a suitable breeding-ground, frequently wet, decaying vegetable material, is important for the pollinating activity of such small insects, which do not travel far.

Kugler (1955a) has shown that many small hymenopters, in most cases beetles and lepidopters as well, frequent the same blossoms as flies. Even if they thus do not possess the exclusivity presumed by the classical writers, fly blossoms form a fairly distinct type well named after its most characteristic pollinator, the unspecialized, usually small fly.

The pollinating effectivity of small dipters — like any small insects — is limited by their size. They cannot carry great pollen loads. In larger blossoms small pollen feeders often do not reach the stigma, and nectar eaters may miss the anthers as well. In small blossoms they can be very effective, and their limited transport capacity is compensated by their number. Literally hundreds of them may visit an inflorescence of, for example *Aruncus* at the same time (K. F.), and in the insect-pollination syndrome of tropical grasses they seem to be of primary importance (Soderstrom and Calderon 1971). Because of their low energy demand they are satisfied with minute quantities of nectar, like those in very small flowers. This ecological niche is exclusive to the small pollinator.

The psychology of flies is less well known than that of hymenopters. However, it is known (Kugler, 1956) that ordinary flies (*Lucilia, Calliphora, Scatophaga*) possess a power of discrimination between some colour groups (at least yellow and blue), and that they show a positive preference for yellow (except in the presence of decaying protein odour, when brown-purplish hues are preferred). Radiating marks (nectar guides) have a certain positive effect, and the positive effect of filiform appendages has also been demonstrated experimentally. That blossom specialists among dipters, such as *Bombylius*, can distinguish between colours, has been known since Knoll's earliest experiments (1921). On the other hand, even a specialized pollinator like *Bombylius* does not exhibit any blossom constancy (Knoll *loc. cit*). van der Goot and Grabandt (1970) found constancy in some syrphid individuals, in others not. The species in question was apparently specialized on pollen of anemophiles (grasses, *Plantago lanceolata*).

Syndrome of fly blossom. Blossom regular, simple, no depth effect. Colours generally light, but dull. Nectar guides frequently present. Odour imperceptible. Nectar open or easily obtainable. Sexual organs well exposed.

Besides this series there is a completely different ecological group of dipters also attracted to blossoms, viz. carrion and dung-flies. Together with the corresponding group of beetles they represent allotropic, but regular pollination. Carrion and dung-flies belong to many different taxa. Some of them are curiously small and may occur in large numbers in one blossom.

In sapromyophily there is naturally no adaptation of the flies for flower visits — the basis for the visit is deceit: substances released by the blossom activate the instincts for feeding or oviposition (for a list of amino acids, amines, and skatoles released by stinking aroids, cf. B. Meeuse 1966). Consequently such adaptations as exist are all found in the blossom. Sapromyophilous blossoms are found in various taxonomic groups, and it is instructive to see how the external appearance of the blossom is similar in all these groups: Asclepiadaceae, Aristolochiaceae, Sterculiaceae, Rafflesiaceae, Hydnoraceae, Taccaceae, Araceae, Burmanniaceae, Orchidaceae. All these families are highly evolved. The type is rare in

Polycarpicae but has been reported from Annonaceae. They may be primarily cantharophilous. Sapromyophily must be a later development, succeeding other pollination principles, in orchids thus coming after melittophily. It is interesting to see that, whereas *Stapelia* in adapting to desert conditions developed sapromyophily and a cactoid habit, the Cactaceae never developed sapromyophily — possibly because of innate morphogenetic limitations or lack of suitable vectors. The constant occurrence of flies' eggs on *Stapelia* blossoms in nature shows the success of deceit in this genus.

Typical carrion or dung-flies are uninterested in blossoms as such and have no incentive for visiting special parts of it, except to go against the gradient of odour, "expecting" to find some protein as its source. Disappointed in this expectation they will usually, after some aimless peregrinations, try to leave the blossom. More than with any other type, sapromyophilous pollination depends on traps: one-way bristles, slipways, see-saw petals contribute to the effectivity and precision of trap mechanisms. In most of the above-mentioned families, traps are known. Typical and well-authenticated examples are those of *Arum maculatum* (Dormer 1960) and various *Aristolochia* species. There are many strong similarities between these blossom traps and the leaf-traps of insectivorous plants belonging to Nepenthaceae and Sarraceniaceae with *Darlingtonia* as the most highly evolved (window openings, light windows, colour, appendages). Even in Phallaceae very similar structures are found in *Clathrus*, *Aseroë*, etc.

Ordinarily, carrion flowers have nothing real, or very little, to offer the imago visitor. In *Aristolochia grandiflora*, Cammerloher (1923), and in *Eucomis*, Pascher (1959), have described the simultaneous presence of nectar and carrion odour. As many insects which place their brood in carrion or dung, eat sugars themselves, this is interesting but not especially unexpected.

Some rather extravagant features recur in sapromyophilous pollination units so frequently as to belong to the syndrome. They must in some way or other be connected with the instinctive reaction of the insects concerned, but it is hardly known how. The first is the frequent occurrence of window openings, through which the flies crawl into the blossom — or the trap. The folded-over tip of the *Arisaema* spatha forms two such openings, but the lantern type blossom of *Ceropegia* (L. Müller 1926; Vogel 1961) is a more elegant example. In most cases — if not in all — the presence of window openings seems to be combined with light windows towards which the flies crawl. The second and even more enigmatic feature, mentioned above, is the frequent occurrence of filiform appendages, sometimes large and stiff, sometimes slender and motile. They recur so regularly, from the drawn-out tip of the *Arisaema* spatha to the thread-like corolla segments of *Aristolochia tricaudata* or the labellum appendages in *Cirrhopetalum*, that there can be no doubt about their significance, perhaps as odour-emitting organs, too. A related feature is the presence of very small, slender, club-shaped trichomes (*Flimmerkörper*); they are also features of this class of blossoms. By their constant flickering they give the blossoms a hairy appearance, and may also "imitate" crawling flies already present in the blossom. Dark, hairy spots, like those found in Cypripedilinae, may have a similar function, both cases activating the aggregation instincts of flies (as proved by Wiesmann 1962).

These features must not be confused with the — sometimes extreme — division of perianth segments observed in moth-pollinated blossoms, nor with the filiform blossom type.

As mentioned in the previous chapter, the instinctive reactions of carrion and dung-beetles are so similar to those of the corresponding flies, that the syndrome of sapromyophily also includes the principle of coprocantharophily even if different animals react to different attractants. Thus in *Amorphophallus titanum* the terrible stench attracts special beetles only, no flies. The question arises whether the more fanciful attractions of sapromyophilous blossoms are also of importance in attracting beetles.

Dung and carrion insects are apparently able to locate their sources of food from a long distance, but it is uncertain whether this is due to an actual, extremely acute olfactoric sense to the corresponding odours, or whether it is based on the fact that these animals are constantly on the move (like cruising taxidrivers), so that after a short time they are bound to come within reasonable smelling distance of any quarry — real or imaginary.

Deceitful attraction is evidently the rule in sapromyophily. But there is an almost infinite variability in pollination, and every theoretical possibility is apparently met with. It is perhaps therefore not more than to be expected that some of these insects actually find a breeding-place and hatch in such blossoms as described in Section 8.5.

Syndrome of sapromyophily. Pollination units generally radial, but frequently with great depth, then developed as guiding traps, frequently lantern types and presence of filiform appendages. Colours dull, dark, brown—purple—greenish; no nectar guides or other leading lines, but chequered with dark spots and frequently with motile hairs or appendages. Transparent windows. Odour resembling that of decaying protein. Generally no nectar or other primary attractant. Sexual organs generally hidden in the interior of the blossom. As mentioned above, Kugler has demonstrated that whereas the dull, purplish, chequered colours have no special attraction for this class of pollinators under ordinary circumstances, the same colours do possess a positive attraction value in the presence of the odour of decayed protein. Their regular recurrence makes it very difficult not to ascribe a definite function to them.

Fungus gnats, Mycetophilidae, usually breed in mushrooms. Some specimens — of unknown sex — are found in sapromyophilous blossoms. Vogel (1973) has described a special blossom type in various genera (*Arisarum*, *Aristolochia*, *Asarum*, Cypripedium, *Masdevallia*), exposed near the ground, imitating mushrooms not only in odour (*Cypripedium debile*: "the odour of super-market mushrooms"; Stoutamire 1967: 168), but also in morphology, especially the lamellae. In *Arisarum proboscideum* the spadix is transformed to a mushroom-like structure which almost fills the opening of the spatha. Both sexes of fungus gnats congregate in this type of blossoms. They carry pollen and do pollinate; eggs are laid, but larvae do not develop.

In addition to these major types of fly pollination, some examples should be mentioned which are at present rather isolated, but which may, in the future, be found to represent more general principles, inadequately known today. For the pollination of *Trollius europaeus* by a fly as elucidated by Hagerup, see p. 175. The interesting thing about this pollination is that closed blossoms are usually unattainable for all except the most accomplished pollinators, but in *Trollius* such pollinators do not seem to be interested. In *Trollius* a very primitive pollination unit (class A) has been closed, belonging to a family in which pollination is usually primitive.

Another interesting but uncorrelated observation is made by Clements and Long (1923). The flowers of *Pachylobus caespitosus* have a corolla tube *ca.* 10 cm long, and are, as might be expected, visited by (hawk)-moths. However, when the blossoms open, anthers dehiscing and stigma receptive, these organs are covered with very small black dipters. That these animals must at least cause self-pollination seems obvious, but perhaps the solution of the enigma as to how they got there might also indicate whether they are pollinating these blossoms in other ways.

The blossom-visiting activities of flies (Muscidae and more specialized dipters) have been known since the earliest days of the study of pollination ecology. More recently has it been realized that blood-sucking diptera are very frequent blossom visitors. In addition to the blood sucked by the females, both sexes of at least many, perhaps all, species feed on sugar. The protein of the blood is apparently of importance for the development of gonads (some utilize the protein of pollen grains, which they slit open; Downes 1971: 244), but carbohydrates form the staple (energy) food of these animals (Haeger 1955; Nielsen 1963). They are allotropic, and will take honey-dew or contend with any other source of sugar as well as with nectar. Their effectivity as mess and soil pollinators in primitive blossoms is not known, but there is hardly any doubt that it has been underrated. Hocking (1968) describes mosquitoes (female *Aedes* sp.) functioning as alternative pollinators in the Arctic, also in small orchid blossoms (*Habenaria obtusata*), which are otherwise moth-pollinated (Stoutamire 1968; Thien and Utech 1970). The contents of amino acids in the nectar is hardly sufficient to maintain egg production.

On the night-blooming *Silene otites* in the Netherlands, *Culex* spp. (both sexes) were regular visitors in addition to the moths. Because of their small demands they were not very effective as pollinators, being able to satisfy their food demands from a single branch of the plant, but they carried pollen also when visiting female flowers (Brantjes and Leemans 1976). In blossoms that are too large to be effectively pollinated by small visitors, these may not only be useless, but their presence may depress the pollination effectively of legitimate pollinators, not least by partial depletion of nectar (Bohart *et al.* 1970).

These feeding habits of ordinary mosquitoes throw new light on some older observations of species that seem to be regular pollinators as testified by Schnuse and Hetschko (quoted from Porsch 1958). The mouth-parts of both sexes of these insects (*Opistomyia elegans* and *Liponeura cinerascens*) are transformed into a sucking apparatus better suited for nectar-licking than for blood-sucking.

The deceitful attraction of midges to the blossom of *Arum conophalloides* has been mentioned already. If the interpretation given is correct, this would represent something completely different from the blossom visits dealt with here. However, the observation of the pollination of *A. conophalloides* may be in need of reconsideration. Blood-sucking Ceratopogonidae are reported as pollinators in *Ceropegia* spp. and *Aristolochia clematitis*, but then the main food intake of many blood-sucking dipters is sugar to much a greater extent than usually realized.

Wilson and Lieux (1972) have found pollen grains in the guts of field-collected horseflies. They interpret this as an indication that these insects also obtain carbohydrates from blossoms, but considering that oak pollen was frequent, that cannot be the explanation. If the observation means anything, it should mean that horseflies obtain part of their protein requirement from blossoms — which would be highly interesting.

In the Faroes, where bees do not exist, Hagerup (1951) has shown that flies have taken over the pollination of a number of blossoms usually considered melittophilous, even including *Iris pseudacorus* and *Orchis maculata*. Some of these flies are ordinary types, but among them are the bumblebee-like *Eristalis* spp., especially *E. intricarius* L. which act as substitutes for the bees.

This replacement of bees with flies is found on a much larger scale in New Zealand, where pollinating hymenopters did not occur (Thomson 1927). Their ecological niche is there filled, amongst others, by various flies.

11.1.3. Hymenopters in General. Wasps

The large group of Hymenoptera comprises some of the most interesting, most highly evolved, and economically most important pollinating insects. However, as the type of pollinating activity within the various subgroups varies considerably, and as we shall have to refer to these subgroups later, we shall give a very short summary of the main ones. The summary has perforce been so arranged as to bring the blossom activities into the foreground (cf. Table 5).

Of the major groups of Hymenoptera, bees (Apioideae) and ants (Formicidae) will be treated separately in the following chapters. In pollination ecology it is customary to treat the remainder of hymenopters under the vague, collective term "wasps". These animals represent the most varied life-cycles, and in their relation to pollination ecology many of them correspond to flies: no brood management, a mixed diet which is only in part covered by nectar and pollen. Their mouth-parts are primitive, and the total length to the tip of the tongue is of the magnitude of 1−3 mm. The tongue is flat, and can be used for lapping up nectar only. The blossom visits of such "wasps" are restricted to allophilic blossoms with open nectar; open extra-floral nectaries are much visited by these insects. At any rate most of the "wasps" are as unreliable and unsteady pollinators as are the flies, and for the same reasons. This does not prevent some of them from being very important. But many of the recorded visits of "wasps" in blossoms − and they are a heterogenous mixture − are futile and are part of a general looking round for food. The instinctive apparatus to build up a systematic utilization of one or very few suitable blossoms is not particularly well developed in these animals. Whereas bees can be trained to distinguish between two or three different colours at the same time, wasps can apparently not be trained on more than one (Mazokin-Poshniakov and Grasvekaya 1966). *Vespa* can be trained to an odour and also has an excellent time-memory. Various ichneumonidae and sphecidae visit open-nectar blossoms and can thereby effect pollination.

The true wasps, Vespidae, and other highly evolved groups are social and nurse their brood; consequently their food requirements are great. However, they are primarily animals of prey and the diet of their larvae chiefly consists of animal protein, which is partly returned as amino acids to the imagines when they lick the larvae. Wasps are interested in carbohydrates chiefly for maintaining their own energy metabolism. This becomes more apparent with age, after the brood-rearing season is over and the colonies have reached full strength. In seasonal climates, this means that wasps are especially active in blossoms towards the end of the flowering season when there is no more brood to be nursed. In a study on the evolution of bees, Grinfeld (1973) postulated a development from carnivorous habit via the use of pollen to nectar utilization. This development does not quite correspond with the picture of social wasps which have retained their character as predators, but take nectar as a supplement. They are not interested in pollen.

In some higher wasps, mouth-parts have been modified into tube-like structures, 5−10 mm long, through which nectar can be sucked up. In their ability to utilize blossoms, these wasps approach bees. Some of them (*Polistes*) are regular pollinators, which even store nectar for the brood in addition to animal food.

Some predacious aculeate hymenopters, e.g. Sphegidae, frequently swarm near flowers, catching the "proper" pollinators. Together with occasional uptake of nectar this may have led to more stable relationships with and utilization of blossoms. For a record of such visits, cf. Kugler (1955b).

Some alleged "wasp blossoms" have since Herman Müller (1873), been characterized by

TABLE 5. The main groups of hymenopters

Group	Social life	Larva food	Imagines food
I. Symphyta, sawflies	Solitary	Phytophagous, a few parasites	Mixed, with some nectar and pollen. No preferences
II. Terebranthes, ichneumon-like wasps			
(A) Ichneumonidae and others	Solitary	Parasitic, generally in or on eggs or larvae of arthropods	Mixed, with some nectar and aphid excreta
(B) Chalcidoideae	Solitary	Generally parasitic, some few phytophagous, gall-producing	
(C) Cynipioideae	Solitary	Generally phytophagous in galls, some social parasites	
III. Aculeata Vespoideae			
(A) Pompilideae and others	Solitary	Carnivorous	Mixed, with nectar as one constituent
(B) Formicidae, ants	Social	Carnivorous and/or vegetarian, great quantities of nectar and other sugar-containing substances used. Social parasitism in some genera	Like that of larvae
(C) Vespidae, wasps (s.s.)	Solitary—social	Carnivorous with some, but generally very little nectar. Nectar and pollen exclusively in some masarids	Carnivorous and/or nectar
Sphecoideae	Solitary	Carnivorous	Mixed
Apioideae, bees		Nectar and pollen. Some cases of social parasitism	Nectar and pollen
(A) Prosopidae	Solitary		
(B) Andrenidae	Solitary, gregarious		
(C) Megachilidae, leafcutter bees	Solitary		
(D) Bombidae, bumblebees	Solitary—social		
(E) Apidae, honeybees	Solitary—social		

dull, brown colours: *Scrophularia nodosa* (cf. Schremmer 1959; Shaw 1962) is one of the classical cases. As many wasp visits are post-floral, and as bees are also busy in these flowers, the rôle of wasps as pollinators has been doubted. On the other hand, *Vespa* immediately "recognized" foreign *S.* species in a Dutch garden (v. d. P.). Wasps have also been found to pollinate orchids with open nectar (*Epipactis* spp., Judd 1971, the glistening of the labellum may serve as a secondary attractant), *Symphoricarpus* etc.

In the Tropics the type may be represented by *Celosia argentea, Jatropha curcas, Manihot glaziovii* (v. d. P.), *Leea robusta* (Heide 1927). However, the adaptive problem is rather obscure and it is hardly possible to establish a syndrome of wasp blossoms, even if it should exist for blossoms adapted to visits by higher wasps. In Kew Gardens, hornets were feeding avidly on *Hedera colchicha* while at the same time flies were swarming round *H. helix* (K. F.). There is not much difference between the two species, but for some reason the wasps preferred one of them and the flies stayed away from it.

In addition to these "regular" wasps it should be remembered that hymenopters are active in a number of cases of "irregular" attraction: *Ophrys, Ficus,* etc.

11.1.4. Ants

If there was ever a scoundrel in the pollination drama, that rôle has been assigned to ants. Ants are notoriously fond of sugar, and will tap whatever source they can get at: blossoms, sugar-bowls, or aphids. As they are brood-rearing, they also need protein, and it is by no means above them to raid any anthers they can get at. On the other hand, ants are so small that they can sneak in and out of many blossoms, without even touching anthers or stigma. In addition, their bodies are hard and apparently not adapted for pollen transport. Ants are also gregarious and notoriously bellicose, and if they have invaded a plant they will take up an offensive position against any other insect trying to land on it — and the insects stay away. Ants therefore have been considered prototypes of nectar thieves, and innumerable are the devices which have been interpreted as designed to bar their access to flowers, with the sticky belts of the *Viscaria vulgaris* stems as one of the most spectacular.

On the other hand, the gregariousness and bellicosity of ants sometimes have a positive side. We refer here to the so-called *ant-guard*, which has been known, but inadequately, for a long time. The question of ant-guards in temperate climates is still rather obscure; von Wettstein (1888) has described ants visiting extra-floral nectaries in composites and keeping away beetles, thus preventing them from laying their brood in the inflorescences. The importance of ant-guards in the life of plants generally has now been established beyond doubt (cf. Janzen 1966; Elias *et al.* 1975a, b), but reliable observations on the possible influence of these ant-guards on pollination are still scarce.

In the Tropics, the carpenter bees (*Xylocopa*), which are indispensable pollinators, have a very strong tendency to rob blossoms of their nectar by stealing. In unadapted species all blossoms are often punctured. In other species, ants are attracted to extra-floral nectaries at the base of the blossom, and it has been proved experimentally that their presence there deters the carpenter bees from robbing the blossom and forces them to enter the legitimate way.* Such a function of extra-floral nectaries and ant-guard has been demonstrated in

---

*Cammerloher (1929) misinterprets the function of the guard in stating that they prevent all visits by pollinators.

Thunbergia grandiflora, Canavallia, and other large Papilionaceae, species of *Iris* and *Ipomoea* (van der Pijl 1954), *Hibiscus henningsianus* (Gottsberger 1972), and *Bixa orellana* (Bentley 1977).

Vello and Magalhaes (1971) describe how the presence of an ant (*Azteca chartifex spiriti*) increases the pollination of cocoa plants in Brazil, although the ants themselves do not pollinate. A suspension of ground-up ants has the same effect.

Considering the number and omnipresence of ants, especially in the Tropics, one may say their absence from any blossom is in reality more remarkable than their presence, and one might ask why they stay out of, for example, *Thunbergia* flowers instead of raiding also the floral nectaries. Experiments have shown that not only are there many ant-deterring structures in blossoms, but many petals (especially) have a direct, probably chemically conditioned ant-repellant effect. Ants refuse to enter such blossoms (van der Pijl 1955).

Even without bringing into the picture the true myrmecophytes, disentangling the question of ants versus plants presents ample difficulties. Many of the interpretations of certain structures as ant-deterring are stretched, to say the least, even if the dense pubescense inside the corolla of *Symphoricarpus racemosus* or the narrow openings to many nectaries certainly do make it difficult for small, short-tongued animals of any kind to get at the nectar. It is obvious that ants crawling around in flowers and inflorescences may cause auto- or geitonogamy and that this will occur wherever they discover an available source of nectar or pollen. The importance of this for pollination may be questioned, especially if the geitonogamy obviously counteracts the principles upon which the pollination mechanism is founded, e.g. dichogamy.

What seem to be genuine cases of ant pollination have been described, for example, by Dahl and Hadać (1940) for *Glaux maritima* in south Norway and by Kincaid (1963) for *Orthocarpus pusillus* in the U.S.A. More pertinent may be Hagerup's observation (1932) of the importance of ants as pollinators in the Sahara;* near the burning hot soil very few other insects are about, and if biotic pollination is to take place, ants must be active. In Timbuktu, where these observations were made, Euphorbiaceae form much of the vegetation. Now this family has extra-floral nectaries, and ants are known to be especially interested in these. In Israel, van der Pijl (unpubl.) found harvester ants covering large distances of desert sand, visiting and clearly lapping up nectar from the blossoms of low cushions of *Polycarpon succulentum* (cf. also Hagerup 1943).

Hickman (1974) has described a well-substantiated case of ant pollination in *Polygonum cascadense*. He points out that ant pollination is a low-energy system: crawling ants spend relatively little energy on travelling, and establishes the following syndrome of ant pollination: dry-hot habitat; nectaries small, quantity of nectar too small to interest larger visitors; blossoms exposed near the ground, sessile, small with minimal visual attraction; few blossoms in anthesis at the same time, gregarious occurrence of several individuals further xenogamy; small quantities of sticky pollen prevents too eager cleaning, number of ovules per flower small.

11.1.5. Bees and Bee Blossoms. Melittophily

Bees are, on the whole, better adapted for blossom visits than the groups dealt with previously; indeed, they are better adapted than any other comparable group. They also

---

*In the South African desert, flies are apparently very important, but Vogel (1954) does not discuss the function, if any, of ants.

possess a whole range of patterns of behaviour from rather simple ones in "parasitic" and solitary bees to the incredibly complicated ones of honeybees and other social bees. In the following, the term bee will comprise all these except when a particular group is specifically mentioned.

Table 5 shows increasing interdependence between plant and pollinator, but it also shows increasing independence of the pollination process of the general synecology of the biotope. Lower bees, dependent on blossoms in general or often on one or a few definite species, are nevertheless also dependent on climate for time of emergence and on soil and climate for nesting opportunities. This imposes restrictions which have been felt by alfalfa growers. The solitary bees are also dependent on the community for nesting material, which is only rarely offered by the blossoms. Some exceptions are *Dalechampia*, the cyathia of which offer resins to *Trigona* (Cammerloher 1931), *Ornithidium* presenting wax on the labellum (Porsch 1905), and *Eria vulpina* a resinous fluid (Kirchner 1925).

In contrast, *Apis* has become more independent, and consequently more reliable as a pollinator, by producing wax itself (however, collecting propolis!), by being active all the year round if temperature is sufficient, and by being polytropic *cum* constancy. This genus could therefore spread to various vegetation types, including open plant communities. The genus belongs to the tropical and subtropical parts of the Old World and is, unlike *Bombus*, not indigenous in N. Europe or in N. America, where it has been a very strong, more or less disastrous competitor to the native solitary bees. In the tropical parts of the New World other social genera play the same rôle (*Melipona, Trigona*).

Among the lower solitary bees in Table 5 the Prosopididae have short mouth-parts, eat the pollen directly, and regurgitate it for the brood. The Megachilidae (leaf-cutter bees) are gastrilegic, i.e. they collect pollen ventrally, with some predilection for sternotribic blossoms of the flag type even if they can handle also other types (Schremmer 1953). Higher bees are podilegic: foot-collectors. Primitive types (Andrenidae) have hairy feet, which in the more specialized forms have developed into the hind-leg corbiculae. Higher bees tend to favour nototribic blossoms, but know how to handle any type. In oligolectic bees the collecting apparatus may be specially adapted to the one pollen type (stickiness, size of grains, etc.). We shall not enter upon the futile discussion which adaptation came first.

For details about the limiting length of the tongue, we refer to Knuth-Loew, to Kugler (1955b), and to handbooks of entomology.

Where they occur, social hymenopters comprise the most versatile, the most active, and, consequently, the best-known pollinators; so well known as to obscure realization of the great importance of other groups. Contributing to this are the facts that they are relatively big animals with a great demand for food both for themselves, and, even more, for the carefully looked-after brood, that they normally take all their food from blossoms, and that they are, if the expression may be permitted, more intelligent than other pollinators and able to operate successfully mechanisms that baffle other ones ("their wonderful instincts (apparently not unmixed with true intelligence)": Webster's New dictionary 1920 edition).

In areas where social bees are scarce, their physiognomic rôle is taken over by large solitary bees, which in tropical countries play a rôle similar to *Bombus* and its relatives in temperate ones. Solitary bees occur also in temperate areas, but are, on the whole, small and insignificant there, and many are rather ineffectual pollinators.

The large bees — social or solitary — are also stronger and can open doors that are closed to smaller and weaker insects, though occasionally their strength turns out to be disastrous for narrow blossoms which may split under the attack of a strong bee (Schremmer 1953).

The strength they exert during a visit is reflected in the marks their claws leave in the blossoms.

Their perception of and ability to remember plant forms also seem to surpass those of most other pollinators (von Frisch 1914; Kugler 1942; Bateman 1951). It should also be noted that apparently bumblebees are able to recognize — crudely — the general form of the whole plant, even without its flowers, as is shown in nature by their attempted visits to plants before the anthesis of any single flower (observed already by Darwin and demonstrated experimentally by Manning 1956). According to Attsatt (1970) pollinators (honeybees and solitary bees) open the closed flowers of *Orthocarpus*, and Kennedy (1973) has described how the explosive flowers of a maranthaceae are opened by bumblebees, and Estes and Thorp (1975) describe how matinal bees, *Hemihalictus lustrans*, tear open the anthers of *Pyrrhopappus carolinianus* and remove the pollen before it is available to other insects. The effectivity of the process is beyond doubt, but the instinct that triggers off the act may be more questionable.

On the other hand, bees do not seem to be able to recognize or discriminate between purely geometrical shapes (squares, triangles) which have no relevance to their practical needs under natural conditions.

There seem to be some subtler psychological differences between different groups of bees, perhaps with regard to colour vision, but also with regard to preference for zygomorphic versus radial, much versus less dissected blossoms, etc. For instance, bumblebees (*Bombus*) prefer the more complicated alternative (cf. Kugler 1930—42; Knoll 1922).

Besides anatomical differentiation, social bees have also developed a communication system permitting them to inform each other about the location and sources of food. In the adaptation of animals to blossom visits this represents the ultimate refinement, not found anywhere else among invertebrates.

The communication system of bees can be subdivided as follows (cf. Kerr 1960):

(1) Odour: returning bees bring with them the odour of the blossoms recently visited and spread it in the hive. The sense of odour in bees does not seem to be very different from man's — size considered.

(2) Odour marks (secretion from the Nasonoff glands) are left in the blossom and on the way home, creating an odour path from the nest to the source of food. The distance between individual marks varies, according to species, between 1—2 and 20 m (Haas 1952 and earlier). The significance of the odour mark in the visited blossom is less obvious — to save followers the work of searching for nectar?

(3) A special buzz tone alerting the other individuals.

(4) Returning bees make zigzag runs knocking into individuals not engaged in field work. This has developed into the fantastic dance communication system of the honeybees by means of which they can indicate direction, distance to, and yield of a source of food (Frisch 1950, Lindauer 1971). The direction orientation is by a sun compass with correction for time-lapse and wind-drift effects. Polarized light (prevalent in the sky region at $90°$ angle with the axis sun/earth) is important. In the Tropics this fails at midday, with the result that honeybee communication breaks down and flights are disoriented or temporarily suspended. Genuine tropical bees, e.g. the more primitive *Trigona*, use simpler signal systems (sound and scent marking).

Thus both the structure and behaviour of these insects show distinctive adaptation to their feeding in blossoms and, incidentally, to their life as pollinators. Most experiments seem to indicate that bees locate blossoms by vision. However, that is not always so; Cumber (1953) relates how bumblebee queens are evidently attracted to *Salix* blossoms by scent during their first flights, but not later. It is well known that in honeybees, and possibly throughout the genus *Apis*, sources of nectar are discovered by a special clan of scout bees, which communicate their findings to the ordinary worker—collector. Odour is easily communicated, but no system is known by which bees can communicate colours.

The hoarding instinct of higher, social hymenopters, especially the genus *Apis*, is of great importance for the effectivity and assiduity of these insects as pollinators. Evidently there is nothing like a "full cupboard" in their housekeeping, and an instinct to stop collecting when a certain point is reached, does not exist. This must be seen as an adaptation originally to a changing, seasonal climate, with dry or cold periods during which collecting is impossible and the colony is obliged to live on its stores. However, when the same hoarding instinct is also active in a non-seasonal climate, it loses all ecologic sense and results in an incessant blossom activity.

The adaptations of hymenopters to blossom visits have their counterparts in the adaptations of blossoms to visits by these insects. Higher bees also utilize blossom of a primitive type — they are frequent on big umbellifers which produce very great quantities of concentrated nectar such as *Heracleum*, *Archangelica* — and they share these with other visitors. But because of their size, strength, and "intelligence" they are the only insects able to utilize the more complicated typical bee blossoms. These blossoms therefore represent an exclusive source of food which cannot be tapped by other visitors and which will therefore generally yield more to the adapted bee, in many cases one caste of the bee species only.

Big bees are strong animals. They secure a good foothold before starting to work a blossom, and they are the only insects able actively to push aside blossom parts barring entrance to the nectar. Together with the distance to the nectar (spur length, etc.), this is the main reason why otherwise relatively simple blossoms like *Aconitum* and *Delphinium* are exclusively dependent on bumblebees for their pollination. Many large-flowered labiatae, Papilionaceae, and Scrophulariaceae with closed blossoms are dependent on bumblebees for the same reason (cf. case histories). Even more than in Europe with the genus *Bombus*, this is the case in the Tropics with *Xylocopa* blossoms, some of which, like *Thunbergia grandiflora* or *Centrosema* conceal their nectar inside hard-walled containers. The negative corollary to this is, of course, that insects steal the nectar by piercing the corolla.

The hairy bodies of bees and, even more, of bumblebees, are well adapted for pollen transport — 15,000 grains have been counted on a single individual (Kendall and Solomon 1973; cf. Witherell 1972). Parts of these loads are groomed away, but especially zygomorphic blossoms often place pollen in such positions that the animal cannot reach it. A certain transfer of pollen from one individual to another takes place when bees brush against each other (Free and Williams 1973), but the importance of this to pollination is unknown. The number of ovules in typical bee-pollinated blossoms is very great, corresponding to the great number of pollen grains carried. Even so, there may not be enough. According to Adlerz (1966) eight bee visits are necessary to ensure good seed-setting in a watermelon blossom.

Syndrome of bumblebee and honeybee blossoms. Zygomorphic with great depth effect; mechanically strong, with adequate facilities for landing and a surface that gives a good foothold, frequently intricate, semi-closed. Colour lively, generally yellow or blue; nectar guides generally present. Odours fresh, generally not very strong. Nectar hidden, but not very deeply, in moderate quantities. Sexual organs concealed, stamens few; many ovules per ovary.

This syndrome should not be accepted without some very important reservations. One is that it concerns nectar-collecting only. Bees, especially bumblebees, will frequently collect pollen from the same blossom, but for pollen collecting specifically they will also visit very primitive pollen blossoms. Another reservation is that social bees — perhaps more than any other group — will visit any blossom type that yields sufficient nectar; the blossom types characterized above are those specialized for bee visits. The only blossoms bees cannot utilize are those in which the nectar is too deeply hidden. The deepest blossoms can be operated in a non-destructive way by butterflies and birds (and some specialized flies) only. But butterflies, on their part, are not sufficiently strong (and "intelligent"?) to open up closed blossoms, and moths in addition do not have a sufficient foothold to do so. In consequence there are many blossoms neither of these pollinators can utilize, even if they could have reached the nectar. The typical bee blossom is found in Labiatae, Scrophulariaceae, Papilionaceae, and Orchidaceae.

In many solitary bees the number of the brood raised per imago is very small, whereas it has been calculated that the quantity of food carried in by a honeybee is of the magnitude of 100 times its own requirements. Size varies a great deal, too, from the small *Andrena* and *Halictus* species to the big *Xylocopas* and Euglossids (the *Abeilhas* or *Hummeln* in publications from the American Tropics).

The small, solitary bees in temperate countries are generally oligolectic (Linsley 1958) or (at any rate within each region) most probably mono to oligotropic (Olberg 1951). They also show the host–parasite relationship to their food plants characterizing such animals, a relationship which must be based upon chemical release of innate instinctive reactions. Their bonds to the host plant, even if very specific, thus represent a rather primitive principle, like the one met with in beetles. The strict monotropy leads to a restriction of the occurrence of these bees, locally to the habitats of their host plants, temporally to the flowering period of the latter. The "flower-calendar" can be matched by a calendar of mono- or oligotropic bees. This restriction obviously limits the possibilities of such bees developing towards sociability and long-lasting colonies.

On the other hand, the much larger "solitary" bees in tropic countries are polytropic and can utilize different food plants. Many are in reality not truly solitary, but breed in loosely integrated colonies. A very important point to take into consideration with some of these animals — the Euglossids — is the difference between sexes. The females feed on pollen and nectar both for themselves and the brood, and collect these foods from various blossoms. That ends their activity in flowers — building material is mainly fetched elsewhere. Males also feed, but sparingly so, and only for themselves; they generally feed in the same blossoms as females. But this activity in blossoms is quite subordinate to their "sexual" activity in various orchids. The latter activity is monotropic — at least locally. In no other group is the principle of sexual attraction so well developed as in euglossids, and in no other bee group do the males live for so long, i.e. for at least half a year (van der Pijl and Dodson 1966: 47).

In addition to these types of melittophily, Vogel (1954) speaks of micromelittophily, into the syndrome of which enters — besides the small size of blossoms — the filiform blossom type with attendant motility, which may form part of the attractant principle.

The filiform blossom is apparently not a class, comparable to the classes A–F in Chapter 10. These blossoms have smallness as a common characteristic, but may otherwise belong to flag, gullet, etc., types. They will attract small visitors; many of these blossoms are micro-melittophilous; but so far we have no reason to suspect that, apart from smallness, there is anything remarkable in the behaviour of these pollinators. The filiform blossom type, discovered in South Africa, may be more widespread than we know; in the European flora, plants like *Vicia tetrasperma, Cicendia filiformis* come to mind, plants which have, perhaps, been a little too readily dismissed as autogamous. This has nothing to do with the filiform appendages of sapromyophily. The small nectar quantities available per blossom excludes larger species with higher energy requirements.

11.1.6. Butterflies and Moths. Psychophily and Phalaenophily

The two groups of Lepidoptera, butterflies (Rhopalocera) and moths (Heterocera), are in typical development so different in their relationship to blossoms that they are generally treated apart. However, the fundamental difference is not one of taxonomy, but of ethology. One generally finds the correlation butterflies: diurnal, alighting, vs. moths: crepuscular or nocturnal, hovering. However, these factors may occur also in other combinations: hovering butterflies, diurnal moths, etc. In the same way there are intermediate types between the easily recognizable typical butterfly or moth blossoms. Even if the typical behaviours and the corresponding blossom types are quantitatively predominant, the intermediate ones serve to make borderlines indistinct.

Some features are common. None of these insects feed their brood, all food collecting is for their own consumption — if they feed at all. Some do not and their digestive tractus is rudimentary. Even among those that can feed, intake of food does not always seem to be a necessity. Many moths and butterflies are therefore rather spurious pollinators and, further-more, their presence is also dependent on those plants on which the larvae feed, which is rarely one of those furnishing nectar to the imagines. Primitive lepidopters still have chewing mouth-parts, and a varied diet, micropterygidae eat pollen from *Caltha* and *Ranunculus* spp., again indicating the primacy of pollen attraction. In the more evolved species, the mouth-parts are represented by a long, thin proboscis, and the food is exclusively liquid, nectar (and water) being preferred. But even higher lepidopters are also known to feed on bleeding sap, on blood, faeces, and urine without showing any tendency towards "sapromyophilous" pollination. Some butterflies cover their (low) nitrogen requirement from amino acids in nectar (cf. p. 67). According to Gilbert (1972) *Heliconius* butterflies feed on the contents of pollen, which is leached out. These nutrients are evidently essential for their reproduction. As these butterflies have the ordinary proboscis, not biting mouth-parts, the utilization of pollen is clearly a secondary development in this case.

Factors in the speciation of butterfly and moth blossoms are, in addition to tongue length, the existence of tongue guides and rough surfaces to be avoided, and also the strength required to insert and withdraw this vital organ. In Asclepiadaceae weak visitors may have difficulties in withdrawing, and there is a whole literature on this, especially about *Araujoa* ("planta-cruel").

Butterflies are diurnal creatures, and butterfly blossoms correspond in their syndrome, which is more or less self-explanatory (Table 6). Ilse (1928) demonstrated the existence, in different butterflies, of inborn preferences for various colours, showing fidelity to colour variations in species of *Lantana, Aster*, etc. This evidently may influence speciation (cf. Levin 1972a). The colour vision, at any rate in some species, seems to include pure red. It is

TABLE 6. Comparison between butterfly and moth pollination

Psychophily		Phalaenophily	
Butterflies	Butterfly blossoms	Moths	Moth blossoms
Diurnal life	Diurnal anthesis, no closing at night	Nocturnal life	Nocturnal anthesis, often closing during day-time
Olfactory sense not very strong	Odour weak, generally fresh, agreeable	Strong olfactory sense with instinctive preferences	Strong, heavy-sweet perfume at night
Visual sense well developed, also for colours, can see red	Vividly coloured, including pure red	Visual sense sensitive to colours at night	Mostly white or faintly coloured, sometimes red or drab, insignificant
Probably not sensitive to deeply dissected contours	Blossom rim not much dissected	Probably sensitive to dissection of outlines	Deeply dissected lobes or fringed petals
Alighting on blossoms	Blossom erect, radial, rim generally flat, but often narrow; anthers fixed	Hovering in front of blossoms without alighting	Blossoms horizontal or pendent, rim absent or bent back; zygomorphy, if present, caused by lower rim bending back; anthers versatile
Long, thin proboscis	Nectar well hidden in tubes or spurs, tubes narrow	Very long, thin proboscis	Nectar deeply hidden in long tubes or spurs, narrower than in bird blossoms
Less active flyer, metabolism not very high	Nectar ample	Active flyers with very high metabolism	More nectar than in butterfly and bee blossoms
Some preference for guiding marks for inserting proboscis	Simple nectar guides or mechanical tongue guide (groove)	Some preference for guiding marks for inserting proboscis	Nectar guides generally absent, guidance by contour of blossom

unknown if nectar guides mean anything to butterflies, or if they are there "for the benefit of" other groups of visitors to the same blossoms.

Butterflies alight on the blossom, generally sitting on the margin of funnel- or trumpet-shaped blossoms. The presence of chemoreceptors on their feet should be kept in mind, even if its ecologic significance is not known. They seem to prefer sucking nectar out of narrow tubes, frequently florets of Compositae.

Like bees, butterflies are also able to utilize other blossom types, including primitive ones, and may even be seen on umbellifer inflorescences, but the typical butterfly blossom is one with a narrow tube and flat rim, e.g. *Lantana* or *Buddleia*. These two genera exhibit the aggregation of flowers into dense masses, frequently found in this blossom class. This has both a visual effect and minimizes travel costs. Probing into deep tubes is a time-, i.e. energy-consuming process.

As a group, moths are ecologically and ethologically diversified. Noctuids do not hover (they may flutter), but use their legs when alighting on a blossom. They may then collect pollen on the legs. For speciation this is of importance. Some noctuids fly also at day-time, ?.g. *Plusia* in the afterseason.

Sphingids, also the day-time flying species, mostly hover while taking nectar, and thus collect pollen only on proboscis and head. They have a high metabolic rate, especially when hovering, and require much food: consequently they are relatively important pollinators. Some of them extend their activities to butterfly blossoms, even to *Bougainvillea*, e.g. *Macroglossa*, the classical object of Knoll's experiments, which among others proved their sensitivity to other colours and its blindness to red. Sensitivity to ultraviolet has been demonstrated in Pieridae (Eisner *et al*. 1969).

Hovering accounts for some of the differences between butterfly and moth blossoms in Table 6 especially the non-availability of a place for landing (labellum absent or bent back). In some blossoms the former landing-place has acquired a new function as visual attractant (split up into narrow fringes) or dispenser of odours or for guiding the insertion of the proboscis.

The long distance between sexual organs and nectar in moth and butterfly blossoms is not only a negative character to exclude bees, but it is also of positive value for the correct use and placing of the proboscis. Capparidaceae have developed from Papaveraceae-like, strictly epigynous choripetalous (and chorisepalous) ancestors, and a tube could not form; instead, the spacing between nectar source, on the one hand, and pollen and stigma, on the other, is established by the elongation of filaments and the placing of the ovary on top of a gynophore or an androgynophore in the blossom.

Olfactoric attraction must play a greater part in moth blossoms than in most others, and they can fill the air of the tropical night with overwhelming fragrances, some of which are well known also in extra-tropical gardens or hothouses (gardenias, tuberoses, lilies, *Pseudodatura*, etc.). In *Cestrum nocturnum** the odour is too strong to be acceptable near the house (see Overland 1960). The strong periodicity of odour production is remarkable; flowers that fill the air with fragrance at night may be virtually scentless during the day (*Pseudodatura*, night-flowering cacti).

One interesting aspect of the night-blooming syndrome has been described by Bhaska and Razi (1974) in some nocturnal *Impatiens* spp. The pollen germinates best in the night and remains viable for some hours after sunset only. This may be important in arid areas.

*See the popular names: *Dama de noche*, or in Sundanese, *Sundel malam* = night-whore, because, it is said, it is unobtrusive at day-time, agreeable at night, and disgusting in the morning!

In phalaenophilous blossoms not only the periodicity of odour production but also anthesis as a whole shows a correlation to nocturnal visits. If anthesis lasts for more than one night, the blossom closes (sometimes imitating wilting) during the intervening day(s) so that it loses its visual as well as its olfactoric attraction. The very rapid opening of some night-flowering blossoms should be noted; *Calonyction bona nox* opens so fast that it can be followed by the eye.

Sphingids that hover produce a more characteristic syndrome than the Noctuids that usually alight or support themselves on the blossoms. Sphingids are sensitive to strong winds, which make landing impossible. Eisikowitch and Galil (1971; cf. also in Heywood 1973) have shown that pollination of the sea-shore plant *Pancratium maritimum*, which has a hawk-moth flower, is imperiled by strong sea winds (above 3 m/sec).

The odours of phalaenophilous blossoms are very similar to each other as they affect the sense of man, and much confusion exists in the literature about their specificity. Visits are not specific, and pollen loads are mixed. In experiments Brantjes (1973) found sensitivity in many species to various nocturnal flower odours, but specific differences were noted, especially in the broadness of the odour spectrum perceived. This may be the basis of discrimination.

Knoll (1922, 1925) demonstrated that hidden flowers were discovered, obviously by their odour alone, but the importance of odour perception for the orientation of some moth species has been doubted. Schremmer (1941) found that scent orientation was important for newly emerged *Plusia (Autographa) gamma* which is not particularly nocturnal. This could later develop into constancy to one scent, but also to colours. Whether this odour-binding should be considered secondary or not, is mainly a semantic question.

A more detailed study of odour attraction (Brantjes 1973) divides the process into various parts: orientation from afar, near orientation, decision to visit, and the final guidance into the blossom. In the sphingidae used for the experiments, the presence of a releaser odour started warming-up by vibration of thorax muscles and, if already flying, a change from random to the special, probably anemotactic searching flight leading to the blossom and decision to visit. When the source of the scent is discovered, the proboscis is stretched out and entered into the blossom. Visual cues enter into the reaction chain in some species, less in others, and can dominate, e.g. in *Macroglossa*.

The question of visual attraction of nocturnal pollinators is a difficult one. The fact that moths see colours in the dark is not evidence against the usefulness of the frequent off-white colour found in these blossoms, nor does it prove that there is optical attraction in dull-coloured blossoms of the *Hesperis tristis* type.

The longest prosboces found anywhere among flower-visiting animals occur in moths, including the famous *Xanthopan morgani* f. *praedicta* that presumably pollinates *Angraecum sesquipedale* (spur 25—30 cm) in its native habitat.

A good example of a temperate moth blossom is *Lonicera periclymenum*. Its lack of a place for alighting makes it very difficult for bumblebees to work even if big bees may be able to reach down to the nectar, and their antics in and around such a blossom can be very entertaining, demonstrating clearly the negative adaptional function of a syndrome. The original type of the basitonic orchid was probably bee-pollinated; some genera have later developed towards lepidopter pollination. There is a good example of a butterfly blossom in the strongly coloured *Anacamptis pyramidalis*; the faintly coloured *Gymnadenia conopea* is visited by moths as well as by butterflies, and the greenish-yellow *Platantheras* chiefly by nocturnal or crepuscular moths. All of these have such long and narrow spurs that bees can hardly get anything out of them.

The most characteristic moths are the nocturnal ones, and the syndrome of blossoms pollinated by them is now fairly self-evident on the background of what has previously been said. As in other groups, small, primitive moths do not conform to the pattern. Some of them are diurnal and approach butterflies in their habits. Others are rather dystropic.

The three main characteristics of moths in relation to pollination are nocturnal habit, long proboscis, and hovering. None of them is exclusive to moths. Nocturnal bees are known; they compete with primitive moths with a short proboscis. The other characteristics are among insects found in some of the more extreme blossom-flies (Bombyliidae and Nemestrinidae) and in *Nemognathus* (flower beetle). These insects (most of them at any rate are diurnal) frequent blossoms of a similar type and may compete with the more highly developed moths.

In addition, the diurnal hawk-moths especially compete with pollinating birds (hovering in hummingbirds), and the syndromes of lepidopter and bird blossoms are rather similar: the strong colours, great quantities of nectar. Porsch (1924) has shown that this similarity goes so far that birds (not always much bigger* but certainly stronger) recognize diurnal hawk-moths as competitors and drive them away. However, in pollination ecology the other alternative always seems to exist somewhere, and so it is not astonishing to learn that in South America a moth (*Castnia eudesmia*) is said to chase birds from its food plant, *Puya alpestris* (Gourlay 1950).

The main differences between the syndromes of ornithophily and of day-lepidopter pollination seem to lie in absence or presence of odour, in the narrower and frequently curved tube of moth blossoms and the versatile anthers (not fixed as in ornithophily). In addition, the corollas do not need the same mechanical strength: there is a great difference between the beak of a bird and the proboscis of a moth. The ordinary butterfly sucks up the nectar through a very narrow and frequently long tube; birds scoop up nectar through a much coarser beak. Consequently, bird blossoms can offer more viscous, i.e. more concentrated nectar, i.e. greater energy offer. Birds are ultraviolet blind. Apart from that the colour difference between butterfly- and bird-pollinated flowers is not clear-cut. *Caesalpinia pulcherrima* (Vogel 1954) represents an intermediate case: odourless, often visited by birds, but classified as psychophilous because of the presence of a landing-place in the stiff filaments, and actually pollinated by large butterflies in America (Cruden 1976).

11.1.7. Other Invertebrates

The visits of many other invertebrates in blossoms have been recorded, but the relationship between blossom and visitor generally remains obscure. A notorious and obscure case is that of malacophily — pollination by snails and slugs. This has been postulated time and again, but even the best examples, such as *Rohdea* (Loew 1895) are very much in need of corroboration (cf. Lorougnon 1973).

Small animals roaming over and into everything will discover any source of food that can be reached and utilized, and they utilize any that can be discovered. Both pollen and nectar furnish excellent sources of food, and so do the soft, succulent tissues of blossoms. No

---

*It is not too uncommon that people who do not know hummingbirds by personal experience confuse them with big hawk-moths like *Acherontia*.

wonder, then, that a great number of invertebrates and unspecialized insects are found in blossoms. Many of them lick up nectar or eat pollen, and their bodies may be completely dusted with pollen. Flying on to another blossom they may cause pollination – geitonogamy if nothing else. That this is not only a theoretical construction is shown by many observations of representatives of various insect families found in blossoms, engaged in such activity (Porsch 1958).* However, even if there can be established a definite relationship between insect and blossom, one hesitates to accept these allotropic animals as anything more than chance pollinators, more or less on a par with the boy in the cherry-tree. If they are to be considered as anything more, blossom food must be established as a regular part of their diet, causing regular, repeated visits, season after season. Very occasionally this has been observed; but there is always the possibility that such "trespassers" have been disregarded because of preconceived notions about "proper pollination" (see Porsch 1958; *Mesocerus marginatus*).

Few of the many insects mentioned by Porsch have been observed so frequently that they can be suspected of being regular pollinators. In some few cases even a single observation may suggest that a more regular visitor is concerned, viz. if there are remains of several pollinia, testifying to several visits to orchid blossoms (*Chloroperla torrentium* Porsch, *loc. cit.*) or of several translators, referring to asclepiad blossom visits (unidentified heteropter on *Asclepias* sp. K. F.). The Heteroptera is a very interesting group in this respect: their frequently observed flower visits may be prompted more by the succulent tissues of the blossom than by the nectar; regular pollination may result nevertheless, but if the ovary has been damaged in the process, the plant gains very little. More investigations are needed here.

So far, Hagerup's observation of the activity of thysanoptera in the blossoms of *Calluna* and *Erica* provides one of the few good examples of pollination by other invertebrates, but there is every reason to presume that both this group of insects and other small and insignificant, unobtrusive animals may prove to be of much greater importance in pollination than hitherto suspected. They have, in fact, been found to be active as alternative pollinators both in *Phlox* and in alfalfa. According to Soria Vasco (1970) thysanoptera (*Frankliniella parvula*) take over as pollinators of cocoa during dry spells.

In this connection can be noted the curious observation of *Tyora* nymphs pollinating cocoa in Ghana (outside the native area of the tree!) by crawling across the flowers and dragging pollen with them in their wax-thread cover (Kaufmann 1973). It is an unusual insect taxon (homoptera), it is an unusual attraction (*Tyora* is a "minor" parasite on *Theobroma*), and it is an unusual developmental stage: regular pollinators are always imagines.

Grinfeld (1959) communicates some very interesting observations on nocturnal pollination of sunflowers by various insects, not usually thought of as pollinators, even less as nocturnal ones: these include moths (various agricultural pests), lacewing flies, earwigs, and grasshoppers. Heads uncovered during the night but covered during the day set 23.6 per cent seed against 0.5 per cent in heads that had been covered continuously. Obviously pollination activity at night may include facts that are not, or not sufficiently, taken into consideration.

---

*It is remarkable how many of Porsch's observations are made on the blossoms of *Listera ovata*, considered by Darwin to be one of the most refined orchids in existence but ecologically a backward ichneumonid wasp-orchid.

11.2. VERTEBRATES AS POLLINATORS

Whereas vertebrates do not pollinate blossoms in Europe and were therefore overlooked as pollinators by the classical writers on our subject, it has become increasingly clear that they are of great importance in other continents.

When making comparisons with invertebrate pollinators it must be kept in mind that vertebrates, and especially the warm-blooded ones, have much higher, more continuous, and more complicated food requirements than have insect imagines, and that they need relatively great quantities of protein in addition to energy food — carbohydrates or fat. The protein requirement is generally covered from other sources utilized by these animals before the establishment of blossom utilization. However, there are substantiated cases of birds and several of macroglossine bats eating pollen to cover their protein requirement partly or wholly.

Pollen has been found in the stomach of hummingbirds in museums. Porsch (1926a) reports a sun-bird, *Anthotreptes phoenicotis*, collecting pollen from *Casuarina*, which is usually anemophilous. Churchill and Christensen (1970) describe a brush-tongued lorikeet (*Glossopsitta porphyrocephala*) using its tongue for collecting pollen from *Eucalyptus diversifolia*. Nectar from the same blossoms was taken as supplementary food during times of nectar flow. In this combination, pollen yields more food than nectar, which would not usually be produced in sufficient quantity for such a relatively heavy bird (*ca.* 50 g).

March and Sadler (1972) report that a North American pigeon feeds on *Tsuga* pollen, albeit only for part of the year. Undoubtedly, many other cases will gradually be discovered and may demonstrate that in the same way as evolution has lead to total blossom-dependence in invertebrates (bees), there are also vertebrates that cover both their energy and protein requirement from blossoms.

There is nothing to indicate that pollen has been a primary attractant for blossom-visiting vertebrates; sugar is the original one, and in practically every case is still present. Incidentally, easily digestible sugars may be a necessity for animals with such record high metabolism as that of hummingbirds, which eat twice their own weight per day.

Energy intake by insect food, as compared with sugar, must be negligible, but insect food is of great importance because of its chemical composition.

Another very important difference between vertebrates and insects is the greater longevity of the former, at least a year or so as compared with the few summer weeks, rarely months of active life of insect imagines. Vertebrates consequently demand food all the year round, and vertebrate pollinators are therefore primarily a phenomenon of the Tropics,† where blossoms are always available. Birds can to some extent compensate for seasonal lack of blossoms by migrations, and hummingbirds move north to the U.S.A., Canada, and even Alaska, following the flowering of plants to which they are adapted. Robertson found that the appearance of *Trochilus colubris* in Illinois coincided with the flowering of ornithophilous species of *Lobelia*, *Tecoma*, *Castilleja*, *Lonicera*, and others. Some hardy migrants may resign themselves to red clover, alfalfa, or even to pecking fruits for sap, thus reverting to a more primitive diet. From the relevant literature one frequently gets the impression that blossom visiting vertebrates prefer nectar as a source of energy, but that they

---

*It is the *evenness* of the tropical climates, i.e. the absence of seasonal variations, not the high temperatures (c. Troll 1943), that is of the greatest importance as shown by the occurrence of vertebrate pollination fairly high up in the mountains (e.g. Vogel 1958), even into the region of regular nightly frosts in African high mountains where insects have to lead a concealed life to protect themselves from the vicissitudes of the climate (Hedberg 1964).

can also use other types of energy food. Some birds (and bats?), apparently not able to substitute, are said to be dependent on a constant supply the year round.

Many small vegetarian or omnivorous vertebrates, especially mammals like squirrels, tree-shrews, and lower primates (cf. Petter 1962) roam in the crowns of trees and eat blossoms or parts of blossoms or suck nectar. In many, perhaps the majority, of these cases the effect is destructive, though even a destructive feeder may more or less accidentally leave some pollinated pistils. Much research remains to be done to establish relationship between possible regular pollinators and the blossoms in which they work. A rather unexpected case, which apparently must be accepted as an established relation, is the present-day pollination of the originally ornithophilous *Freycinetia arborea* by rats in Hawaii. At night rats (*Rattus hawaiensis*) climb into the trees to eat the succulent bracts and while feeding transfer pollen (Degener 1945). Coe and Isaac (1965) have described the pollination of *Adansonia digitata* by a small primate, the bush baby (*Galago crassicaudatum*). Undoubtedly, more similar cases will come to light among primitive primates, rodents, and especially marsupials. Their inability to fly restricts their movements from one plant individual to another, and thus also their effectivity as cross-pollinators. To some extent this may be compensated for by the great quantities of pollen which they may transport for a long time in their fur.

Many of these prospective or suspected pollinators are omnivorous and show no special adaptations to blossom visits. Others are more or less specialized with the small south-west Australian marsupial *Tarsipes spencerae*, the "honey-mouse", as an extreme type (see Glauert, 1958). The animal resembles a shrew-mouse, is about 7 cm long with a 9-cm long, grimpant tail. Its snout is strongly projecting, most of the teeth reduced or absent, but the tongue very long, extensible, vermiform, and the outer part brush-like, thus well adapted for collecting nectar from narrow blossom tubes. Its chief source of nourishment seems to be nectar from different Proteaceae. The source of protein is not clear.

In addition to *Tarsipes*, Morcombe (1969) describes from the same area another anthophilous marsupial, the rediscovered, "lost" *Antechinus apicalis*. There also is an endemic rat, *Rattus fuscipes*, which visits the inflorescences of *Banksia attenuata* and possibly other Proteaceae. It is apparently not an exclusive nectar-feeder, and also shows relatively little morphological adaptation to blossom visits, in contrast to the marsupials. This is not unexpected, as the history of the rats in Australia is much shorter than that of the marsupials.

Whereas at any rate *Tarsipes* exhibits very distinct adaptations to blossom visits, no blossom is recognized showing adaptation to visits, either by this animal or by the others mentioned above.* As far as is known today only two classes of vertebrate pollinators correspond to a definite syndrome in the blossoms, viz. birds and bats; these will be treated separately. The other vertebrate pollinators are of great theoretical interest also, because among them are very distinct examples of adaptation by animals to what we must presume were previously existing types of blossoms. This shows the adaptibility of animals in this respect and may in the more evolved types serve as an argument for the mutuality of adaptation. Obviously, the animal adaptation must here be rather recent, and Baker and Hurd rightly point out (1968) that pollination by vertebrates must have developed out of insect-pollination syndromes.

---

*It should be added that Porsch (1936) is inclined to interpret some structures in Australian plants as adaptations to pollination by non-flying vertebrates.

11.2.1. Pollination by Birds. Ornithophily

As flying animals with a rough surface, birds possess good external prerequisites for becoming pollinators. Whereas everybody apparently takes for granted the fact that various insects find their food in blossoms, the corresponding habit of birds seems to have caused a great deal of astonishment and speculation about how birds "got the idea" of utilizing the nectar of blossoms (the lack of bird pollinators in Europe may have contributed to that attitude!). Among the ideas proposed has been that pollination is a development from destructive eating of flowers, perhaps primarily by fruit-eating birds.* It has also been proposed that woodpeckers or sapsuckers (*Sphyrapicus*) may have been tempted to change their diet under the influence of the sap flowing from their holes (some of them peck fruits, too; *Dendrocopus analis* the fruits of *Cassia grandis*). A third group of "explanations" suggests that birds were chasing insects in the flowers, and happened to find the nectar or to puncture succulent tissues; or that they originally drank water collected in the blossoms to quench their thirst since water is difficult of access to many tree-living animals in the tropical forests. That hummingbirds are primarily chasing insects in the blossoms is sometimes advanced even today. The fast resorption of nectar makes it difficult to identify it in the stomach of the bird, whereas more or less accidental, indigestible insect remains are easily observed. However, ornithological literature abounds in observations of digestive systems filled with nectar. The occurrence of nectar-thieving by puncturing the basis of the corolla is another proof that nectar is the object; insects cannot be obtained that way. Nor are they obtained from the closed flowers of javanese Loranthaceae, flowers that do not open until they explode under the impact of a nectar-seeking bird (Docters van Leeuwen 1954).

Hummingbirds need much energy, especially when hovering (215 cal per hour and gram body-weight). The more advantageous rate of expenditure between hovering and flying (plus resting) may explain the small size of these birds. After periods of non-feeding reserves may run dangerously low in spite of low metabolic rates during sleep.

The efficiency of nectar uptake and nectar metabolism differentiates between visitors with different energy budgets (cf. Schlising *et al*. 1972). A concentrated occurrence of blossoms with much nectar is a prerequisite for territorial defence in hummingbirds (Grant and Grant 1968; Stiles 1971). Reference could here be made to the migrations of hummingbirds to areas of concentrated bloom, especially during breeding periods.

Anybody who has witnessed the way in which sparrows can completely demolish a bed of crocuses in spring will know that birds will eat food they can utilize wherever they find it and, naturally, those birds that "like" sugar cannot help discovering sooner or later the sources represented by flowers — just as the sparrows discover them in some springs and not in others.† What is remarkable is the way plants and birds have adapted themselves to each

---

*Some of the actual examples of this are furnished (i) by bulbuls (*Pycnonotus*) which devour the fleshy bracts of *Freycinetia funicularis* and function as legitimate pollinators. Characteristically this is a species with fiery red, odourless, diurnal blossoms; (ii) by semi-dystropic birds observed drinking from less specialized blossoms like *Bombax (Gossampinus)* or plucking petals of *Dillenia* spp., (iii) the fantastic case of *Boerlagiodendron* (Beccari, 1877), which is said to attract bird pollinators (pigeons) by means of fruit imitations (sterile flowers) between the flowers; (iiii) by birds pollinating nectarless *Calceolaria uniflora* when snapping off food bodies from the flowers (Vogel 1974: 153).

†McCann (1952) notes how sparrows and chaffinches (*Passer domesticus* and *Fringilla coelebs*) in New Zealand learn to rob bird blossoms by puncturing them at the base; see also Swynnerton (1915) and Iyengar (1923). In south European gardens local, unadapted birds have frequently been observed to rob imported ornithophiles (*Abutilon*, *Erythrina*), mostly just damaging the blossoms, but sometimes also pollinating. Cf. the remarkable way in which blackbirds in the Islands of Scilly have adapted to taking nectar from the Chilean *Puya* cultivated there (Ebbels 1969).

other, but again, this is no more – nor less – remarkable than the way plants and insects are mutually adapted.

From the point of view of pollination it is quite irrelevant whether birds visit flowers for the sake of nectar or of insects as long as the visits are regular. Whether nectar or insects is the object is a problem of adaptation only, not a functional problem. In Java, *Zosterops* visits the non-ornithophilous *Elaeocarpus ganitrus* to collect the mites which are very numerous in the flowers (v. d. P.).

There is little doubt that birds may have approached blossoms for all the reasons mentioned, and the example of the sparrows shows that this happens today, too. Even if the blossoms have been demolished from a gardener's point of view, the crocuses may have been successfully pollinated. Destruction of the flower is in itself irrelevant as long as the pistil is not affected. Explosive blossoms are self-destroying.

Sparrows were observed to pollinate pear-trees (K.F.).

Other such cases of accidental blossom visits by possibly dystropic birds have recently come to light in ornithological observations of migrating birds arriving in Britain dusted with pollen from southerly regions (Ash *et al*. 1961). Campbell (1963) observed various birds in Britain chasing insects in blossoms and getting dusted by (very little) pollen.

From these examples of dystropic blossom visits there is a gradual transition through many allotropic birds with a mixed diet in which nectar constitutes one of the ingredients (Porsch 1924) to eutropic ones, to establishment of the true ornithophily.

Obviously, observations of blossom visits by hummingbirds must have been made long ago. As a scientifically recognized phenomenon ornithophily was established by Trelease during the last two decennia of the last century, by Johow (1900), Fries (1903), and, decisively, by Werth (1915) in a more comprehensive study. However, it was not till Porsch during the 1920s (see references) had collected a huge amount of data and had given a convincing summary of the now familiar characteristics that ornithophily was unanimously accepted as a phenomenon, even if its origin is still, partly, under discussion.

The habit of nectar collecting is obviously a polyphyletic one, which has arisen in different groups of birds in different regions. The most famous example, reaching the peak of adaptation, are the hummingbirds (Trochilidae) of both Americas. Hummingbirds must originally have been insect-eaters, but have later largely switched to nectar; their young are still reared on insects in addition to nectar (higher protein requirements of growing organisms). Also the imagines take insects.* It is remarkable that birds so rarely have switched to pollen as a source of protein.

Another American group of more or less eutropic flower-birds are the much less important sugar-birds (Coerebidae, "honey-creepers", "quits"). In the Old World other families developed the same characteristics as the hummingbirds, even if the adaptation is generally less extreme. In Africa and Asia the sun-birds (Nectarinidae) are found, in Hawaii the honey-creepers (Drepanididae) so intimately linked with native lobelias, in the Indo-Australian region the honey-eaters (Meliphagidae) and the brush-tongued honey-parrots or lorikeets (Trichoglossidae).

Less-specialized blossom visitors with a mixed diet (allotropic pollinators) are also active, but less so, as pollinators especially in lower bird-flowers (*Bombax*, *Spathodea*),

---

*Marden (1963) tells a marvellous story about flies being attracted to (late anthesis) flowers of *Stanhopea graveolens* by the odour, being preyed upon by a camouflaged spider, which in its turn is preyed upon by a hummingbird, *Glaucis hirsuta*, which, in its turn, pollinates the flower.

demonstrating that flowers and their birds may have developed together, with mutual influences. They are found in many other families, e.g. some tropical bulbuls (Pycnonotidae), starlings (Sturnidae), orioles (Oriolidae), and even amongst tropical woodpeckers (Picidae), where a fringed tongue tip shows the first sign of morphological adaptation.

The flower-peckers (Dicaeidae) visit diverse flowers but show a curious specialization in favour of one plant group, viz. tropical Loranthoideae, where they not only visit the ornithophilous flowers, but are also adapted to digest the fruits and to spread the seeds (see Docters van Leeuwen 1954). The oldest references to bird pollination we have come across are by Catesby (1731–43) from the New and by Rumphius (1747) from the Old World.

The regions in which at least some kind of ornithophily may be found thus cover practically the whole of the American continents and of Australia; further, tropical Asia and Africa south of the deserts. Israel used to be a northern outpost of this area with a *Cinnyris* visiting a red *Loranthus* (according to Werth, 1956b), but is now, Galil informs us (in litt.), teeming with these birds on introduced flowers in gardens.

In Central and South American mountains, the number of ornithophilous species in the flora seems overwhelming. When bees are present at high elevations in Mexico they are equally effective pollinators as the birds except that birds are superior under bad flying conditions (Cruden 1972b). However, the species of *Bombus* are more independent of climate, and their presence may change the picture, as shown by Docters van Leeuwen (1933). Similar results are indicated by Stevens (1976) with regard to *Rhododendron* pollination in the Papuasian mountains.

In Australia and New Zealand also the number of eutropic pollinating insects is apparently low, and the function of, especially, higher bees in other continents is taken over by birds, cf. the dominant rôle of the predominantly ornithophilous genus *Eucalyptus*. We have more precise data on the percentages of plant families with ornithophilous tendencies in certain regions only.

The scattered occurrence of blossom feeding in various groups of birds, its great geographic distribution, and the scattered occurrence of ornithophilous types of blossoms in many groups of plants, all indicate that ornithophily is a young phenomenon.

The ability to hover is well developed in hummingbirds (cf. analysis by Greenewalt, 1963), but rare in other groups; it occurs, for example, in the meliphagid *Acanthorhynchus*, and it is imperfectly developed in the Asiatic *Arachnothera*. Some other birds may hover in a strong counter wind.

The brilliancy of plumage, resulting in a great similarity of colouring between flower birds and bird flowers, may seem rather queer. There is good reason to see it from the point of view of protective colouring. van der Pijl thus saw a dangerously conspicuous flock of red-and-green *Loriculus* (flower parakeet) disappear out of sight when alighting on a flowering *Erythrina*. Obviously, these animals are most vulnerable when immobilized during feeding.

Grant (1949b) has maintained that blossom "constancy" is but weakly developed in birds, and that their feeding habits are too complex to be grouped in a linear treatment. The evaluation of blossom constancy by other authors varies a great deal. Much of the variation may be due to lack of distinguishing properly between imposed and preference fidelity. Birds seem to take their food wherever it is available. Naturally, if there is rich flowering of a plant with much nectar, the apparent preference of birds for this plant is a matter of statistics, not of food preference. If there is no such dominance of flowering, they may flutter between species or even turn to alternative food; any observable constancy will be imposed, even if length of blossom tube versus length of bill, composition of nectar, etc., may play a rôle in selection of blossoms. Under extraordinary circumstances (migration and breeding) the birds accept miscellaneous blossoms. Johow (1900) noted in Chile that hummingbirds may even switch over to European fruit trees or *Citrus* species. Hemitropic birds will more frequently switch to fruit (sometimes devastatingly). In a tropic environment birds apparently have a special preference for freshly blossoming trees as compared with other attractions. The ecological value of this is, of course, not absolute, but relative and competitive, and may constitute selective pressure.

Tropical plant species and higher groups show a directed phylogenetic development, leading to a definite and easily recognizable syndrome of characteristics pointing to birds, to the exclusion of other pollinators.* Accidental combination is out of the question. The mutual interdependence was very obvious in Hawaiian Drepanidids and their flowers, which later had to find an escape into autogamy when the birds were exterminated (see Porsch 1930; Amadon 1947).

Some bird blossoms are of the brush type (*Eucalyptus*, heads of Proteaceae and Compositae; *Mutisia*), others belong to the gullet type with oblique profile (*Epiphyllum*), the tube type (*Fuchsia fulgens*), and some papilionaceous blossoms (*Mucuna* spp., *Erythrina*) are also typical ornithophiles.

This multitude of types shows that the phenomenon is a late development on top of pre-existent ecomorphological organizations, defying "Bautypen" and the like, but leading to a secondary convergence of style. Some cases of similarity between unrelated blossoms cited by morphologists as mysterious "repetitive pairs", by others as orthogenetic, easily prove to be parallel adaptation in relation to pollination. Considering the phylogeny of these convergent changes we may say they frequently arose independently in several phylogenetic lines.

The syndrome of ornithophily is given in Table 7, illustrating how the points correspond to the ethology of the birds concerned (cf. also the discussion of the genus *Salvia*, p. 147).

Some comments on the table may be useful. The characters are partly positive (attractive), partly negative (excluding competitive visitors). The neglect of bird blossoms by hymenopters demonstrates the exclusion, easily observed with regard to *Mimulus cardinalis*, *Monarda* spp., and *Salvia splendens* in every botanical garden. Darwin already observed that *Lobelia fulgens* was neglected by bees, standing among melittophilous species in a garden.

The effectivity of this syndrome is demonstrated by the way in which typical bird blossoms, when grown in European botanical gardens, attract the attention of short-billed, unadapted dystropic birds and also by the way in which flower birds immediately recognize and try to utilize foreign bird blossoms (Porsch 1924). The size of the blossom does not enter into the syndrome. Many bird-pollinated blossoms are comparatively small. Apart from the depth effect, bird blossoms do not belong to any definite blossom class, even if the brush and the tube are the two most characteristic classes.

Flower birds are not always confined to blossoms exhibiting this syndrome. In time of dearth of nectar, flower birds will utilize "unadapted" blossoms, too, as already mentioned.

One refinement should be added to the table. There are regional differential characteristics between flowers for hummingbirds and for other birds. In the former

---

*For the differential diagnostic characters between the class of ornithophiles, and that of diurnal lepidopter flowers, see p. 116. The differences are rather indistinct, especially in American plants.

TABLE 7. The syndrome of ornithophily

Bird flowers	Flower birds
1. Diurnal anthesis	Diurnal
2. Vivid colours, often scarlet or with contrasting parrot-colours	Visual with sensitivity for red, not for u.v.
3. Lip or margin absent or curved back, flower tubate and/or hanging, zygomorphy unnecessary	Too large to alight on the flower itself
4. Hard flower wall, filaments stiff or united, stiped or otherwise protected ovary, nectar stowed away	Hard bill
5. Absence of odour	Scarcely any sense of smell
6. Nectar abundant	Large − and great consumers
7. Capillary system bringing nectar up or preventing its flowing out	
8. Possibly deep tube or spur, wider than in butterfly flowers	Long bill and tongue
9. Distance nectar − sexual sphere may be large	Large, long bill; large body
10. Nectar-guide absent or plain	Intelligent in finding an entrance

(American) the flowers stand out or hang down, exposing their organs towards the free space, ready to be pollinated by hovering visitors (cf. *Pedilanthus, Quassia*). It is said that hummingbirds do not readily visit blossoms that face upwards (Frankie 1975: 202). In the latter (Asiatic and African) there is a perch near the flowers and the flowers point towards it (*Spathodea campanulata, Protea, Aloë*). We might analyse the species of *Fuchsia* and *Erythrina* (Toledo 1974) on this point to verify their American or Old World exterior − imitating Linnaeus: *Hic flos facien americanam habet* (or whatever he might have said). Even so, there are American bird blossoms provided with a perch, e.g. *Heliconia rostrata*

In the Chilean *Puya* (subgenus *Puya*) the outer part of each partial inflorescence is sterile and forms a distinct perch* that is utilized by the legitimate pollinator, an icterid, "tordo" (Gourlay 1950) and by the blackbirds in England (Ebbels 1969). In the African flora, *Antholyza ringens* offers a fine example of a specially formed perch. In the absence of possibilities for perching, American bird blossoms cultivated in Java proved to be inaccessible to nectarinids and to be punctured (van der Pijl 1937a). According to Johow (1901) the African *Aloë ferox* is neglected by hummingbirds in Chile, but well pollinated by a tyrannid, *Elaeina*. Cruden (1976) describes other instances (e.g. species of *Eucalyptus* and *Leonotis*) in which the adaptation to perching birds have a negative effect in relation to hummingbird visits when these plants have been introduced to America.

It is no objection against point 2 that many bird-visited blossoms are white. The association of bird and colour is not absolute. In some geographical regions bird blossoms are predominantly non-red (e.g. in Hawaii). Against the denial of the general importance of red we may, however, refer to statistics proving the relative preponderance of red in the Tropics, especially the Andes (see Porsch 1931a; for South Africa see Vogel 1954). Also reference is made to the colour preference (known to every observer) of trochilids, moreover to general sense-physiological researches on birds proving their high sensitivity to red and their much lower sensitivity to blue. With true red colour being invisible to most, if not all, pollinating

*There is a very good illustration in J. Roy. Horticult. Soc. 87 (1962), fig. 13.

insects, red blossoms, visible to birds (and to man!), represent an empty ecological niche, which has been open for utilization by various ornithophilous blossoms (cf. K. Grant 1966). To migrating American pollinating birds — both seasonal and irregular — the red colour would have the effect of a common signal, indicating the availability of a suitable source of nectar (like a roadside inn sign), which again would increase the effectivity of visits. A further discussion, with illustrations and also information about the birds, is given by Grant and Grant (1968, cf. Raven 1972).

The spectral sensitivity of birds' eyesight varies between species. In a hummingbird Huth and Burkhardt (1972) found a shift towards the short-wave end as compared with man (363 to *ca*, 740 nm against *ca*. 390—750).

In *Columnea florida* birds are attracted to red spots on the leaves, whereas the flowers themselves are hidden. As the colour pattern does not simulate a flower, this presumes a high degree of mental integration on the part of the pollinating birds (Jones and Rich 1972).

For the "parrot-colours" reference may be made to *Aloë* spp., *Strelitzia*, and many Bromeliaceae.

To point 3 we remark that zygomorphy (usually a sign of entomophily) in ornithophiles results from doing away with the now in two respects dangerous lower margin. This typical shape is formed even in ornithophilous Cactaceae, a family the rest of which has regular, actinomorphic flowers (see below). This "avoiding" of a part of the blossom that would represent a landing place for insects and an obstacle for birds can be admired in the red *Corytholoma* in flower shops.

To point 5 (smell) we may add that odour is in itself no obstacle, but that its absence is characteristic for ornithophiles. It is still present in transitional flowers like *Bombax* and *Spathodea*. von Aufsess (1960) found that both pollen and nectar in ornithophilous blossoms have subnormal odour production and bees could not be trained to these odours.

To get an impression of the quantity of nectar in ornithophiles (point 6) one should, in temperate botanic gardens, look at the nectar in *Phormium* spp. or at the dripping *Aloës* and *Proteas* from the Cape. Even though more concentrated than butterfly-blossom nectar, that of ornithophilous plants cannot be too viscous, otherwise capillary feeding systems may not function (Baker 1975).

Tubes (point 8) arise most easily in Sympetalae but they are "improvised" in many Choripetalae as *Cuphea*, *Cadaba* spp., *Tropaeolum*, *Fuchsia*, and *Malvaviscus*. In the ornithophilous *Iris fulva*, discovered by Vogel (1967) there is a long, strong-walled tube in contrast to the short tubes of melittophilous species. Birds can stretch out their tongue and exploit legitimately flower tubes longer than their bills, but short-billed hummingbirds have a tendency to puncture blossoms and rob them.

On point 10, the absence of nectar guides, it should be remarked that the strong reduction and deflexion of the corolla limb makes the placing of a nectar guide difficult anyhow.

We said that the transition to ornithophily is mostly recent and peripheral, but in some groups the tie seems older. Porsch (1937a), alas without substantiation by field-work, found a supra-generic group in Cactaceae, the Andine Loxanthocerei, where ornithophily apparently has become fixed in the tribe.

Amongst Euphorbiaceae with condensed cyathia, *Poinsettia* has enlarged glands and red bracts for hummingbirds. The genus *Pedilanthus* (cf. Dressler, 1957) represents a higher specialization, independent since the beginning of the Tertiary, with the glands enclosed in a spur, the flowers sticking out and the whole blossom zygomorphic.

Even in the orchids, a recent group and a bee family *par excellence*, some species have switched over to ornithophily in the unending probing of the sexual environment, typical for the family. In the South African genus *Disa* some species have probably become ornithophilous (Vogel 1954). In connection with its butterfly pollination the flowers within the genus were already red, provided with a spur, and with reduced labellum. We suspect the same for *Cattleya aurantiaca* and some New Guinean mountain *Dendrobiums* (van der Pijl and Dodson, 1966: 95—96). Bird visits have been observed by Dodson in *Elleanthus capitatus* and *Masdevallia rosea*.

Dressler (1971) gives a list of bird-pollinated orchids and maintains that their pollinia are dark-coloured (in contrast to the usual yellow), therefore not contrasting with the colour of hummingbird beaks, which prevents the release of brush-off movements.

11.2.2. Pollination by Bats. Chiropterophily

Like birds, bats possess a rough surface with great pollen-carrying capacity. They also move swiftly and across great distances. Pollen in bat faeces could be proved to come from plants at least 30 km distant. It is therefore no wonder that bats have been used as pollinators.

The first actual observations of the flower visits by bats, with understanding of the issues, were made by Burck in 1892, in the Buitenzorg (now Bogor) Botanic Garden. He saw fruit-eating bats (probably *Cynopterus*) visit the inflorescences of *Freycinetia insignis*, now known to be fully chiropterophilous, in contrast to its ornithophilous sister species (see p. 123).*

Later authors like Cleghorn, McCann (India), Bartels, Heide, Danser, Boedijn (all in Java) described other isolated cases, of which *Kigelia* (the sausage tree) became classical. Since 1922 Porsch has been the prophet of chiropterophily, analysing its characters and predicting many possible instances. After a visit to South America he published (1931b) the first investigated case in its native land (*Crescentia cujete* in Costa Rica, cf. also Porsch 1934—36).

Through the work of van der Pijl (1936, 1956) in Java, Vogel (1958, 1968—9) in South America, Jaeger (1954) and Baker and Harris (1959) in Africa, bat pollination has now been established in many plant families. Some plants, which were earlier considered ornithophilous, have proved to be bat-pollinated instead, e.g. *Marcgravia* spp.

Bats are "normally" insect-eaters, but vegetarians have developed independently in both the Old and the New World. Possibly the evolution went via fruit-eaters to blossom-feeders. Fruit-eating bats are known from the two suborders of bats, inhabiting different continents, and a mixed diet is found, e.g. in the African Pteropinae. The possibility that nectar feeding has developed from insect hunting in flowers, as presumed for hummingbirds, has been suggested.

The relation between megachiropteran fruit-bats and blossoms is still partly dystropic, in Java *Cynopterus* was found to eat flowers of *Durio* and parts of the inflorescence of *Parkia*. In the eastern part of Indonesia and in Australia many flowers (*Eucalyptus*) are squashed by *Cynopterus* and *Pteropus*; an as yet unbalanced condition.

In the Macroglossinae there are more exclusively flower animals than even hummingbirds. Animals caught in Java showed only nectar and pollen in the stomach, the latter in such

*The often-cited observations by Hart in Trinidad in 1897 on *Bauhinia megalandra* and *Eperua falcata* were confused by incorrect conclusions.

quantities that accidental uptake is out of the question. Without pollen the animals would obviously not obtain the protein their ancestors possibly obtained from fruit juices. In Glossophaginae pollen consumption, though actually found, seems more incidental.

Howell (1974) maintains that *Leptonycteris* covers its whole protein requirement from pollen, which is sufficient both in quantity and quality. She also states that the chemical composition of pollen of bat-pollinated blossoms is adapted to the utilization by bats, and different from the composition of pollen of related species, which are pollinated by other animals. This may be considered as the blossom's part of the coevolution of the chiropterophilous syndrome. So far no report has come to light about African fruit-eating bats ingesting pollen.

Within the class of bat blossoms there is an early sideline forming a subclass of its own, within which Pteropineae have been found to be the only pollinators. In these blossoms solid food only (with the typical smell) is presented in specialized structures. We find here no nectar and no large masses of pollen. *Freycinetia insignis* offers sweet bracts, *Bassia* and *Madhuca* species a very sweet and easily detached corolla. Perhaps another Sapotaceae, viz. the African *Dumoria heckelii*, belongs to this subclass.

The New World nectarivorous bats are more or less tropical, but some migrate in the summer to the southern states of the U.S.A., visiting some cacti and agaves in Arizona. In Africa there are no records of bat pollination from north of the Sahara, whereas *Ipomoea albivena* in the Zoutpansbergen of South Africa is just inside the Tropics.* In Asia the northern limit of bat pollination runs north of the Philippines and Hainan, with small Pteropinae taking over in the latitude of Canton. The eastern Pacific limit runs with a sharp salient through the Carolines up to Fiji. Flower visits by Macroglossinae are known in North Australia (to an introduced *Agave*), while the indigenous *Adansonia gregorii* has chiropterophilous characters, indicating that chiropterophily must exist in this continent, too.

The knowledge of pollination by bats may help in solving puzzles about the origin of plants. The chiropterophilous flowers of *Musa fehi* indicate that the species must be introduced in Hawaii, where bats do not occur. It would be more at home in New Caledonia, where some botanists have already placed its origin.

Nectarivorous bats show various adaptations to their mode of feeding. In the Old World Macroglossinae there are adaptations to the life on blossoms, viz. decrease in dimensions (*Macroglossus minimus* weighs 20–25 g), reduction of molars, long snout, very extensile tongue with long, soft papillae on the tip (not a hard brush, as mentioned in old publications). Our description is based on live observations; denial of extensibility (as found in the literature) may be based on an examination of animals preserved in alcohol.

Similarly, some New World Glossophaginae possess a longer snout and tongue than their insectivorous relatives. In *Musonycteris harrisonii* the tongue length is 76 mm for a body-length of 80 mm (Vogel 1969a: 311). Also according to Vogel (*loc. cit.*) the hairs of *Glossophaga* are especially adapted for pollen transport, being equipped with scales corresponding in size to the scales on the hairs of bumblebee abdomens.

The sense-physiology of Megachiroptera deviates from what is "normal" in bats. The eyes are large, sometimes with a folded retina (allowing rapid accommodation), with many rods and no cones (causing colourblindness). Night photographs of live, fruit-eating *Epomops*

---

*The possibility of bat pollination in the white-flowered, arboreal strelitzias (*Strelitzia nicolai*, etc.) from the eastern Cape area should be investigated.

franqueti (Ayensu 1974) show enormous eyes, almost lemur-like. The sense of smell seems more important than usual (septate, large nasal cavity) and the sonar apparatus is less developed. According to Novick (quoted by Vogel 1969a: 323) sonar location organs are present in *Leptonycteris* and other pollinating Microchiroptera. In American bats with a mixed diet: nectar and fruit/insects, the sonar apparatus is intact. They make extended flights with fleeting contacts to sometimes poorer blossoms with less firm corollas (hovering visits are more frequent here).

The Macroglossinae have a power of flight which reminds one at first of swallows. Some species may also hover almost like hummingbirds. This has also been reported for Glossophaginae (Heithaus *et al.* 1974).

The existence of a definite harmony between blossom and animal in structure and physiology makes it possible to establish "bat blossoms" as a definite concept. Secondary self-pollination may interfere as in *Ceiba*, or even parthenocarpy as in cultivated *Musa*.

It is remarkable that, though the development of chiropterophily in America took place independently and probably much later than elsewhere and though the bats concerned developed late as an independent line, the syndrome of chiropterophilous characters became the same all over the world. Flower bats and bat flowers from all regions fit in mutually. This points to general characters in the physiology of all bats as involved. Sometimes also general properties of plant families may underly the development of chiropterophily in different lines.

We shall again list the adaptive syndrome in a comparative table (Table 8), again partly positive, partly negative.

TABLE 8. The syndrome of chiropterophily

Bat flower	Flower bat
1. Nocturnal anthesis, mostly only one night	Nocturnal life
2. Sometimes whitish or creamy	Good eyes, probably for near orientation
3. Often drab colour, greenish or purplish, rarely pink	Colour-blind
4. Strong odour at night	Good sense of smell for far orientation
5. Stale smell reminiscent of fermentation	Glands with stale odour as attraction
6. Large mouthed and strong single flowers, often strong (brush) inflorescences of small flowers	Large animals, clinging with thumb claws
7. Exceedingly large quantity of nectar	Large, with strong metabolism
8. Large quantity of pollen, large or many anthers	Pollen as sole source of protein
9. Peculiar position outside the foliage, flagelliflory, cauliflory	Sonar system less developed, flying inside foliage difficult

To point 1. Nocturnal anthesis is easily observed in the banana where the large bracts covering the flowers unfold every night.

Many of the flowers open shortly before darkness and fall in the first hours of the morning. As there is some time overlap between the diurnal birds and the crepuscular bats and so also between bird and bat blossoms it is no wonder that bird visits are on record for some chiropterophilous plants. Werth (1956a), who obviously never observed at night, therefore maintains *Musa paradisiaca*, *Ceiba*, and *Kigelia* on his list of ornithophiles, though birds are nothing but robbers in these flowers.

4 and 5. To an investigator with some experience the smell of bat blossoms is most definite. It has much in common with the smell of the animals themselves, which may have some social function in their clusters and may have some stimulatory effect. A strong effect was found on *Pteropus* specimens reared in captivity.

The same kind of odour, reminiscent of butyric acid, is found in fruits that are dispersed by bats (e.g. the guava). This circumstance, as well as the way in which the fruits are exposed, offers a convenient starting-point for the development of chiropterophily in a taxon with bat-dispersal, an older condition widely spread in the Tropics (see van der Pijl 1957). In many Sapotaceae, Sonneratiaceae, and Bignoniaceae this smelling substance may have helped in establishing relations. Vogel (1958) found a pronounced bat smell in the fruits of a *Drymonia* sp., whereas other Gesneriaceae (*Campanea*) have bat-flowers.

A bat smell is still or already present in some ornithophilous species of *Gossampinus*, *Mucuna* and *Spathodea* related to species with bat flowers.

The change from nocturnal sphingophilous odours seems relatively easy. Porsch (1939) suggested this chemical change for some Cactaceae, where the nocturnal anthesis, the useful cauliflory and the multitude of anthers were already present as organizational characters. This prediction was confirmed for the giant cactus, *Carnegiea*, in Arizona by Alcorn *et al.* (1961). The pollen had already been found in the bat *Leptonycteris nivalis* and the authors confirmed its visits, although as yet under artificial circumstances.

The smell is sometimes musty (*Musa*) or reminiscent of cabbage (*Agave*). A chemical investigation seems indicated.

6. The typical claw marks usually betray the nightly visits in flowers that have been shed. In banana inflorescences the number of marks on the underlying bracts allows a count of the number of visits. Incidental hovering may account for absence of claw marks (*Carnegiea*).

7. The nectar is still more abundant than in bird blossoms. In *Ochroma lagopus* (balsa) 7 ml was found, in *O. grandiflora* as much as 15 ml. We have no data on a possible special composition. In the banana the nectar forms a solid jelly on cold mornings. Heithaus *et al.* (1974) describe two strategies in nectar feeding on *Bauhinia pauletti*. Large bats come in groups, land, and take a relatively long time to drain the nectar from the blossoms. Small bats hovered in front of the blossoms and lapped up the nectar during repeated, very short, visits. Evidently there was no marking of flowers that had been visited. Sazima and Sazima (1975) describe a strategy more like the trap lining.

8. The enlargement of anthers is obvious in *Ceiba, Bauhinia, Agave, Eugenia cauliflora* and the Cactaceae, their multiplication in *Adansonia* which has 1500–2000 anthers.

9. The necessity for open space when alighting and departing and the relative inefficiency of echolocation in Megachiroptera have been proved by experiments with obstacles placed before the flowers, by observations of collisions and the fact that bat hunters catch Megachiroptera much more easily than Microchiroptera.

The exposure of bat flowers is similar to that of hummingbird flowers, but is much more pronounced; it often takes the form flagelliflory (penduliflory), with the flowers dangling on long, hanging stalks (*Adansonia, Parkia, Marcgravia, Kigelia, Musa, Eperua*). It is most obvious in some *Mucuna* spp., where stalks of 10 m or more sink the blossoms to open places.

A pincushion type with stiff stalks projecting flowers upwards also occurs (*Markhamia, Oroxylum*). The giant inflorescence of *Agave* speaks for itself. The "pagoda-structure" of some Bombacaceae is also favourable.

Chiropterophily also makes clear why cauliflory, as mostly adaptive to bat visits, is practically confined to the Tropics with its estimated 1000 cases. Good examples are *Crescentia*, *Parmentiera*, *Durio*, and *Amphitecna* In many genera (*Kigelia*, *Mucuna*) flagelliflory and cauliflory occur together or alternate in different species.

All known theories of tropical cauliflory have been discussed and proven as superfluous in previous papers (van der Pijl 1936, 1956). Cauliflory is a secondary phenomenon. Its ecological nature agrees with the results of investigations into its morphological bases. Numerous cases had no taxonomical, morphological, anatomical and physiological common basis.

In most examples of cauliflory, where the flower was not chiropterophilous, another bond with bats was found, viz. chiropterochory, dispersal of seeds by fruit bats (van der Pijl 1957). Fruit bats have left on tropical fruits (and consequently on the position of their flowers) an earlier and more widely spread stamp, including colour, position, and odour. This older syndrome shows an exact parallel with the newer one of chiropterophily. Basicaulicarpy may also be connected with the syndrome of saurochory (dispersal by reptiles), which is more archaic than angiospermy.

A succession of flowering periods is necessary to make life possible for both plants and bats. In large Javanese plantations of *Ceiba*, which has a definite flowering period, the flowers were bat-visited only in the parts next to gardens with *Musa*, *Parkia*, etc., on which bats could feed when *Ceiba* was out of bloom.

The relatively recent nature of chiropterophily as a whole is reflected in the distribution of bat blossoms among the plant families. In the *Ranales* bat fruits are known, but bat blossoms not. Bat blossoms occur in more highly evolved families from Capparidaceae and Cactaceae onwards, with concentrations in the Bignoniaceae, Bombacaceae, and Sapotaceae. Many cases are entirely isolated.

Some families (Bombacaceae and Bignoniaceae) include chiropterophiles apparently independently evolved in Old and New Worlds, obviously on some pre-adaptive basis, as already discussed in the foregoing parts. This may also have happened within some genera like *Mucuna* and especially *Parkia*, which was considered in this light by Baker and Harris (1957).

The same Bignoniaceae and Bombacaceae, also *Mucuna* and *Musa*, have some species which are intermediate between bird and bat pollination. *Bombax malabaricum* (*Gossampinus heptaphylla*) is ornithophilous, but incompletely so with open, red, cup-shaped, diurnal flowers. It has, however, the bat-smell of its chiropterophilous sister-species *B. valetonii*. In Java bats neglect it, but in border regions of South China, Mell (1922) found it eaten by Pteropinae. The direction of the change seems to be from birds to bats in Bignoniaceae and probably reverse in Bombacaceae and in *Musa* where the subtropical species are bird-pollinated. The switch-over from sphingid flowers in Cactaceae has already been considered.

It is too early to attempt a more quantitative analysis of the relations and their genetical consequences. Sometimes the bats (especially the slower Pteropinae observed by Baker and Harris) remain restricted to one tree, causing self-pollination. The Macroglossinae, swift flyers, make the rounds of distant trees with an obviously fine memory for spatial relations. There is, however, no blossom constancy, as examination of the pollen on the fur and especially of the big pollen lumps in the stomach reveals. How, if at all, genetical purity is maintained between related chiropterophilous species, as for instance the wild *Musa* species, is not clear.

CHAPTER 12

"RETROGRADE" DEVELOPMENTS

12.1. "REVERTENCE" TO ABIOTIC POLLINATION

If we presume that the blossom developed from primitive towards more refined types under the influence of selection with regard to pollination, a development back to an allegedly more primitive type constitutes true revertence. On the other hand, one frequently meets the terms "revertence" and "retrograde" development for any wind-pollinated group, based on the idea that wind pollination is a more primitive means of pollen dispersal in angiosperms. We have shown previously that this idea is hardly tenable; why should ovules be enclosed in carpels as long as plants were anemophilous? Today there seems to be growing agreement between phylogeneticists that angiosperms are monophyletic. Should one presume a polyphyletic origin of angiosperms, one might question if there might be primary anemophily in some lines, e.g. in Amentiferae (see Lam 1961), but this now seems less likely. Also, the group can hardly be considered natural. Tropical species of *Quercus* and *Castanea* are insect-pollinated with pollen as attractant and spermatic odour — recognizable also in *C. vesca*, the inflorescences of which still show the syndrome of biotic pollination: nectar, colour, brush type. *C. vesca* is to a great extent pollinated by insects; chestnut honey is well known among south European bee-keepers. During the first part of anthesis the pollen grains are sticky; only later do they dry up and are dispersed by wind. The pollen grains of Fagaceae are not typically those of anemophilous plants, and unisexuality in Fagales is still only half-evolved.

Shift to anemophily can be facilitated by "pre-adaptation", i.e. the existence of characters which, whatever their origin, also fit into a syndrome that manifests itself later in the evolution of the group, in this case the wind pollination syndrome. Change of habitat, e.g. from forest to plain, would then give these characters a new usefulness connected with transition to anemophily. The evolution of pollination systems is a dynamic process, and there is reason to believe that evolution in this sphere goes relatively fast.

We have already pointed out that anemophily is found in many families (1), that are highly evolved, and (2) the ancestry of which from zoidophilous ancestors can easily be traced: Gramineae, Cyperaceae, Juncaceae, and Plantaginaceae.

It is therefore in these families no simple *revertence* to abiotic pollination. Anemophily is a further development in pollination, and it is tempting to see it as a response to poor climates, in which the possibilities for biotic pollination are smaller. In themselves, many of the morphological characteristics of wind-pollinated blossoms represent further development of tendencies seen already in the development of biotically pollinated blossoms, e.g. reduction in number of members of the blossom.

According to Kugler (1975) there is no direct correlation between the number of insect species and the percentage of anemophilous plant species in various biocoenoses. However, mutual competition factors, the effectivity of individual pollinators, etc., make the picture far too complicated to be explained by species statistics alone.

The interpretation of anemophily as an adaptation to lack of suitable pollinators (like autogamy, etc.) explains the scarcity of anemophilous species in the most lush coenoses, above all in the tropical rain-forest, where there is a great and constant supply of pollinating animals, but little wind. Many genera that are in temperate regions typical anemophiles are represented by entomophiles there; even grasses may be entomophilous, e.g. *Pariana* (cf. Soderstrom and Calderon 1971). In addition to the American olyroid grasses dealt with *loc. cit.*, we also suspect the Asiatic *Ischaemum muticum* and wonder about the function of the extranuptial nectaries between the spikelets of many *Eragrostis* and *Andropogon* spp., often combined with sticky pollen early in the morning. According to Porsch (1958), several species of beetles are involved in these blossom visits, and Neal (1929: 25) describes "pili honey" from Hawaii produced from the grass *Heteropogon contortus*, cf. also Karr (1976).

In many species the possibility of wind pollination occurs together with entomophily: *Calluna*, *Salix* spp., produce nectar, and the blossoms are regularly visited by and pollinated by insects. In addition, great masses of their pollen are spread by wind, and additional wind pollination must be inevitable. On the other hand, uptake of anemophilous pollen by pollen collectors is inevitable. The great quantities of pollen produced by regular anemophiles attract pollen collectors. Honey bees may bring home pure grass pollen pellets (Sharma 1970). These visits may also cause pollination, but they are hardly effective because of the mutual positions of anthers and stigmas. Frequently the flowers are unisexual and female flowers are not visited. On the other hand, the typical pollen blossoms with their surplus pollen production are better adapted for the transition to anemophily. The idea of revertence only holds if one projects backwards beyond the origin of angiospermy, to hypothetical ancestors, the spores of which are presumed to have been wind-dispersed. In the case of hydrophily, the development of abiotic pollination is a response to edaphic rather than to climatic conditions. Interesting cases of real revertence, but the opposite way, are found in Cyperaceae. There can be no doubt that the family is derived from entomophilous plants with a six-membered perianth of which only rudiments, if any, are left in the present anemophilous representatives. Species of *Dichromena* and *Sickmannia* have again turned to entomophily – pollen attractant – but the perianth being lost, the involucral bracts have to function as advertising organs as should be expected according to the rules of irreversibility (K.F. cf. Leppik 1955).

Other examples of entomophiles in this otherwise anemophilous family are given by the South Africa *Chrysothrix capensis* – a blossom-like inflorescence with blue filaments and orange anthers. *Mapania*, which is a forest-bottom plant like *Pariana*, has white capitula with vivid red bracts. Porsch (1956) mentions that one species has an agreeable scent. Entomophily has been corroborated by Lorougnon (1973). The presence and possible importance of pollen odour in such blossoms should be investigated.

Another example of return to entomophily by forest plants is provided by the genus *Ficus* in which the pollination syndrome is much more complicated and effective than in the grasses referred to above.

12.2. AUTOGAMY

In spite of all the structures observed in blossoms, the *effect* of which must be to counteract self-pollination (whatever their origin), there is always some chance that self-pollination takes place nevertheless; even in orchids pollinia have been observed to fall out of their loculi and down on to the stigma. The effects of these perhaps omnipresent accidental self-pollinations vary both with compatibility conditions and the time factor – the interval, if any, between allo- and autogamous pollination, and the difference in growth rate between pollen tubes. With the exception of strongly self-incompatible species it must be assumed that a rather large proportion of the yearly seed output is the result of this kind of autogamy.

In some blossoms self-pollination is spontaneous: even if no external pollinating agent (biotic or abiotic) touches the blossom, its own structures cause the transfer of pollen from anthers to stigma in the same blossom. In some plants this is combined with strong self-incompatibility and is thus part of an allogamous syndrome, e.g. in Papilionaceae or Proteaceae. In blossoms where there is no such incompatibility, this is an autogamous syndrome. In other blossoms autogamy has to be induced: the pollen transfer is caused by a pollinating agent, an animal messing around in the blossom, or wind shaking the plant.

Darwin (1876) established a range of compatibilities from species the seed output of which after selfing was equal to that after out-crossing to such species in which the seed output after selfing was negligible. Such series have some value, but it should not be forgotten that they are highly dependent both on external circumstances and on racial characteristics. There is no doubt that the plant material of gardens, especially botanic gardens, has been subject to positive selection for this kind of self-compatibility.

There is also a more regularly occurring autogamy. The development towards this kind of autogamy, like that towards abiotic pollination, may be looked upon as a compensation for poor chances of allogamous pollination. We may postulate a whole series of stages in the degeneration of the pollination mechanism. A list of autogamous species is given by Fryxell (1957).

Weeds and aliens present special problems with regard to autogamy. There is "Baker's law" (Baker 1955): autogamy is a prerequisite for the successful establishment of long-distance migrants, the functional definition of which must be that the offspring is so far removed from the original population that no genetic feedback is possible. It has escaped its "normal" visitors – pollinators and predators, and if establishment is to succeed, alternative pollination processes must exist. Weeds are by definition plants occurring outside their "natural" habitat (many of them have no known habitat that is not made by man), and their success depends on their ability to adapt to different habitats, including different breeding systems. There are two possibilities: either predominant autogamy (or agamospermy) with reduction of attractants (cf. Mulligan and Kevan 1973), or the establishment of a simple, generalized pollination syndrome with showy or odoriferous – or both – blossoms attractive to any insect that might happen to be there. Many of the old field weeds belong to this type: poppy, cornflower, etc. Macior (1971) has pointed out that the existence of the same pollination syndrome in the new station may provide an immigrating alien with a readymade pollination system from the beginning. The similarity between pollination syndromes in *Dodecatheon* and *Solanum* has facilitated the establishment of *S. dulcamara* in North America. In *Melochia* increasing weediness goes hand-in-hand with break-down of distyly and self-incompatibility (Martin 1967).

A first stage in the evolution of autogamy is represented by the breakdown, during anthesis, of external and internal factors preventing self-pollination or self-fertilization, permitting autogamy to occur *late* during anthesis. Many pollination mechanisms include bud self-pollination as a regular part of the process, e.g. in Papilionaceae (p. 181; extreme self-incompatibility is a prerequisite if this is not to result in autogamy, and a shift towards self-compatibility is the only step necessary to make such a flower autogamous. As pointed out by Stebbins (1957), the genetical processes involved in the breakdown of self-incompatibility are much simpler than those involved in its establishment. Dichogamy may become obliterated by the prolonged activity of the agent of the first phase (anthers in protandry, etc.). Especially in protogyny this would be a very effective way of securing autogamy if allogamy fails. Examples of the breakdown of herkogamy with ensuing autogamy are found in *Commelina coelestis*, and in small-flowered Labiatae and Scrophulariaceae. A prerequisite for the effectivity of the breakdown is that genetic self-incompatibility breaks down at the same time, and we may again refer to the observation by Lamprecht (1929) that the concentration of the impeding substances may decrease during anthesis (cf. also Ascher and Peloquin 1966). This would then permit a late- or post-anthesis self-fertilization. In ephemeral flowers, like those of *Commelina* referred to above, or *Nicandra*, this type of autogamy is important.

Autogamy after the main anthesis very markedly stands out as a first emergency measure in blossoms that have for some reason not been properly pollinated. If incompatibility factors are still at work, xenogamous pollen, if present, will have arrived at the stigma before the autogamous one, and the ensuing pollen tubes will also grow faster. A further development of emergency measures, then, is autogamy *during* regular anthesis. This presumes inactivation of any incompatibility genes (or pre-anthesis breakdown of the substances involved) as well as immobilization of any external mechanism unless they are replaced by other mechanisms, compensating for herkogamy. For rain pollination to be effective, dichogamy must break down, as it does, for example, in *Nymphaea*. Autogamous flowers of this kind are found in many species belonging to entomophilous genera or families. Small-flowered alpine representatives of Orchidaceae (Hagerup 1952) and Oenotheraceae (Resvoll 1918) may be mentioned as examples, and many other plants with small and insignificant flowers. Small-flowered shore plants are capable of self-pollination in an airbubble even under water (*Subularia, Limosella*), although transitions to cleistogamy are frequent.

Autogamy is sometimes brought about more or less accidentally, frequently by pollen simply falling down from the anthers (of the same or a neighbouring flower) on to the stigma, the so-called gravity pollination; or it may take place when a deciduous corolla with anthers slides past the stigma (Hagerup 1954). In other cases the filaments bring anthers into contact with the stigma, with the result that pollen is deposited directly there and then. Sometimes the anthers have dehisced before they touch the stigma; sometimes they do so subsequently. In other cases pollen grains start growing in the anther, and the pollen tubes are the organs which reach the stigma.

A very interesting group are the self-fertilizing cultivated plants: wheat, barley, oats, beans, etc. It is easy to see how they may have been created by unconscious selection, which would be more severe as soon as crop plants were removed from their original area (especially with regard to entomophilous species). In *Lycopersicum* self-pollination is achieved by minor adjustments in the mutual position of anthers and stigma. The stubborn resistance towards self-compatibility of some very old crop plants like rye and maize is very interesting in this connection, showing the limitations of selection.

One characteristic effect of adverse conditions is the break-down of heterostyly in alpine or arctic species. This occurs, for example, in *Primula* and *Armeria* (Iversen 1940), and many others: the temperate species are heterostylous, the arctic-alpine homostylous with break-down of the self-incompatibility factors. The equilibrium between heterostylous and homostylous forms (with self-compatibility) may also be influenced by population size and the distance to the nearest colonies of the same species (Ganders 1975). This must not be confused with a primary homostyly, preceding the heterostylous state (Ernst, 1953, 1955).

In flowers at this stage of development of autogamy, allogamous pollination is still possible and does take place, although in different degrees, depending on the species and also on external conditions. Thus Lamprecht (1929) reports that in ordinary years there will be about 0.3 per cent outbreeding in *Phaseolus vulgaris*, but the percentage may rise to 1 per cent in hot summers when many insects are about. If some of the internal anti-self-compatibility factors are still at work, xenogamous pollen may have an advantage, even if it is deposited later than the first autogamous grains.

The next step in the development of complete autogamy is formed by flowers in which autogamy takes place *before* anthesis, with *Lobelia dortmanna* as the classical example. In this flower, all the paraphernalia of allogamy are − still − present, but completely inactive. Even if the flowers should receive visits, there is hardly a chance that the pollination will produce any result. Other examples are quoted by Hagerup (1951) from the Faroes. Again, these may be autogamous ecotypes of otherwise allogamous plants produced under the selective pressure exerted by the absence of suitable pollinators.*

The morphological establishment of self-pollination is in itself highly interesting, but even if the growth of pollen tubes from autogamous pollen has been established, this is not sufficient to prove that xenogamy does not take place concomitant with and perhaps superseding autogamy. Only experiments can decide this − and simple isolation is indecisive. Isolation may show if autogamous pollination can lead to seed-formation, but tells nothing about the possibility for xenogamous pollination.

In gynodioecious species bud autogamy in hermaphroditic flowers may secure seed-setting under adverse circumstances, whereas allogamous seed-formation is secured by the existence of female flowers (e.g. in *Silene noctiflora*: see Halket 1936).

Myosurus minimus represents an aberrant, possibly unique case (Stone 1957), in which self-pollination regularly takes place during the first part of the anthesis, whereas it is impossible during the later part both because of developing herkogamy and because of dichogamy (exhaustion of anthers). The plant is self-compatible, and the long receptable will therefore carry nutlets produced by apogamy at its base, by autogamy further up (lack of integration in the gynoeceum).

Now, if a flower is already pollinated before it opens, and there is no likelihood of later pollination having any effect, the whole anthesis is functionally redundant, and could be done away with. *Cleistogamous* flowers with reduction of organs, especially affecting those serving as advertisement and producers of attractant, represent the final stage in this series of degenerative phenomena. Generally, the anthers and number of pollen grains are also reduced, corresponding to the greater effectivity of the pollination process. A list of cleistogamic plants is given by Uphof (1938).

The classical example of cleistogamy, already known to Linnaeus, is *Viola mirabilis*, which produces, more or less simultaneously, both cleistogamous and chasmogamous ("normal") flowers, the latter generally sterile. Similar conditions prevail in *Oxalis acetosella* and many grasses, white *Lamium amplexicaule* produces cleistogamous flowers at the beginning and end of the season (Schoenichen 1902), and fertile chasmogamous flowers

---

*While fully recognizing Hagerup's great achievements in showing the presence of bud autogamy in many unsuspected cases, we feel that perhaps a kind of iconoclastic joy has led him to un-warranted generalizations about the relative merits of auto- and allogamy. The same applies also to some of the "anti-teleologists" who in their justified disbelief of the more extreme applications of Knight − Darwin's law have also rejected perfectly good examples.

during the summer. This is an example of an environment-induced cleistogamy (in *Oxalis* conditioned by humidity — and low temperatures? — cf. Vereshchagina 1965). Other examples can be shown to be due to lack of nourishment, inundation, etc. They are generally not periodical. Cleistogamous flowers are subterranean in some species such as *Cardamine chenopodifolia* (Troll 1951). Examples of facultative subterranean cleistogamy are cited for some mycotrophic orchids, but there is a possibility that *Cryptanthemis* may have open subterranean flowers.

It is easy to picture cleistogamy also evolving from the phenomenon of closing of flowers during periods of unfavourable weather. Self-pollination is frequently the result of these repeated opening and closing movements, and if there is self-compatibility the result may under adverse climatic conditions be that this becomes the most effective pollination method. From facultatively open flowers the step is not far to permanently closed ones.

The development from biotic to abiotic pollination and further or (as the case may be) directly to autogamy may be considered an autonomous orthogenesis; but mostly it will represent also an increasing adaptation to adverse conditions. Such conditions may be represented by climates in which biotic vectors do not occur in sufficient numbers to safeguard pollination, e.g. in too dry or too cool climates. An example is afforded by the predominant autogamy within the floras of the alpine summits of Java (Docters van Leeuwen 1933), rising out of a very rich vegetation with dominating biotic pollination. On the other hand, this rule does not always hold true, Swan (1961) insists that there are enough prospective pollinators as far as the highest vegetation zones in the Himalayas. And Kevan (1972: 667) maintains that there is ample pollination in the Arctic, where, incidentally, there are many species, especially with low chromosome numbers, dependent on allogamy.

The genetic effect of autogamy in isolated populations is increasing homozygosity, even the splitting off of microspecies. Autogamy is rare in perennials. Nevertheless, there have been some misunderstandings with regard to autogamy in tropical rain-forests because of the distances between individuals of the same species. The discovery of the trap-lining strategies of various tropical pollinators (cf. p. 57) has changed the basis for these assumptions, and it is now known that devices for out-crossing are predominant in the rain-forest (Ashton 1969) and also in drier tropical lowlands (Bawa 1974).

According to Frankie (1976) the shift of pollinators from one tree to another in the tropical forest may be induced by (temporary) depletion of the attractant below the acceptable level, or by aggressive territoriality of a competing pollinator.

So far, adverse climatic conditions have been considered as the main possible cause of autogamy, probably working through the effect of cold, dryness, etc., on the occurrence of possible pollinators. Levin (1970, cf. 1972b) sees competition for pollinators as a possible cause for the development of autogamy in any climate. The adverse climate effect should be considered a special case of this. However, there are other factors that may come into the picture as well. One of them is active in those plants in which the pollination mechanism has apparently become too difficult, and where only small external changes may lead to a maladjustment that can only be remedied by a complete break-down of the whole mechanism with autogamy as a consequence. It is no surprise that at least one *Ophrys* species, *O. apifera*, seems to belong here (C. Darwin 1890; Schremmer 1959).

Levin (1968) describes a case of heterostyly *cum* cleistogamy in *Lithospermum caroliniense*. In the population studied, chasmogamic flowers produced 16.9 (thrum) or 13.9 per cent (pin) seeds as against 90 per cent in cleistogamic flowers. Legitimate pollen grains per stigma numbered 4.9 and 2.8 respectively, illegitimate 27 and 650. In spite of the enormous quantity of illegitimate pollen on pin flowers they produced less seed than the thrum ones. In this case cleistogamy proves to be an emergency measure.

The extermination of the "proper" pollinators seems to have been the cause of the development of autogamy in Hawaiian ornithophiles (Lobelias), adapted to pollination by Drepanididae, which have later become extinct.

In taxa, which grow within a wide range of climates, one subordinate taxon may be autogamous while another, occurring where biotic pollinators are available, is allogamous and even self-incompatible (*Potentilla glandulosa* subspecies: Clausen and Hisey 1960).

Stebbins (1958) has pointed out that autogamy is linked with certain habitat factors. They are concentrated in unstable habitats, or they are frequent in annuals, being less frequent in climax vegetation or among long-lived perennials.

Genetically, geitonogamy is equivalent to autogamy* and may also, if effective, be considered a degeneration of the pollination mechanism. The "measure against" such a degeneration is, above all, self-incompatibility even if secondary dichogamy and ephemeric anthesis may also be effective. Geitonogamy may be indirect, i.e. dependent on a a vector (biotic or abiotic), but it may also be direct, haptogamy, through the flowers touching each other, more or less permanently as in gregarious *Galium* spp.

12.3. APOMIXIS AND VEGETATIVE PROPAGATION

Autogamy is still a sexual process retaining the possibility, though limited, for new gene recombination and production of new genotypes of higher selective value. In an apomictic population this faculty of variation has been very much reduced, and the higher degree of efficiency in reproduction has been paid for by a loss of adaptibility which will, one may surmise, seriously restrict the power of colonization of new habitats. By apomixis all that was gained by the introduction of the diploid generation and the sexual propagation has been lost again.

Apomixis may therefore be considered the final step in the reduction of pollination dealt with in the preceding section; the male nucleus has eventually lost all genetic functions. Correspondingly, we find apomictic genera in which the flowers and the stamens are almost vestigial, e.g. *Alchemilla,* where one or more stamens may have been lost, or, if present, have empty anthers. And when pollen is produced a high proportion, often most, of the grains are highly irregular and have probably lost all sexual potentiality.

It is curious that many apomictic plants have large, open pollination units, apparently functioning quite normally: *Taraxaca, Hieracia, Rubi, Potentillae,* etc., and to some extent they do function normally; one of the main reasons for taxonomic trouble is that every now and then a normal fertilization does, in fact, take place, starting another group of apomicts. Besides, as is well known, many apomicts need pollination to start the apomictic development of seed, even if fertilization does not seem to occur. This must be considered a relict of secondary pollination effects (production of hormones and/or auxins) known in many species both of gymnosperms and angiosperms. Thus, in cycads pollen exerts a non-specific action on the growth of ovules, in *Ginkgo* on the growth of the female

---

*Laczynska-Hulewiczawa (1958) insists that geitonogamy gives better results than autogamy s. str. and Alcorn *et al.* (1959: 40) maintain the same with regard to *Carnegiea.* The genetical basis for this, if corroborated, would be difficult to explain.

prothallium, in orchids on the differentiation of the ovule.* As is well known, parthenocarpic fruit development may start after a more or less specific pollination, cf. Yasuda (1939).

It has been suggested that the attractive blossoms of certain apomictic species may have some competitive value by attracting pollinating insects from obligate insect-pollinated species in the same plant community. In this way, these latter species should reproduce themselves less effectively. Actual tests of the idea are not known. The tetraploid *Antennaria alpina* may have arisen in response to adverse climatic conditions, and apomixis in consequence of tetraploidy.

The final break-down of sexual propagation under adverse external conditions is not apomixis, but vegetative reproduction. In many higher plants, as well as in many lower ones, the sexual process is all but eliminated, and the plant lives on vegetatively. *Hedera helix* near its northern limit is a good example (Fröman 1944), and some aquatics may also be quoted with Lemnaceae as the better known ones. This type of vegetative reproduction should, of course, not be confused with those cases that are due to the fact that a single, self-incompatible clone occurs alone as mentioned earlier, represented among aquatics by *Acorus calamus* in Europe.

Heslop-Harrison (1959) pointed out that a combination of asexual (apomixis, vegetative) and sexual reproduction should give a plant optimal versatility; the sexual reproduction produces new ecotypes, the asexual perpetuates the successful ones. As compared with ordinary vegetative reproduction, apomixis has the great advantage of using the seed with its greater resistance (the dormancy period) and its dispersal mechanism. Environmental selective pressure may in the end be decisive for the equilibrium between sexual reproduction with its diversity and asexual reproduction with its effectivity, and pollination is one of the environmental factors.

*We shall not here enter upon various theories about secondary effects of pollination.

CHAPTER 13

THE DEVELOPMENT OF FLOWERS
IN RELATION TO MODE OF POLLINATION

In the following chapters we shall follow two characteristic flower types as they occur within a given taxon, exemplifying each type by a central typical blossom. We shall try to trace the development of the blossom from simpler form and show its further development into a more specialized or derived type. Our examples start with zygomorphic flowers, the pollination unit classes A–C, and to a great extent F, being considered rather self-explanatory. Compare the Case Histories.

13.1. THE FLAG BLOSSOM IN LEGUMINOSAE

The pollination and function of a typical flag blossom without any complications may be seen, for example, in the genus *Astragalus*. The operative parts are the following:

1. The calyx forms a longer or shorter tube which protects the nectary and supports the inner part of the unguiculate petals.

2. The median petal, *vexillum*, is symmetric, its outer part is broadened and forms an angle with the axis of the blossom. It is the chief advertising organ of the blossom.

3. The four other petals clasp together and are parallel to the axis of the blossom, all four are distinctly unguiculate with very narrow, pedicel-like claws, and are asymmetric.

4. The two upper lateral petals, the wings, *alae*, are free (see below). Besides having folds which will be discussed later, each possesses a projection pointing towards the centre of the blossom and inwards towards the axis of symmetry.

5. The two lower lateral petals coalesce along their lower edges, thus forming a boat-like structure, the *carina*. The upper edges of the carina may be free or connate.

6. There are ten stamens, the filaments of nine of which form a sheath surrounding the pistil. Only the outer ends of the filaments are free. Anthers and style are found in the tip of the carina. The tenth stamen is free and lies as a lid on top of the filament sheath.

7. Nectar is produced at the base of the pistil and collects in the inner part of the filament sheath. At the base of the loose tenth filament there is a hole at each side, through which insect probosces can penetrate to the nectar.

As entrance to the nectary from the sides is closed, the only practicable way in (apart from biting through) is from the front. Insects land on the alae and force their way to the nectar along the top of the filament sheath and the tenth filament, under the middle line of the vexillum. During this process they depress the alae and carina, the claws of which form a flexible system. In contrast, the pistils with the surrounding filament sheath are rigid and do not follow the petals in their downward motion. As a consequence, anthers and style emerge from the carina and touch the under side of the abdomen of the visitor, which follows the petals in their downward motion.

The motion of the various parts of the flower in relation to each other is *not* due to the weight of the pollinator. Both nature observations and experiments show that in many species the pedicels are much too weak to keep the flower in the horizontal position, which would have been a prerequisite for the weight of the animal being the force opening the blossom. Flowers drop to a more or less hanging position when the insect visitor lands on them. The petals are therefore actively forced apart by the pollinator, which must consequently be (1) strong and (2) able to get a secure grip on whatever part of the flower it has alighted. In other words, the insect must be a bee. The typical flag blossom is adapted to being operated by bees.

After the visitor has left the blossom, the petals resume their original position. This is partly due to stresses establishing themselves (through differential motion) between the claws of alae and carina and partly to the above-mentioned backward projections of the alae, which act like springs and pull the rest of the petal back.

The zygomorphy of a flag flower in Papilionaceae is due to position and size effects only. The flower diagram is radial.

The development of this remarkable blossom type is easily followed within the order. In primitive Caesalpiniaceae, the flower is almost radial, also in position, only the median petal differs from the others (frequently being more vertical) and the pistil and stamens are found in the lower part of the flower. Pollination is sternotribic; in *Cassia* and *Bauhinia* insects land on the stamens and "milk" them, at the same time throwing pollen up on their own backs. To progress from this completely open flower to an incipient flag type all that is needed is a concentration of the morphological elements of the lower part of the flower. An example of this is *Camptosema nobile*, in which the four lower petals form a kind of open tube, containing the sexual parts of the flower which protrude slightly.* The red-flowered *Camptosemas* are evidently ornithophilous, and *C. nobile* represents an intermediate stage between the brush types so frequent in Caesalpiniaceae and Mimosaceae, and the true flag types, which do not seem to be primarily suited for pollination by birds. We shall later see examples of modification of the flag flower in response to bird pollination.

Within typical flag flowers, there are a number of minor modifications, which chiefly modify two points, viz. the connection between alae and carina, and the presentation of pollen. In some − primitive − types, the alae and carina are completely free and move independently of each other. In these types the alae do not form part of the mechanism proper, for example in *Wistaria*. In other blossoms the alae and carina are connected, with the result that movement in one set of petals is transferred to the other. Because of the unguiculate shape of the alae, these petals will generally move outward—downward when depressed, and thus contribute to the opening of the upper part of the carina.

In some flowers the alae and carina are connate inasmuch as the epidermis of corresponding spots at the base of the broad part of both groups of petals adhere to each other. The epidermal cells at these spots generally possess verrucate protuberances that increase the effectivity of the connation. Such a system is rigid, and the alae and the carina are absolutely fixed in relation to each other. Typically, the alae and carina are flat, without distinct buckles or folds. Generally, there is a combination of this type of connection with the following, but a rather pure case of adherence is found in *Melilotus*.

The second type is characterized by the presence of a fold in the alae, fitting into a

*Without further study it can hardly be decided whether in *C. nobile* this structure is truly primitive, as presumed by Lindman (1900), or secondarily reduced.

corresponding fold in the sides of the carina. The axis of the fold is turned downward and, when depressed, the alae will therefore take the carina with them.

In comparison with the former type, this kind of a joint is more flexible, and the parts may move a little in relation to each other. The fold is situated in the inner end of the broad part of the petals, but with varying relative length of the claws the joining may occur far out or far in. In *Genista*, it is situated almost at the base, in *Coronilla emerus* very far out. In some genera, for example *Vicia*, there are two such interconnected folds on each side.

Since Delpino's days three or four main types of pollen presentation have been recognized within the family. In the simplest case, the upper edge of the carina is not closed, or the two petals adhere so loosely that the carina opens when pressed down under the influence of a pollinator. Filaments and style are stiff and do not bend, consequently anthers and stigma protrude during the visit, and are concealed again when the pollinator leaves the blossom. This is the type found in *Trifolium, Astragalus* and others.

In the second type, the carina is rostrate and the upper edge is closed, with the exception of the outermost tip, where there is a small opening. The anthers open and shed their pollen in the rostrum. The pollen is prevented from falling down into the carina by the swollen ends of the filaments. When the carina is pressed down, these ends function as a pump piston, pumping the pollen mass out of the end of the rostrum, like a thin sausage.* This is the type found in *Lotus* or *Coronilla.*

The third type is also characterized by secondary pollen presentation, but here the pollen is brushed out of the carina by the upper part of the style, which is sharply bent and densely strigillose, generally on the inner side only. This type is found in *Vicia, Lathyrus,* etc.

All these types can be worked several times, since more pollen is available than is spent during the first visit and this is stored in the carina for future use. In contrast to them are the explosive blossoms which occur in some genera and constitute a fourth type. The principle of explosion is always the same: pistil and filament sheath are confined in the carina under pressure. As long as the blossom is left alone, the adherence between the upper edges of the carina petals is strong enough to withstand this pressure, but the additional force exerted from outside by the pollinator is sufficient to cause the upper edge to rip open, and the pistil rushes out, spreading a cloud of pollen. In *Medicago* the pressure is exerted by the pistil, which after explosion is pressed against the vexillum. In *Genista* pressure is exerted by the carina and alae which after the explosion take a downward position at right angles to the primary one. In *Desmodium* both these movements take place. In *Cytisus scoparius* stamens and style curl up. The explosion mechanism is more complicated inasmuch as five short stamens hit the pollinator under the abdomen, while five long ones and generally also the style hit it on the back.

These blossoms have pollen as their only primary attractant, and the thorough dusting with pollen caused by the explosion may ensure that some pollen is left for pollination even after most has been collected for other purposes by the visitor. The explosion mechanism as described here functions once only, and it is easy to observe how bumblebees are not attracted to exploded blossoms of *C. scoparius* as they are to fresh ones. On the other hand, secondary visits by syrphids to exploded blossoms are frequently observed. However, such visits are of no consequence from the point of view of pollination. In *Cytisus racemosus* there is also an explosion mechanism, but the explosion is rather weak, and the changes of

*To Delpino it suggested macaroni production.

position so insignificant following the first visit that anthers and stigma are again concealed in the carina and may later function as in an ordinary flag blossom of Type 1.

Another minor modification of the ordinary flag blossom in the family concerns the filament sheath. In nectar blossoms the tenth (top) stamen is free, and two openings at its base give access to the nectary at the base of the pistil. In pollen flowers this access is redundant and the tenth filament is firmly connected with the others, forming a closed tube. In *Coronilla emerus*, the claws of all petals are freely exposed as there is no nectar to conceal, whereas in nectar blossoms the petals, especially the vexillum, clasp each other to form an outer tube that can be entered from the front only. An elongate calyx may contribute to this as in *Anthyllis*. In *Trifolium pratense* the petals have fused together and with the filament sheath, forming a narrow, 9—10 mm long tube, which leaves the outermost parts of the petals free for carrying out movements of the kind described above. In *Arachis hypogaea* even longer tubes exist, but their function is obscure (see Heide 1923), apart from exposing part of the flower while keeping the ovary in a more protected place near the ground. From Africa Vogel (1954) mentions sphingophilous *Camoensia* with a 7-cm long tube.

In spite of all these modifications all taxa mentioned still have typical flag blossoms. However, within the family are also found more radical deviations in the morphology and ecology of the flower. The form of the flower degenerates in some cases, the differentiation between its members disappears. In *Petalostemon* the flower is almost regular, short tubular, and with sexual organs protruding out of the tube. The dense inflorescences form secondary brush blossom types, and pollination is of the mess and soil type, caused by a variety of allotropic to eutropic insects. This corresponds to the *Mentha* type in the development of the gullet blossom. Theoretically, the type may also be primitive. Degeneration occurs also via autogamy (*Vicia lathyroides*) to cleistogamy (*Lespedeza*). The corolla has been lost in *Hardwickia* (Caesalpiniaceae) and the blossom is anemophilous.

In *Centrosema*, *Clitoria*, and others the flower is turned upside-down. The flat vexillum forms a large landing-space for big bees and the rest of the flower is hanging over the entrance to the nectarium like the upper lip of a gullet blossom. In such a blossom the two functions of landing and of deposition, reception of pollen, typically combined in the flag blossom, have been separated. The result is a gullet-like blossom. However, there is a flag flower pollen deposition mechanism, type 3, like *Vicia*. It requires a great deal of force to push the vexillum and carina apart, due to different reinforcements at the base of the petals (van der Pijl 1954). Autogamy does not occur.

A more curious change of direction of symmetry plane is found in *Phaseolus*, in which pistil, filament sheath, and the closed, long-rostrate carina are spirally wound. In *Ph. vulgaris* the spiral makes one turn, with an attendant slight asymmetry of the vexillum. The species is generally autogamous. In *Ph. caracalla* the principle is taken to the extreme, the carina spiral forming no less than four turns. At the same time the symmetry of the flower is completely lost, and the petals have taken up new positions – the vexillum in the lateral and the alae in the median plane. Except that pollen is deposited on the side of the visitor, this fantastic structure functions like that of *Vicia*, etc., the pollen being brushed out with the end of the style. The blossom is visited by a large bee, *Xylocopa augusti* (Schrottky 1908); containing great quantities of nectar, the blossom keeps the bee busy for a long time (30 sec). A structure like this defies any attempt at typification. It is of great interest that spiral carinae occur also in other genera not closely related to *Phaseolus* (*Lathyrus rotundifolius*, K. F.), and the same principle is found in rostrate *Pedicularis* (Wendelbo 1965).

As mentioned initially, the flag blossom is adapted for visits by bees which creep into the flower and force the petals apart. Pollen is collected on the ventral side. Varying sizes of blossoms correspond to different-sized bees. Other pollinators, behaving like bees, can also be effective. Butterflies may penetrate to the nectaries with their probosces, but generally without forcing the petals apart; they are therefore nectar thieves.

Some Papilionaceae are ornithophilous, and their blossoms are modified accordingly. Hovering birds in particular are unable to force a blossom open. *Camptosema nobile* has been mentioned already, having an open carina—alae tube. In *Clianthus puniceus* the flower forms a vertical strip, the vexillum turning up and the other petals down. Both parts of the blossom are narrowly conical. The alae are very small and free from the carina, having no function in the pollination mechanism. In their original home (New Zealand) the blossoms are visited by a perching meliphagid, "tui" (*Prostemadera novaeseelandiae*). Pollen is deposited on the head of the visitor (McCann 1952). Cultivated specimens are also visited by hummingbirds in countries where these birds occur.

Further modifications in response to ornithophily are seen in *Erythrina*. *E. crista galli* (tropical America) has very stiff blossoms that are resupinate, like those of *Centrosema*. However, here the vexillum is bent down and even a little back, so that it cannot be used as a landing-place. The alae are reduced almost to absence, whereas the carina forms a very stiff, closed tube, the position of which in relation to the flower axis cannot be changed without destruction. Anthers and style are freely exposed at the end of the carina tube, and pollen deposited nototribically on the visiting birds. Bumblebees may also pollinate the blossom, but the lack of adaptation to these visitors is shown by the awkward behaviour of bees which land on the carina and laboriously climb along it towards the centre of the blossom. Nectar production is very rich, and nectar seeps out of the filament tube at the base of the carina. The deposition of pollen on the top of the head of visiting birds exclude this blossom − like that of *Centrosema* − from the flag blossom group, and means that it must be grouped together with the strongly modified gullet blossoms, adapted for ornithophily (e.g. *Salvia splendens*). All species of the genus *Erythrina* are apparently ornithophilous; those occurring in other continents, being visited by perching birds, are not always resupinate and thus still keep within the limit of the flag blossom. In *E. variegata* var. *orientalis* the inflorescence forms a second order flag blossom (Docters van Leeuwen 1931) whereas according to Scott Elliot (1890) *E. caffra* thus forms a brush blossom.

13.2. THE GULLET BLOSSOM IN TUBIFLORAE

In this order, gullet blossoms are very frequent, especially in the two families Labiatae and Scrophulariaceae, with which we are chiefly concerned here. As the discussion refers to blossom types only, examples are drawn from both families indiscriminately.

A typical example of a gullet blossom is given by *Galeopsis speciosa* (p. 188). The characteristic features of the blossom are the following.

1. The lower part of the blossom forms a tube, divided into a distal (lower) division comprising three petals and a proximal (upper) division comprising the other two petals. The distal division is turned out and down forming a landing platform (lower lip) generally with some irregularities that have been interpreted as assisting visitors in getting a foothold. The proximal division, the upper lip or *galea*, continues in the general upward direction of the tube, and then contracts into a hood.

2. Nectar is produced at the base of the ovary and fills the bottom part of the tube.

3. Anthers and stigma are placed under the outer, hooded end of the upper lip. They are protected against rain, and they touch the back of an insect that forces its head down into the upper, wider part of the tube. The zygomorphism of the bilabiate gullet blossom is mainly one of relative size. The only diagram zygomorphism is due to lack of the fifth, median, stamen, which would have been in the path of visitors, and also stand in the way of the style.

The type, as evidenced by *G. speciosa*, is found with small variations in a great number of related genera: *Stachys, Glechoma*, and others. The covering of anthers and stigma by the upper lip may be more or less complete; in *Teucrium*, there is no upper lip left. The morphology of the anthers and the character of the pollen varies from regularly opening thecae with coherent pollen to box-like structures formed by all four thecae, out of which falls a dry pollen powder on contact with the visitors.

However, *G. speciosa* is already a rather highly evolved type. A more primitive one is represented in *Digitalis purpurea*; the flower is bell-shaped, almost radio-symmetric, without a pronounced difference between upper and lower lip. The blossom is sufficiently wide for pollinators (bumblebees) to creep into it, as in a bell-shaped blossom. But the position of the anthers is the same as in *Galeopsis*, and pollination is nototribic. In *D. lanata* the greater length of the distal part of the flower produces a lower lip, on which pollinators land. Still more primitive types are found in the subfamily Pseudosolaneae of Scrophulariaceae, in which the blossoms are bowl-shaped, and do not show any very strong tendency towards the placing of anthers in the upper or lower part. Similar types, bell-shaped and frequently with exserted stamens, are found in Mentheae; they are not truly primitive, but reduced in connection with the development of second-order (brush) blossoms in dense inflorescences. These types are pollinated by flies and lower hymenopters, frequently by mess and soil.

Refinements, as compared with the *G. speciosa* blossom, are found in butterfly- or moth-pollinated species with a long, narrow corolla tube and in some blossoms with versatile and non-didynamic anthers (*Orthosiphon, Catopheria*). The occurrence of hairs in the tubes has been interpreted as a protection against small nectar thieves (*Stachys, Salvia, Scutellaria*, etc.). Bracts enter into the attraction unit, sometimes even replacing the corollas as such (species of *Salvia, Lavandula, Castilleja*).

More profound differences are found in the species of *Salvia*, of which *S. patens* is a classical example. Thanks to Correns (1891) and later authors (Werth 1956b; see Vogel 1954), many types of blossoms are known within the genus. The most primitive types have a short bent connective giving a Y-shaped stamen with two fertile thecae (Hedge, 1960). Of those with short filaments and long (actually broad) connective, *S. officinalis*, in which both thecae are fertile, is a more primitive type than *S. pratensis*, where the lower thecae are sterile, but the mechanism is similar in both species. Obviously, the pollen of the lower thecae cannot be very effective in pollination. An intermediate stage is represented by *S. glutinosa* in which the original character of the lower part of the anther is still recognizable, even if sterile. The development from these two species through *S. pratensis* culminates in *S. horminum*, in which the lower thecae form one inseparable unit, making any progress into the blossom without working the mechanism absolutely impossible. In other species this remarkable mechanism has apparently degenerated again. The anther of *S. verticillata* is immovably fastened to the filament; however, the absence of the lower half of the anther suggests a derivation from forms with movable anthers, like *S. pratensis*. In *S. verticillata* the blossom is operated by bumblebees which push back the upper lip.

An interesting pair are the sympatric *S. mellifera* and *S. apiana* described by Grant and

Grant (1964): the former is normally built, and is pollinated by small, solitary bees. The latter lacks the upper lip and has an explosive mechanism built into its lower lip, tripped only by heavy *Xylocopa* bees.

Whereas bumblebees are the preferred pollinators of European *Salvia* species, other groups of animals are operative elsewhere. In South Africa we find the section *Nactosphace*, mostly with violet, short-tubed bumblebee blossoms. One species, *S. aurea*, kept the "see-saw" mechanism, but became golden orange. It has a larger disc, the tube is 45 mm long, and the lower lip curves back. Nectar fills the tube. Honey-guide and odour have disappeared. Typically this represents a direct functional change of the whole pollination unit towards ornithophily. Bird pollination was observed already by Scott Elliot in 1890.

In quite a different section (*Eusphace*) we find in America a number of more or less distinctly ornithophilous species: *S. coccinea, S. gesneriaefolia, S. heeri*, and *S. splendens*, of which the latter shows the most extreme adaptation. The visual attraction is the intensely red corolla, and even the calyx has the same colour. The flower has become tube-like, 35 mm long, with a lateral entrance. The lower lip is reduced almost to absence, and there is no landing-place for visitors. There is no odour, but every local child knows that the tube contains much nectar. The ornithophilous syndrome could hardly be more distinct, and pollination by hummingbirds was described as early as in 1881 (by Trelease).

The intricate "see-saw" mechanism of the *S. pratensis* type has degenerated in *S. splendens*: the filaments are more parallel to the general direction of the flower and the anthers than in the *S. pratensis* type, and there is very little room for movement within the tube-like flower. The observations of pollination are not quite satisfactory on this point. Anthers seem to dip down (see *Galeopsis*) simply as an effect of the dilation of the corolla, the lower part of which is flattened with many ridges. Shaking by the fluttering birds will be sufficient to ensure that pollen is deposited on their heads. The difference between this and *S. aurea* which is visited by perching birds, is significant. However, the *S. splendens* blossom can also be utilized by perching birds (sun birds) in Java (v. d. P.). *S. involucrata* is an example of an ornithophilous species in which the see-saw mechanism still works, thanks to an inflation of the corolla tube below the entrance.

In the mountains of Java, van der Pijl observed a mixed planting of *S. splendens*, other species, and *S. coccinea* (pink, long tube, but still bilabiate, yet without lever, hence semi-ornithophilous). Bumblebees swarmed around. Some clung to the lip of *S. coccinea*, though the tube was too long for them and a sharp bend made it impossible to reach the nectar. The tube was often punctured by bumblebees. *S. splendens*, however, was in that case completely neglected.

Finally, the predominantly blue colour of the melittophilous species should be compared to the glowing scarlet of the ornithophilous ones.

There are probably also psychophilous species and perhaps even myophilous (*S. lasiantha*).

The corolla limb of *S. splendens* is so narrow that the blossom approaches very closely the tube type (free petal parts reduced). Other examples of ornithophilous gullet blossoms that have become tube-like are those of *Leonotis leonurus* (Vogel 1954, pl. V) and of *Castilleja coccinea*. The corolla of the latter is reduced and almost included in the calyx. The lower lip is represented by three small teeth only, while the upper one includes anthers and stigma. There is no place for alighting; introduction of an object in the corolla will cause the upper lip to bend down and expose pollen and stigma.

Development toward closed blossoms is found in various series. One of them is the masked flower of Anthirrhineae, as exemplified in *Linaria vulgaris*. This genus shows a special development by possessing a spur, which extends the corolla tube by 10−13 mm. A more regular type is found in the genus *Anthirrhinum*, in which the spur is insignificant. These flowers are interesting because they show two of the tendencies prevalent in Rhinanthoideae: guiding structures and closing of the flowers.

The simple blossom type in Rhinanthoideae is represented, for example, by *Melampyrum* or *Bartsia*; a more or less horizontal, narrow, tube-like flower in which the free parts of the corolla are small but sufficient to afford a foothold for bee visitors. The upper part of the flower is compressed and pinches the anthers. The thecae are vertical and turned towards each other. They also open towards each other, forming four boxes in which pollen is contained. In order to reach the nectar, bees are obliged to force their heads into the outer part of the corolla. This is therefore dilated and the pollen boxes open. Projecting points, hairs, etc., safeguard the contact of insects with the anthers, and the dry pollen is shaken out and sifted on to the back of the visitor. The stigma generally projects beyond the position of the anthers and frequently also outside the corolla.

In the genus *Pedicularis* the type is refined in various directions (cf. Nordhagen in Lagerberg *et al.* 1957 and following studies by Macior in a number of publications between 1967 and 1977). The simplest *Pedicularis* flowers are symmetric, like *P. hirta*, or very slightly asymmetric, like *P. silvatica*. They distinguish themselves from the *Melampyrum* type by the erect position of the corolla tube, by the greater angle between the horizontal lower lip and the rest of the blossom, and by the occurrence of echinate ribs that guide the insect proboscides into definite paths.

A *Melampyrum*-like, possibly primitive type, is *P. semibarbata*, which is pollinated by solitary bees (*Osmia* sp.).

The asymmetry is stronger in *P. palustris*, which forms an intermediate type, and, above all, in *P. lapponica*. The flower here has the same horizontal position as that of *Melampyrum*, but the asymmetric lower lip makes it so difficult to handle that the visitor can only penetrate by first opening the blossom by force. *P. canadensis* apparently belongs to the same type, but in this species the blossoms are echinate on the right-hand side only, whereas the left-hand "entrance" side is glabrous (K. F., cf. Macior 1968a).

Within the genus there are two further development trends (cf. Li 1948−9).

One leads to extremely rostrate blossoms in which the galea forms an elongated, spirally wound tube, as in *P. groenlandica* (Macior 1968b). In the other the corolla is, in addition, elongated into a long, straight tube with the lower lip as a salverform rim, and a rostrate galea (possibly placed under the rim). About the pollination of the latter type nothing is known (moths are probably the only pollinators that can use these blossoms). In the rostrate type only the stigma has followed the growth of the upper lip into a rostrum; the anthers are to be found much further down, and stigma and anthers are much more distant than in the types with hooded upper lip.

An independent evolutionary trend is the one leading to the closed flowers of *P. sceptrum-carolinum* and a few other species. *P. densiflora* is ornithophilous with reduction of the lower lip and red flowers (hummingbird syndrome).

Within some genera of Labiatae the gullet blossom is reversed and functions like a flag blossom. This reversal may be a simple resupination either through torsion of some part of the blossom, or, e.g. in *S. nutans*, because of the hanging position of the inflorescence. In the Ocymoideae (cf. illustration in Tanaka 1972) the reversal is not a resupination, but is due to transference of anthers and style from the upper to the lower part of the flower. They are more or less included in the carina-like lower lip. Although there is a stiff staminal sheath that has a similar function as the sheath in the papilionaceae flower, the mechanism is much simpler. However, as in the other family, there are also explosive flowers in Labiatae (*Hyptis* and *Eriope* — American — and *Aeolanthus* — African: van der Pijl 1972). In many Ocymoideae the individual flowers are rather reduced, forming compound, second-order blossoms of dense inflorescences.

These examples may serve to demonstrate some of the variations within the gullet blossom in these two families. Only one more tendency will be mentioned, viz. that of autogamy ultimately leading to cleistogamy (e.g. *S. verbenacea*, Willis 1895). Because of the very close juxtaposition of stigma and anthers in regular gullet blossom, self-pollination is likely to take place unless there are special structures providing against it, e.g. pronounced dichogamy or — perhaps more frequently — herkogamy. Thus in *S. pratensis* or *Pedicularis palustris* the change necessary to cause self-pollination is rather small. Self-incompatibility has not been fully investigated, but seems to be weak in Labiatae. In a plant like *Linaria minor* self-pollination occurs as a matter of course, and is concomitant with self-compatibility.

A more complex type of self-pollination is found in small-flowered Rhinanthoideae, e.g. many *Euphrasia* species. The blossom on the whole follows the *Melampyrum* pattern with pronounced herkogamy, with the stigma projecting beyond the tip of the corolla. Later, the corolla increases in length, bringing with it the partly adnate stamens whereas the stigma remains in place, as the style does not increase in length. Consequently, the anthers are brought out to the stigma which becomes enclosed in the upper lip, and self-pollination takes place.

CHAPTER 14

POLLINATION ECOLOGY AND SPECIATION

Some of the variations of blossoms within major taxa concomitant with variation in pollination ecology were discussed in Chapter 13. A discussion of which came first — structural changes of the blossom or the change in pollination — can only lead to a war of words. The one change is unthinkable without the other; the structure of the blossom and the breeding system and, especially, the habits (if not also the structure) of the pollinator must have developed together (cf. Haskell 1954; Baker 1960) even if one-sided adaptation also occurs (viz. in cases of deceit).

Many of the structural changes concomitant with variation in pollination should by ordinary taxonomic judgement be considered rather insignificant, and diagnostic on the species level only. Other ones are easily considered more important, e.g. the way in which filaments form an open or closed sheath in Papilionaceae. Greater taxonomic value than is really warranted may easily be ascribed to this unless we realize that this is a structural response to the availability, or non-availability, of nectar at the base of the pistil. As the flower is primarily an organ of pollination there is, of course, nothing improbable in the assumption that major taxa may be characterized by a common pollination mode and the structural characteristics belonging to it. On the other hand, a deviating pollination may, even in close relatives, cause comparatively large differences as shown in the chapter mentioned above. To achieve a good taxonomic judgement it is mandatory that one does not forget problems of pollination when changes in the floral region are to be interpreted (cf. Pennell 1948).

On the other hand, similarity in pollination may be the cause of structural similarities that may mask taxonomic differences. As examples of the same structure occurring in taxonomically widely separated groups may be mentioned the see-saw mechanism of *Salvia*, which is also found in *Roscoea* (Troll 1929), and the flag blossoms which return in, among others, *Corydalis*, even with an explosion mechanism similar to that in *Genista* (Müller 1939). In these cases, the functionally similar blossoms belong to such widely separated taxa that a confusion could not arise — especially as the structures are not homologous. But it is easy to see that corresponding similarities within a smaller taxon might lead to confusion unless the functional aspect is clearly realized.

The species being considered an interbreeding population (Poulton 1938), it is evident that the nature of breeding and pollination systems also deeply influences the speciation process. As long as the cross-pollination system is uniform within a taxon, there will be gene exchange, and further speciation is counteracted, unless, of course, incompatibility or geographical barriers come into existence. On the other hand, small shifts in the pollination system may be sufficient to start speciation or to keep compatible taxa separate, even if they are ecologically and geographically sympatric. Some of the apparent genetical coherence in a taxon (referred to by Clausen and Hiesey 1960) may simply be due to pollinators having a more restricted distribution than the taxon as a whole, or being more discriminating than taxonomists, e.g. discriminating between forms with slightly varying floral dimensions (cf. Bateman 1968). It has been discussed whether a micro-geographical separation came first, resulting in the development of different pollination systems and consequent later breeding isolation, or whether the shift in pollination had taken place by some mutation without preceding geographical isolation. Most probably, both types of development may have occurred (see Grant 1963).

Whereas in most taxa, whether geographically split or not, differentiation in pollination ecology will be concomitant with other ecological differentiation, either as a cause or as an effect, the highly aberrant pollination system in *Ophrys* has apparently led to a state indicated loosely as a "luxury differentiation" of the pollination system, isolated from other ecological differentiation ("diversification simply for the sake of diversification"; Heslop-Harrison 1958). Not all ecologists agree.

Grant and Grant (1968) have described how one subspecies in the *Ipomopsis aggregata* complex has facultative humming bird and hawkmoth pollination, the other has bee blossoms, and they maintain that this may be a starting point of evolutionary divergences, probably both between and within subspecies.

As another example we may mention that differences in the dimensions of the bills of blossom-feeding birds and, as a consequence, of flower tubes may form the origin of speciation through genetic isolation.

Besides specific odours or shapes (leading to oligotropy) and flower constancy other characters may help to set up ecological barriers promoting speciation, for instance a pronounced daily rhythm. Bees with "time-memory" adapt quickly. Bumble bees have been seen to wait, even hovering around the flowers, until they open (*Bombus lapidarius* at *Cichorium intybus*, Thijsse 1934). Ponomarev (1966) and Ponomarev and Rusakova (1968) have demonstrated that daily rhythms may also play a part in speciation in anemophiles.

Due to their preference for using floral characters, taxonomists will easily recognize speciation caused by differences in pollination systems. It should be noted that the theoretical basis for such a recognition, viz. the presumed greater conservatism of flowers in relation to external factors as compared with that of vegetative organs, is in reality fallacious.

An indiscriminating pollen vector (and abiotic pollination is the most frequent) will tend to keep large populations uniform. Even if the average distance of massive wind pollination is not very great, the general effect will be to equalize any differences tending to establish themselves, and to counteract speciation (see Heslop-Harrison 1959 and references). This equalizing may not inevitably lead to a uniform population, but within a large population it will lead to the formation of clines (cf. the conditions in temperate forests, in which the majority of species are wind-pollinated). Given the same habitat conditions, a population pollinated by different oligotropic pollinators, or by discriminating, constant ones will tend to break up into smaller units, leading, if conditions are favourable, to species formation.

Following a suggestion by Heslop-Harrison (1958) we may indicate the possibility that the restriction of some pollinator(s) to particular habitats within the total range of the species may tend in the first instance to isolate ecotypes, which may later perhaps develop into more decisively differentiated taxa. It is remarkable that Mayr (1947), while stressing that "The hypothesis of sympatric speciation is unnecessary . . . The species (is) . . . an aggregate of ecologically different populations". fails to include breeding systems (and especially pollination in plants) among the pertinent ecological factors.

It should be especially emphasized that the very strong constancy of bumble bees and other higher pollinators may, in its genetic effect, almost equal autogamy in establishing homozygosity and genetical diversification. But only almost, as there will be occasional outcrossings, which tend to hold at any rate smaller populations together and prevent them from disintegrating into the multitude of micro-taxa characterizing autogamous and, especially, apogamous groups, in reality representing a morphological poverty (Nilsson 1947).

Charles Dawin (1890) had already realized the importance of breeding systems in the case of the two *Platanthera* species, *P. chlorantha* and *P. bifolia*, which are mainly separated by the angle of diversion between the pollinia and, consequently, by the mutual distance of the viscid discs. Many taxonomists of the day considered this difference too insignificant to warrant separate specific status, but Darwin points out that the differences between the two pollination systems are sufficient to keep the taxa apart.

A recently described example which has already become a classic is the case of pollination in *Aquilegia* (Grant 1952). *A. formosa* has a red flower, and is pollinated by humming-birds, wheras *A. pubescens* is whitish and pollinated by hawk-moths. Colour and position of the blossom and length of nectar spurs differ accordingly. The areas of the two species overlap to a certain extent and, being interfertile, they produce a fertile hybrid by pollination, probably by non-discriminating pollen-collecting bumble bees. The hybrids suffer from the usual back-crossing effects (see Anderson 1939), and the difference in pollination systems will tend to accentuate this and keep the back-crosses apart. Hawk-moths will pollinate the hybrid and *A. pubescens*, while hummingbirds will visit the hybrid and *A. formosa*. Sub-intermediate hybrids will be preferred and so the foreign genes are again sorted out. Consequently the hybrid population is split again. Grant (*loc. cit.*) thinks that the five major divisions of the genus are separated by their pollination mechanisms and can exist side by side, whereas the species within a division will merge if occurring together. This leads to the somewhat disquieting general conclusion that one of the most important conditions for "good" species, if compatible, is that they differ with regard to pollination, and, further, that hybridization experiments in gardens are meaningless as measures of the validity of species delimitation in nature unless accompanied by a study of pollination systems. See also the case of the five sympatric *Papaver* spp. described by McNaughton and Harper (1960), which are kept apart in nature by the constancy of their *Apis* pollinators.

Later, the *Aquilegia* problem was discussed by Chase and Raven (1975), who found that pollinators (hummingbirds and bumblebees) were indiscriminate in relation to a similar species pair in *Aquilegia*. Nevertheless, the species remained separate. Miller (1973, also in litt.), for another similar species pair in the same genus, concludes that "pollination systems within each species pair may reinforce basic ecological differences between species, but pollination . . . should not be considered the primary mechanism of species isolation, at least not in the sense of Grant (1952)". Cf. also the mathematical treatment of the problem by Straw (1972). As has been pointed out later by Grant (1976), the speed and effectivity of such processes must perforce depend on the exclusivity of preferences, and also on effects of site and habitat in addition to the incidence of the unavoidable indiscriminate pollination by other pollinators. Also, the feeding of various blossom visitors depends on the availability of blossoms. If *A. formosa* is not there, hummingbirds will have to visit other plants, e.g. *A. pubescens*, for which they are less adapted, and which they may avoid if food is more plentiful elsewhere.

On the other hand, if hybridization is caused by disturbance of equilibria, and two compatible species are brought together which were previously (geographically) isolated, a common, non-discriminatory breeding system may lead to the extinction of the parent species (Morley 1971), cf. the well-known case of *Symphytum officinale, S. asperum*, and their invasive hybrid *S. "uplandicum"*.

Levin and Anderson (1970) have analysed mathematically the situation when two plants flowering at the same time compete for the same pollinator, and deduce that this must lead to the elimination of the minority species "because it suffers a larger percentage of heterospecific pollinations" (p. 465). This is nothing but the old dictum that two species cannot occupy the same ecological niche. As niches are usually not absolutely identical, the competition pattern will be different in actual practice and will not lead to complete elimination.

Mathematical treatments of hypotheses of evolutionary specialization predict that greater densities of "prey", in our case blossoms to be visited, lead to greater "predator" specialization (Pulliam 1974: 73). Such a hypothesis is not corroborated in pollination ecology: some of the most specialized, i.e. restricted, pollination strategies are found in plants of low densities and small number, of which the *Ophrys* species are good examples, but many others might be quoted as well. Gregarious and conspicuous plants more often than not have an unspecialized pollination strategy, and, on the other side, a strong pollinator constancy — specialization will cause that — will maintain the minority species in a competitive situation, as shown by Levin and Anderson (*loc. cit.*).

The development of the pollination syndrome in *Aquilegia* is also interesting inasmuch as the formation of the spurs distinguishing between primitive and advanced flowers may be due to a single (repeated?) mutation (cf. Prazmo 1960; Baker and Hurd 1968).

On the other hand, the chances of a hybrid of entomophilous plants maintaining its own existence in nature and not being swamped by back-crossing to parents seems, above all, to be dependent on its "finding" a pollinator of its own. Thus, Straw (1956) has described two *Penstemon* species, *P. centranthifolius* and *P. palmeri*, of which the former is hummingbird pollinated (*Calypta anna* and *C. costae*), the latter pollinated by bees (*Xylocopa californica* and *X. orpifex*). An intermediate species, *P. spectabilis*, may be interpreted as a hybrid, which has become stabilized because it has its own pollinator, viz. a wasp, *Pseudomasaris vespoides*. Similarly, the atypical *Stanhopea tricornis* may perhaps be a hybrid between a *Sievekingia* and a proper *Stanhopea*, again stabilized because of its special pollination system (Dodson and Frymire 1961a). Cruden (1972a) maintains that three subspecies of *Nemophila menziesii* are developing towards separate species because they have different pollinators.

Speciation in plants is, therefore, not only a question of genetic instability, but also, and perhaps equally, of breeding systems. The greater the number of prospective discriminating pollinators, the greater the chance a new form (of whatever origin) has of finding a separate pollinator and of establishing itself as a species in its own right. The very vigorous speciation in the humid Tropics may be seen as a function of the teeming insect life of that part of the Earth (van der Pijl 1969).

Invading plants in a new area depend for their success on the presence of suitable pollinators — if they cannot take recourse to autogamy. This explains the prevalence of autogamy — facultative autogamy — in invading plants. Plants with a very specific breeding system may therefore be restricted in their colonizing abilities.

Incidentally, the occurrence of hybrids in nature is one of the best proofs that allogamous pollination does take place. For example the *Cypripedium* hybrids described by Mandl (1924) demonstrate that, whatever the attractant may be, *C. calceolus* must be regularly visited by not too discriminating pollinators, and the hybrids of *Ophrys apifera*   prove that this habitually autogamous plant is sometimes cross-pollinated (Schremmer 1959).

On the other hand, pollination systems influence speciation in pollinators, too. Linsley and MacSwain (1959), discussing this, conclude that closely sympatric pollinator species utilizing the same pollen source could have arisen only in geographical isolation and must have come together afterwards. Especially in oligolectic species, speciation would be dependent on geographical and/or ecological isolation. It is also obvious that the balance between seeds and nectar as the preferred food will also strongly affect speciation in relevant groups of birds, e.g. in cockatoos.

CHAPTER 15

POLLINATION ECOLOGY AND THE BIOCOENOSE

Pollination ecology has, on the whole, been interested in the plant—pollinator relationship on a one-to-one basis: the individual pollinator on the individual plant (species). Even so, various studies have gone beyond this simple relationship and dealt with the total activity of pollinators also on other plants than the one immediately under consideration, e.g. the comprehensive studies by Linsley and collaborators (1963—73) on the bees visiting various species of *Oenothera*. Only recently has the community aspect attracted attention. Negative biocoenosis effects have long been considered in applied pollination ecology: the competition of non-crop plants for the same pollinators which are needed for the crop plants. The opposite case is also known: long-lived pollinators may need additional food sources to maintain a population large enough for pollination of the crop that will have a short period of mass flowering. It may be necessary to build up a strong pollinator population in the community before the onset of flowering of the crop plant. In the more diversified natural plant communities, conditions are much more complicated.

The monolectic pollinator represents no problem. There is a real one-to-one relation, and in a periodic climate the pollinator will follow the rhythm of the blossom. If the blossom is also monophilic, things are even simpler: as pollination goes, these species, both animal and plant, have separated from the community of which they form part.

Usually, the life-time of pollinator and blossom do not coincide so closely; the former may need more than one plant for feeding, and the latter is exposed to more than one pollinator during its flowering time. It is then important that a constant energy flow through the community is maintained at a minimum level to maintain the pollinator community. An insect which is necessary for the pollination of a certain plant species may have a greater energy demand than this plant can provide, e.g. because of shortness of anthesis. The pollen of six different anemophiles in a salt-marsh are utilized for the maintenance of a bumblebee population (Pojar 1973) even though the pollen donors themselves, being wind-pollinated, do not benefit from this. A more direct case is that of pollen and nectar thieves. The maintenance of their food-plants depends on the activity of some other insect, which pollinates. If proper pollination does not take place, the food-plant will disappear from the community, and the thieves will have to find another source or be doomed.

In a biocoenosis context all orchids are parasitic inasmuch as their pollen is not utilized by pollinators. For those orchids that offer no reward (primary attractant) this is even more pronounced, and in order that the orchid-pollinator population be maintained it must always get its protein from other sources, in many cases also energy food. The ultimate parasitism is represented by blossoms that, like *Arisaema* spp. or *Pinellia*, kill the pollinators (cf. Vogel 1965b).

In some communities there may be a surplus of pollinators, and the problem for them is to find sufficient food. The ensuing competition will have greater evolutionary consequences for the pollinator than for the plant. In other communities there are more plants than necessary to feed the insect population, and the strategies must accord with this situation. One of the strategies is staggered flowering: species with different flowering time occupy different time niches and therefore do not compete. This is especially important for the weaker competitors, who lose their visitors when a more proficient source of food becomes available. Staggering of flowering may be both diurnal and seasonal. From a study of four plant communities in British Columbia, Pojar (1974: 1830) concludes that the presence of insects is the "limiting factor in sexual reproduction of many entomophilous species". In a stable community there must be an equilibrium between availability of pollinators and of food for them — whether this food be exclusively presented by the plant to be pollinated or also by other members of the community.

Seasonal staggering is of greater importance in climates with permanent or long pollination seasons. Towards the Arctic, seasons become shorter, and staggering is, in the extreme, no longer possible. Staggering of presentation of rewards and of emergence of pollinators must be synchronized. This is equally important for diurnal as for seasonal staggering.

Macior (1971) has pointed to the existence of pollination symbiosis between plants growing in the same habitat. *Erysimum amoenum* and *Primula angustifolia* are sympatric and flower synchronously. Both share the pollination activity of nectar-foraging bumblebee queens at a time when few other flowers are available for them. In a similar way *E. nivale* and *Polemonium viscosum* share pollinators. When — for spatial or other reasons — individual populations are too small to sustain pollinator interest, this symbiosis enlarges the functional size of the plant population. If one species is already established and the other is an immigrant or simply a minor species, the latter is thought of as a mimic, and it will benefit from the pollinator population maintained by the first species.

Pedicularis groenlandica and *Dodecatheon pauciflorus* form a similar symbiontic pair, but in this case pollen is deposited on different parts of the bodies of the pollinating bumblebee. Several other examples have come to light, and more will undoubtedly be discovered. If one visitor species is a more effective pollinator than another, with which it competes, it may lead to adaptation of the blossom towards the more effective one because of the better seed-set. An example is given by Colwell *et al.* (1974) describing the competition between coerebids and hummingbirds on the same species. The coerebids perch and are nectar thieves (pierce the flower), the hummingbirds hover and are effective pollinators. The seed-set obtained from flowers out of the reach of the coerebids is the greater, and evolution towards exclusive hummingbird pollination is conceivable. The introduction of foreign pollinators may change the picture completely, and so may also the introduction of foreign plant species. The introduction of the honeybee to middle and northern Europe and later to other continents, must have changed the pollination picture completely and influenced the competition between plant species. Ayensu (1974) has shown some of the effects on the local bat fauna (chiefly fruit-eaters) of introducing foreign food-trees. A striking example of the introduction of a foreign pollinator is the immigration of *Xylocopa* to the Galapagos Islands, where it is the only bee. There are some other pollinators in the islands: moths, flies, birds, etc. The appearance of the bee completely changed the competitive status of many plant species. Linsley *et al.* (1966) found that the old element of the Galapagos flora was autogamous — not necessarily obligate — but could also sustain a bee population. The main importance of the bee lies in its rôle as pollinator of later immigrant plant species. The "general attractiveness [of adventives] to carpenter bees indicate that the bees may have helped materially or accelerated the process of establishment" (*loc. cit.*: 16).

Any disturbance of any part of the ecosystem is likely to affect the system as a whole, and may throw it out of balance. Relations among the pollination syndromes are important to the integrity of the total community (Frankie 1976), and the loss of one component may mean that the pollinator of another plant is lost at the same time. Another community factor that comes in is the availability of larval food, e.g. in butterflies, the larvae of which usually get their food from other plants than those delivering nectar for the imagines. Even if this special problem does not arise, many lower-grade pollinators — and especially those attracted by deceit — are dependent on other elements of the biocoenose for real food. The independence of bees from anything but the food delivered by the blossoms is the main factor in making them such superior pollinators. The food delivered by the biocoenosis is not alone in being of importance in the community pollinator syndrome. A factor like the occurrence of nesting sites (and nesting material) may regulate the presence or absence of pollinators.

With major and minor pollinators taking part in blossom—pollinator interaction, conditions in natural communities are extremely difficult to analyse, and apart from the very simple arctic communities studied by Kevan (1972), no attempts at a synoptic analysis have been published. Nevertheless, one has to bear in mind Macior's words (1974: 766) that "it is imperative that studies of particular pollination mechanisms be based upon critical observation not only of the mechanisms themselves, but of the entire ecological context in which they operate". The realization of this would need a multidisciplinary botanical—zoological team, which so far, unfortunately, has not operated anywhere.

One of the more fanciful, but perhaps not completely improbable suggestions that a biocoenosis effect is at work, is the idea that the showy, nectariferous blossoms of some apomicts, like *Taraxacum*, may serve to attract pollinators from other plants, reducing their seed-set and as a consequence thereof their competitive power.

CHAPTER 16

APPLIED POLLINATION ECOLOGY

As stated before, the oldest available references to pollination concern agricultural practices. The classical case is the well-known relief found by Layard at Nimrud, dated *ca.* 1500 B.C., showing two divine, winged creatures each holding a male inflorescence above a female date palm. There are other, similar pictures expressing the same idea and indicating very clearly that, whether understanding the real function of pollination or not, the Assyrians knew how to safeguard their date production by artificial pollination (cf. Roberts, 1929).

The second classical example of artificial pollination is the caprification in order to safeguard fruiting of the fig-tree. Both these practices have been mentioned in the writings of classical authors. Nevertheless, the introduction of fig culture to other countries did not always take place without pollination difficulties, as the "unproductive" caprificus was not included.

In modern Europe the first stimulus was given by the father of floral ecology: Sprengel's essay *Die Nützlichkeit der Bienen* (1811), reprinted and annotated by Porsch (1934). Its impact was insignificant.

Even if applied pollination ecology thus has very old traditions, it has remained for plant-growers of our age — with its profounder knowledge of the mechanism of pollination and fertilization — systematically to influence pollination in plant production. The present status is summarized by McGregor (1976).

Some of the applications of pollination ecology are on the border between pollination and fertilization, e.g. the various techniques to circumvent incompatibility barriers (cf. p. 30) and avoid self-incompatibility barriers. Till now, many of these techniques have been empiric, but a better understanding of the chemical actions taking place may increase their effectivity. Many of them are so difficult and expensive that they have no place in agricultural production. This also includes artificial pollination, which can, however, be important in actual practice, e.g. in the production of hybrid maize. Mostly, these techniques are restricted to breeding practices, but sometimes also production may depend on very delicate artificial pollination techniques, e.g. in *Vanilla* (removal of the cover over the stigma).

In anemophilous plants, a knowledge of the average transport distance of pollen grains is of importance for evaluating the possibility of fertilization, e.g. in forestry (Wright 1953) and in obligately outbreeding commercial crops. It is also important for evaluating the chances of gene transfer, whether this is, in the individual case, desirable or undesirable, e.g. in the production of seeds of strains of agricultural crops. According to Omarov (1973) the effective pollen dispersal distance in hybridization experiments with barley (usually autogamous) was only 3 m. Incidentally, similar reasoning is valid in the cultivation of *Claviceps* for medical purposes: the fields must be isolated, so that regular crops are not contaminated. However, the more involved applications of pollination ecology are usually reserved for plants with biotic pollinations, like the fig, already mentioned. In the date-palm an anemophilous pollination system was influenced by agricultural practice.

Applications of pollination ecology are important under two different conditions. One, as already mentioned, is the safeguarding of crop production, the other is the control of the work of bees to ensure maximum honey production. Little need be said about the latter theme. It is a very old observation that certain plants are good "honeyplants" producing great quantities of nectar during a long period of flowering and occurring in great numbers, often gregariously. Some of these plants are allophilic (e.g. *Heracleum*), while others are more or less euphilic (*Salvia*). Use is frequently made of this observation by cultivating the particular plant in sufficient quantity or by moving the hives when and to places where such plants flower, e.g. to the flowering *Calluna* heath and similar mass occurrences for the utilization of which thousands of hives are temporarily moved each season. The problem is to utilize abundant sources of nectar when and where they occur. As bees are generally very adept at finding such sources, further manipulation is generally unnecessary; but it is of course possible to use odour control and similar measures (see below) to safeguard the results.

In the cases related above the major objective was to ensure a maximum honey yield. If pollination is the main objective, hives are moved in among crops especially into orchards at fruit-blossoming time. Cautwell *et al.* (1970) describe air-dropping containerized bees into cranberry bogs. A prerequisite for success is the availability of compatible pollen. Recent studies (Free 1962a) have shown that in self-incompatible strains the percentage of fertilized flowers drops very fast with distance from the pollen parent. If possible, there should be a pollen parent adjacent to each fruit tree. On the other hand, more fruit is frequently set than can be successfully matured; consequently optimum pollination is not always necessary.

The practice of moving bee-hives does not always lead to the expected results. If moved too early, bees may become constant to other blossoms, or even to the extra-floral nectaries of *Vicia faba* (Free 1962b). Spraying crops with scents that attract bees has also been proposed (Waller 1970), but most such techniques remain at an experimental or pilot project stage. Bees do not travel unnecessarily far, especially in bad weather. In Norwegian orchards honeybees are less efficient owing to their climatic sensibility, and pollination is mainly carried out by bumblebees. These work at lower temperatures and light intensities, in more rainy weather, and keep longer hours (Løken 1958). The relative importance of honeybees will naturally depend both on climate and on the frequency of bumblebee nest sites in the surroundings of the orchard. In view of the great potentiality of bumblebees as pollinators (see also below) cultivators are increasingly interested in preserving their habitats, especially nesting sites near or within cultures that are dependent on them (e.g. Dorr and Martin 1966), and attempts are made to domesticate them on a large scale (Stephen 1961).

There is also a negative side to this, viz. to remove plants that may compete with the crops for the attention of pollinators: dandelions are a nuisance in orchards because bees and bumblebees may collect from them rather than from the fruit-trees. However, things are not always that simple: Palmer-Jones and Forster (1972) found no effect on alfalfa pollination by eliminating competing pollen sources in the (New Zealand) neighbourhood (cf. Free 1968).

Feeding bees with sugar syrup may lead them to collect relatively more pollen. Pollen gatherers are considered more effective pollinators (Free 1962a), provided the blossoms are not too distinctly dichogamous (Løken 1950).

Some crops are autogamous, wheat being the most important. They present no pollination problem. Anemophilous crops like rye or maize are from this point of view almost equally simple. Crop failure because of failure of pollination is almost unknown, even if very uniform wind direction during flowering or constant rain may cause some difficulties. Some zoophilous crops have rather simple blossoms that can be utilized and pollinated by a variety of animals. The problem is to ensure that some pollinating agent is present − as in the orchards mentioned above. The other extreme in this series is formed by oligotropic plants, the pollination of which depends on a single pollinator or on a small group of pollinators.

The classical case is that of seed production of the self-incompatible red clover (*Trifolium pratense*), an important forage plant. Its blossoms are melittophilous and rich in nectar, but so long-tubed that ordinary honeybees do not reach the bottom of the tube (tube length 9−10 mm). Pollination of the red clover is therefore dependent on the activity of bumblebees,* or rather of long-tongued bumblebees like *B. hortorum* (Hawkins 1961) even if pollen-collecting honeybees are of importance in some regions (Skovgaard 1956). When red clover has been introduced to countries where it did not previously grow, e.g. New Zealand (Cumber 1953), it proved necessary to introduce European bumblebees simultaneously to secure pollination. To safeguard clover seed production one might breed for (1) autogamous clover, (2) shorter clover flowers, or (3) longer bee tongues, thus enabling the more useful and easier-to-handle honeybees (present tongue length *ca*. 6 mm) to collect the clover nectar and carry out pollination. These breeding programmes have now succeeded so far that the bumblebees in many countries have become more or less redundant (Gubin 1936). A fourth breeding objective would be for bees with preference for clover pollen. According to Mackensen (1969) this has been done successfully for alfalfa, where problems are similar (cf. below). Formerly, bees would take their pollen from white clover, where it was more accessible. Such breeding programmes are not invariably successful.

A case similar to that of the red clover is represented by introduced alfalfa (*Medicago sativa*) which is also melittophilous and the seed-setting of which frequently fails because of lack of pollinators. As a general rule it must be kept in mind that conditions may vary greatly from place to place, and that measures effective in one place may be completely useless in others, as is easily seen in the extensive literature dealing with seed-setting problems in clover and alfalfa, or in orchards. Often the crop plant is not the only one flowering, and a competition for pollinators may arise. The nature of the competitor plant is important. According to Wafa *et al*. (1972) only 7 per cent of the honeybees visited alfalfa in competition with *Trifolium pratense*, which is a source of both nectar and pollen, against 31 per cent in competition with *Gossypium*, the pollen of which is not acceptable to honeybees.

Some of these situations are easily dealt with: dandelions in an orchard are readily cut. Other cases are more difficult, and applied insect psychology may be necessary to ensure the right results. Thanks to the communication system of bees this can be done. Baits are placed in such a way that foraging bees must find them − especially the honeybee scouts to whom is entrusted the task of finding new sources of nectar. These baits consist of a sugar solution perfumed with the scent of the flower of the crop plant of which the pollination is desired. Waller (1970) obtained a similar effect by the use of secretion from the Nasonov glands of bees (citral and geraniol). The scouts localizing these scents will bring back a message to the

*Compare the famous Darwinian paradox: Old maids keep cats. The more old maids, the more cats. Cats take mice. The more cats, the fewer mice. Mice dig out bumblebee nests. The fewer mice, the more bumblebees. Bumblebees are necessary for the production of red clover seed. The more bumblebees, the better seed-setting. In other words: The more old maids, the more clover seeds.

hive about the specificity of the nectar source (the scent) and the way leading to it (dances or odour paths). By convenient manipulation of the bait the interest of the pollinator is then transferred to the crop plant. This system of scent direction is not yet worked out to perfection, but has stood its first test (von Rhein 1952; cf. Hawkins 1961). An even more refined technique (so far not out of the laboratory stage) consists in making artificial dances by means of an artificial (scented!) bee that can be oscillated at will. In this way a message to go to a specified place could be delivered directly to the bees.

Honeybees are not particularly effective in pollinating alfalfa. Even when collecting from the blossom they avoid "tripping" it, i.e. make the explosion mechanism work, which is a prerequisite for successful pollination. Other bees are more effective, especially megachilidae and alkali bees (*Nomia*). More or less successful attempts have been made to domesticate some of these bees for alfalfa pollinations (cf. Bohart 1971; Bohart and Youssuf 1972).

Unfortunately, bumblebees (*Bombus*) have proved difficult to domesticate. The queens are solitary almost to exclusivity and present a multitude of problems in wintering, rearing, etc. On the other hand, a single successful colony may contain several hundred hard-working individuals (cf. Medler 1962).

As defined by Holm and Haas, successful domestication would entail controlled hibernation of mated queens and the establishment by such queens of new colonies near the crop to be pollinated. This may easily lead to completely artificial rearing of queens for dispersing to the fields. The possibility of cooling down the queens and waking them up again by higher temperature presents itself as a possible treatment. However, so far the attempts have not been successful outside the experimental field and interest in bumblebee domestication has decreased during the last 10–15 years.

In the development of cultivated plants very little attention has been paid to pollination. Many pollination systems are extremely delicate and are easily destroyed by the change of the blossoms due to breeding for other characters. Especially in hybridogeneous plants this may play a rôle, and the possibility that *Medicago sativa* is a crypto-hybrid may explain some of the difficulties of alfalfa seed production. Removal of a taxon from its place of origin may also deprive it of its pollinators.

Some incompatibility problems also need attention in cultivation, especially when the cultivated plants are heterostylous (*Cinchona*). Another problem is the (minimum) number of male plants necessary to secure seed-setting in dioecious species (*Carica papaya, Ilex*). The number of pollen plants and their distance from seed plants vary with the activity level of the pollinator. For anemophilous plants it is different altogether.

In addition to the legumes mentioned above (cf. Free 1970), carrots (Bohart and Nye 1960), cotton, cucumber, onions (Bohart *et al.* 1970), and many other crop plants are more or less dependent on insect pollination, and honeybees are usually the preferred pollinators (Todd and McGregor 1960). Very much work has also been done on the pollination problem in orchards as many commercial "varieties" (i.e. clones) are completely self-incompatible. For many of these the "effective bee population", i.e. the ratio of the number of bees to that of flowers, is a concept of great importance (see Fryxell 1957; Free 1962a).

Artificial pollination has been resorted to for many crops, from the simple hanging of caprificus twigs in the crowns of fig-trees to the careful hand-pollination of the individual stigma in vanilla. Apart from specialized breeding and some kinds of flower production, the latter technique is generally restricted to expensive crops grown under glass – but insects may also be introduced in the greenhouses for pollination purposes. Automatic dusting of honeybees with the "right" pollen has been tried in North American orchards, but the results are not unequivocal. The same applies to some other more spectacular large-scale attempts, like collecting pollen from flowers or, by means of traps, from homing bees (cf. the deleterious effect of packing pollen in the corbiculae!) and scattering it (generally diluted with *Lycopodium* spores or in aqueous suspension — sensitivity of pollen to water!) from aircraft or even by bombs.

Sometimes artificial pollination is resorted to even in large commercial crops, e.g. the oil palm (*Elaeis guineensis*), in which the artificially enforced xenogamy is said to improve the quality of the fruit. Outside Africa anemophily does not seem to be sufficiently effective to give a consistent xenogamy, and the introduction of the pollinating beetles from Africa has been considered too hazardous (cf. Heusser 1912; Devreux and Malingraux 1960). Vanilla is pollinated by hand, even in its country of origin, Mexico. The "natural" vector is still unknown, and neither in America nor in the Old World has any other insect adapted itself to its very specific requirements. In Java *Passiflora quadrangularis* is also mostly pollinated by hand, even if *Xylocopa* sometimes takes over, like the big solitary bees usually do in South America, its country of origin.

Obviously, it is of great importance to have a precise knowledge about the time when conditions for fertilization are optimum. Hayase (1963) has described a cross between two *Cucurbita* species, which succeeded only at 4 p.m.

A special type of pollination is the vibration by so-called "artificial bees" of greenhouse tomatoes to ensure autogamy; this copies the whirring technique of bees when collecting pollen and the effect of wind in causing autogamy (cf. Good and Saini 1971).

A very important practical application of pollination studies is the study of the incidence of allergogenic pollen types in relation to hayfever, about which a great literature has grown up (Wodehouse, 1945). In many countries more or less regular surveys are conducted by trapping airborne pollen, and daily or weekly hay-fever forecasts are given in some places.

There is also a negative side of applied pollination ecology. A bee-hive should not be kept near cucumber cultivations since cucumbers, being parthenocarpic, should not set seeds, which detract from the quality of the fruit. The introduction of male sterile cultivars makes precautions less necessary.

Obviously, pollination, causing premature decay of the flowers, is highly undesirable in orchid cultures, whether in the open or in greenhouses.

The question of the spread of pollen, by biotic or abiotic vectors, has also been very much discussed in relation to the problem of avoiding contamination, especially of seed crops (e.g. Haskell 1943; Bateman 1947; Pedersen *et al.* 1961).

CHAPTER 17

CASE HISTORIES

In this chapter we have collected examples to illustrate some of the principles expounded in the preceding ones. Obviously, it is very far from being complete. Some of the examples are rather ordinary textbook material that has been included because it is particularly well suited to illustrate the point in question. Most of the examples are less well known, and one or two have been included because the usual textbook presentation is less satisfactory.

17.1. POLLEN PRESENTATION TYPES

Saxifraga aizoides L.

(Saxifragaceae. N. hemisphere)

Protandrous with deciduous anthers. Entomophilous. Allotropic. Dish blossom

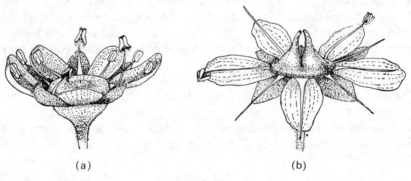

(a)   (b)

FIG. 4

At the outset of anthesis petals are slightly curved, and stamens are adpressed to them and to the sepals underneath. Gradually, one by one, the filaments bend, placing the anther directly above the centre of the flower. After having exposed the anther, the filament bends back again. During the first part of anthesis, several anthers are usually on their way up or down. The gynoecium is undeveloped and chiefly consists of a large discus-like nectary which produces great quantities of nectar lying in the open. No stigmas can be discerned during this phase. Flies and other allotropic animals constitute the majority of visitors.

In the second, female phase of anthesis, the petals open up a little more, and the stamens remain adpressed to them. Most anthers drop off. The upper part of the gynoecium now develops rapidly, and stigmas become receptive.

The figures show (a) an early male phase; (b) a flower in female phase.

Ruta graveolens L.

(Rutaceae. Mediterranean)

Protandrous. Autogamous. Entomophilous. Allotropic

 R. graveolens resembles *Saxifraga aizoides* in its gradual presentation of the anthers, which are, at the outset of anthesis, bent back and half concealed in the hood-like petals, usually two, sometimes one or three in each. When the first stamens bend upwards, the style is very short and the stigmas undeveloped (a). During the male phase the style grows, and it apparently becomes receptive some time before the last stamen has risen (b). At the basis of the ovary there are eight large, eye-like nectaries which produce great quantities of nectar. Being open and unprotected, the nectar is utilized by allotropic visitors.

 So far, the main difference between *R. graveolens* and *S. aizoides* consists in the less distinct division of phases of antheses, and in the fact that anthers do not drop off after stamens have bent back. However, after the female phase in *R. graveolens* stamens bend up again, and autogamy is inevitable (c). During this phase the nectaries are dry.

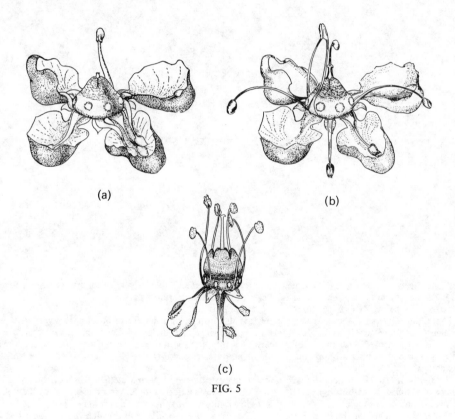

(a)   (b)

(c)

FIG. 5

Polygala chamaebuxus L.

(Polygalaceae. European Alps)

Protandrous. Melittophilous. Flag blossom with secondary pollen presentation

 Flowers *ca.* 1½ cm long, single in upper leaf axils, but frequently 2–3 in simultaneous anthesis at the same branch. Calyx "phlox purple" (637/1, Horticultural colour chart), with the upper median sepal forming a short spur. The upper lateral sepals are large, slightly unguiculate, erect, forming together an advertising organ comparable to the vexillum in Papilionaceae. The lower lateral sepals are scalelike.

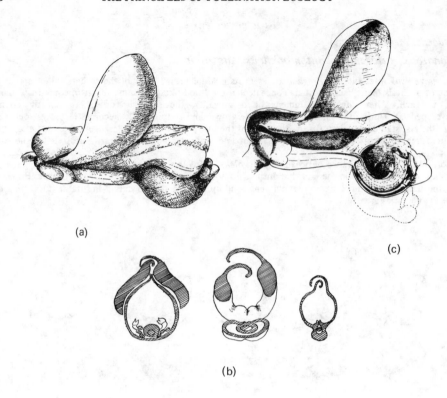

(a)

(c)

(b)

FIG. 6

Petals three, forming a sheath, their outer one-third free. Tube purple, outer parts sulphur yellow (1/1). Filaments connate to the corolla tube except for their outermost part.

Apart from colour differences there are no nectar guides.

The upper edges of the two lateral petals are rolled into each other. Their outer part is very thick and succulent, the outer edges running down towards the hinge and enclosing the style. Lower petal forms a bowl that is connected with the corolla sheath by a fold, functioning as a hinge. Outer part of the bowl is contracted into a channel around the style and the outer edges of the bowl are reinforced, with irregular thickenings.

Filaments are short, anthers open in bud and deposit their pollen on a shelf-like part of the style. The style is very thick, its outermost part (outside the hinge) forming a circle. After deposition of the pollen, the style stretches so much that the pollen is removed out of the sphere of the anthers.

Nectar is produced at the base of the ovary. Access to the nectar is only through the upper part of the corolla tube, which opens backwards. The opening is constricted by two pockets, one in each upper petal. (These pockets, forming blind alleys, may delay the insect and safeguard pollination.)

Visiting bees can only reach the nectar via the upper part of the corolla tube. Trying to penetrate this, they lodge their legs between the irregularities of the edge of the lower petal, and press this down. Movement of the style is prevented by its stoutness and by its fixation in the lower part of the corolla tube. The style plus pollen therefore emerges when the outer part of the lower petal is pressed down, and hits the visitor. After the visit, the lower petal returns to its former position because of its own stiffness.

The figures show (a) a whole flower and (c) a longitudinal section with the lower petal in normal and depressed position. In addition they show (b) three transverse sections, viz. (from left to right) through the outermost petal bowl, through the corolla tube immediately beyond the hinge and near the nectary.

Campanula species

Protandrous. Secondary pollen presentation. Late anthesis autogamy. Bellshaped blossom

The four semi-diagrammatic drawings show the main stages in the anthesis of most *Campanula* flowers. Pollinators are usually large hymenopters that creep into the flower and are able to penetrate with their probosces between the bases of the filaments.

(a) Bud stage. The anthers, which form a tube around the style, open introrsely, so that pollen is deposited on the middle and upper part of the style, which is densely covered with bristles. Nectar is produced by a ring-like nectary at the base of the style.

(b) Early anthesis. The style has stretched, and the filaments have curled more or less. Consequently, the pollen deposited on the style is now freely exposed within the flower. The stigma lobes are close together, therefore no pollination is possible.

(c) Middle anthesis. The anthers have now (if not earlier) withered, and are found as shrivelled remains in the bottom of the flower. Only the bases of filaments, guarding the entrance to the nectar, are still turgid. Stigma lobes have opened and the stigma is receptive.

(d) Late anthesis. The stigma lobes have continued to grow, the lobes curve further and contact the pollen left in the upper part of the indumentum of the style. Autogamy takes place. This last stage is not realized in all species. Degeneration of the pubescence of the style may increase the effectivity of the process (Vogel 1975b).

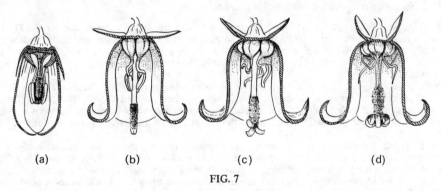

| (a) | (b) | (c) | (d) |

FIG. 7

Lagerstroemia indica L.

(*Lythraceae. East Asia to Australia*)

Heterantheric pollen blossom

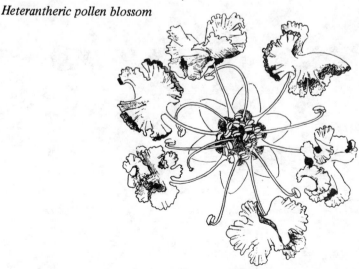

FIG. 8

Lagerstroemia (cf. Schrottky 1908; Harris 1914; Heide 1927) is given here as an example of heteranthery, i.e. differentiation within the androecium. The feeding stamens are conspicuously yellow, have shorter filaments, and are collected near the centre of the flower. The fertilizing ones are sombre and more or less concealed under the pink petals. There is no nectar, and bees visit the flowers to collect pollen from the central stamens. Pollination takes place in a haphazard manner – mess and soil pollination – while the visitor collects pollen. The petals are curious in their high degree of morphological complexity.

17.2. EXTRA-FLORAL NECTAR

The occurrence of extra-floral nectar is widespread, but usually the importance of this nectar for pollination is obscure, if there is any connection at all. In the example quoted below extra-floral nectar plays an important part in the pollination syndrome.

Thunbergia grandiflora Roxb.

(Acanthaceae. South-east Asia)

Homogamous. Melittophilous. Extra-floral nectaries

The large, sky-blue flowers of *T. grandiflora* have been described several times (see van der Pijl 1954 with references). With exception of the upper distal part, the corolla is succulent and stiff. The lower rim is distinctly corrugated. There is a large lower lip that affords ample landing space for pollinators – in S. Asian gardens *Xylocopa latipes*. The proximal parts of the flower are white, giving a positive light gradient towards the nectary.

The calyx is vestigial; its protective function has been taken over by two deciduous bracts that are easily broken off after anthesis has begun. The flower is ephemeral and drops off after the first day. However, owing to its heavy production of nectar, it is eagerly sought by bees, and all flowers are visited very soon after opening. In addition to colour, the flower possess a distinct but not very strong odour attraction.

The flower is divided into two chambers by a constriction. An outer distal one with stigma and anthers, and a smaller inner one with ovary and nectarium. The style is trumpet-shaped with a folded stigma. Anthers are large, opening downwards. Hairs on the anthers serve as arresting mechanisms, spurs projecting downward from the inner end of the anthers serve as triggers. On entering, bees must contact the stigma and also shake pollen down on their backs.

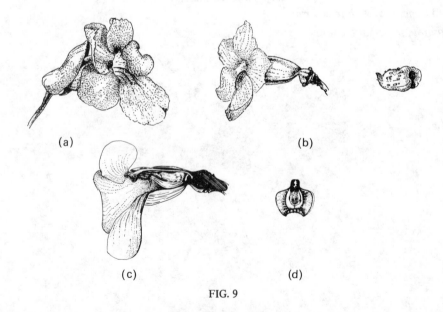

(a)   (b)

(c)   (d)

FIG. 9

The longitudinal section shows that in the region of the constriction the corolla is extremely thick. In fact, it cannot be opened without being broken, which certainly over-taxes the strength even of a *Xylocopa*.

If one looks into the flower, one sees that the constriction is also closed laterally by the widened bases of the filaments. The only negotiable way is in the channel along the style, so there is no doubt that the insects have some difficulty in negotiating the passage to the nectary. Presumably, they must push against the roof of the blossom, thus securing contact with the anthers. The ordinary entrance to the nectar being so difficult, and *Xylocopa* being a habitual nectar thief, which prefers to pierce the base of the corolla instead of using "the proper approach", the *T. grandiflora* flowers would ordinarily run the risk of being robbed. That robbing is the preferred *Xylocopa* technique is easily seen in unprotected flowers. However, unprotected *T. grandlifora* flowers hardly ever occur in nature. Both the outside of the bracts, the pedicel and the calyx are covered by very active nectaries. Like all extra-floral nectaries, these are very popular with ants which swarm over them and carry away all the nectar produced. The glistening drops of sticky nectar – so conspicuous in European hothouse specimens – are in the Tropics only seen on ant-free flowers in a vase. The ants are, as usual, bellicose and occupy a defensive position if a *Xylocopa* approaches. In spite of the difference in size and strength, the "ant-guard" keeps the bees away from the base of the flower. *T. grandiflora* flowers with an "ant-guard" are never robbed, those of related species without one are invariably bitten through at the base and robbed.

In Fig. 9b the extra-floral nectar is indicated. The cross-section (Fig. 9d) shows the longitudinal corrugation and the general thickness of the corolla, and also the way in which the filaments block the entrance to the nectar chamber. In the upper part are seen a section of the anthers with conspicuous hard spurs and the channel through which the style passes (not drawn).

17.3. HYDROPHILY

Vallisneria spiralis L.

(Hydrocharitaceae. Tropics)

Ephydrophilous with free male flowers. Dioecious

The pollination of *Vallisneria* is so generally known and so frequently quoted that one might expect to find good representations in the relevant textbooks. However, the much-reproduced drawing from Kerner suffers from serious defects. The drawings presented here are based, partly, on Chatin's monograph, and

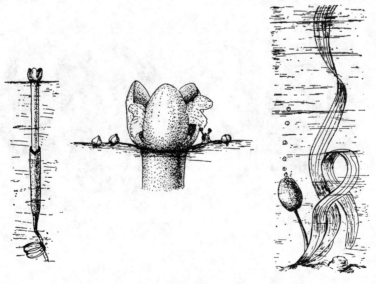

FIG. 10

partly on personal observation. Reservation must be made for the position of the female flower (left), which in nature probably is less stiff than indicated.

Right: A male plant with an inflorescence (cf. Kaul 1970), the spatha of which has opened at the tip. Male flowers are liberated and rise up through the water. On reaching the surface, they open and their two (or three) anthers are exposed. Tepals are bent completely back and support the floating flower.

Middle: Female flower in higher magnification, showing the large outer and very small inner perianth leaves with the large stigmatic branches exposed outside the flower. Male flowers float towards the female one and eventually the anthers will touch the stigma. After fertilization the female flower is pulled down under water by the spirally contracting pedicel.

17.4. TRAP BLOSSOMS

Calycanthus occidentalis Hook. et Arn.

(Calycanthaceae. California)

Protogynous. Beetle-pollinated. Food-bodies. Trap flower

The flower is rather primitive with a great number of spirally arranged members. Apocarpic, but perigynous. Fragrance strong, wine-like. Colour of flower dark red. No distinction between calyx and corolla, perianth members thick, broadly linear. At the beginning of anthesis the outer tepals fold back, whereas the inner ones form a hollow cone. The innermost ones are at this time bent abruptly inwards and downwards and form an entrance similar to that of a lobster-pot. Stamens and styles protrude into the inner chamber formed by tepals. Odour is produced from the central part of the flower and is emitted through the apex of the cone (in Vogel's terminology the petals are osmophores). The innermost tepals, stamens, and staminodes (innermost, sterile stamens) carry at their tips white, granular food-bodies. Hairs and bristles on stamens and staminodes prevent penetration of animals down to the ovules.

C. occidentalis is pollinated by a 3-mm long beetle, *Colopterus truncatus*, which enters the floral chamber and eats the food-bodies. Attraction is probably mainly by odour. No nectar is present. During the early phases of anthesis exit is barred by the reflexed inner petals, the narrowness of the chamber, and by stiff, downwards-pointing bristles on the inside of inner tepals. This first, female part of the anthesis lasts for 1–2 days, and 8–10 beetles may be trapped during this period. If the flower is cut open, beetles fly away, indicating that their stay in the flower is enforced.

In the second, male phase of anthesis staminodes cover the stigmas, preventing self-pollination; also, stigmas wither very soon. Anthers dehisce, powdering the beetles with pollen, and stamens and inner

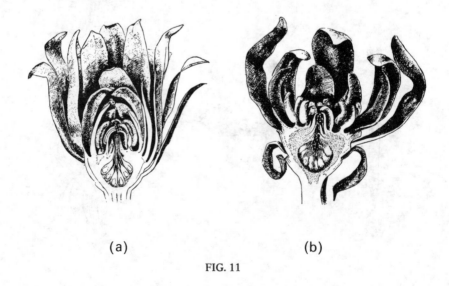

(a)   (b)

FIG. 11

tepals reflex, opening up the flower and setting the beetles free. Experiments have shown that they enter another flower and carry out pollination there. It is not known if the beetles feed on pollen, too.

The illustrations show (a) a flower in female phase with stigmas projecting into the inner chamber and food bodies present; (b) a flower in male phase: stigmas withered, staminodes closing the entrance to the ovaries and food-bodies eaten. (Mainly from Grant 1950c.)

Nymphaea citrina Peter

(*Nymphaeaceae. Central Africa*)

Protogynous. Liquid trap

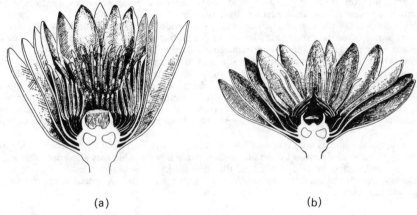

(a)   (b)

FIG. 12

In Schmucker's studies (which were carried out in a hot-house!) the flowers of *N. citrina* had two completely different aspects. During the first day of anthesis they were in a female stage: all anthers were directed straight upwards, and their tips formed a sliding surface. Visitors landing on them lost their foothold and tumbled down into the bowl-shaped stigma, containing a thin sugar solution. The effect was that many of the visitors were drowned. This was a mixed group of insects including bees. It is known that in nature beetles play a prominent part in some *Nymphaea* species. Possibly, only small and unadapted pollinators are drowned?

During the later part of anthesis stamens form a closed cone over the stigma surface, which is therefore no longer accessible. The stamens develop gradually, starting from the time when the flower opens again on the second day. As each stamen ripens, it turns outward, forming, together with the inner tepals, an arena-like space on which visitors land and crawl. Thecae open inwards, i.e. upwards on the arena.

Colour and scent are secondary attractants. As no nectar is available during the male stage, and as the liquid in the stigma bowl is too dilute to attract bees, pollen must be the primary attractant, if there is any.

The stigmatic liquid disappears gradually during the male stage of the anthesis, whether by evaporation or by other means is not known.

Liquid traps are known from other plants, too, e.g. the orchid *Coryanthes*, the labellum of which is filled with a watery liquid. However, the Nymphaeas are unique inasmuch as visitors are drowned. Pollination therefore presumes that the visitor has been in a male flower first – there is no chance of going back and forth. It is thus a precision mechanism, the adequacy of which may seem questionable. The well-known difficulties of germinating *Nymphaea* pollen *in vitro* may have something to do with the remarkable pollination mechanism.

(a) Flower in the female stage. (b) Flower in the male stage. (Chiefly after Schmucker 1932.)

The pollination ecology of other Nymphaeaceae varies between species. In *Victoria* the flower behaves like that of *N. citrina*, as detailed above, but the beetles (chiefly *Cyclocephala hardyi*) are released the following day, after the flower has entered the male stage. In various other night-blooming species, flowers open on two (rarely more) consecutive nights, but do not trap visiting beetles (*Cyclocephala castanea*). In *N. ampla* flowers open for three or four consecutive days and are pollinated by bees (cf. Cramer *et al* 1975; Prance and Anderson 1976).

Arum nigrum Schott

(Araceae. Middle Europe, Mediterranean)

Second-order protogynous. Sapromyophilous. Sliding-trap blossom

This is Knoll's classical object of study (1926 cf. 1923).

The inflorescence is enveloped in a large (in this species) blackish-purple bract, the spatha, the lowermost part of which is brighter and forms a chamber. From below, the stem of the inflorescence carries the following flowers: female; sterile, bristlelike; male; sterile, bristle-like. The topmost part of the stem is sterile and projects out of the spatha: the spadix. It is odoriferous. Flowers are extremely reduced, no perianth being present. Stigmas are rather large, swab-like.

During early anthesis the spadix produces a very strong odour, reminiscent of decaying human faeces. The production of odour is accompanied by strong production of heat, during which great quantities of starch in the outer parts of the appendix are consumed. Carrion and dung-flies are attracted by the odour.

The epidermis of the inner side of the spatha and of the spadix is slippery to the feet of insects 1, because of oil drops on the surface, 2, due to turned-down papillar surfaces, or 3, through very flat surfaces formed by closely fitting cells with no possibility for insect claws to find support. Insects that alight on the upper part of the spatha and start crawling around to localize the source of odour, lose their foothold and fall into the chamber forming the bottom of the blossom. In the right-hand figure the route of an insect is indicated by a dotted line, changing to hatches from the point where it slips and falls into the chamber.

Once trapped, the insects are prevented from escaping, partly by the character of the epidermal cells and partly by the obstacles formed by the bristle-like, sterile flowers, which are also unscalable.

Insects trapped during the female phase of the anthesis lick up liquid secreted by the stigmatic hairs, thus being led towards the stigmas. The female phase lasts for one day. A special tissue provides for air exchange into the chamber.

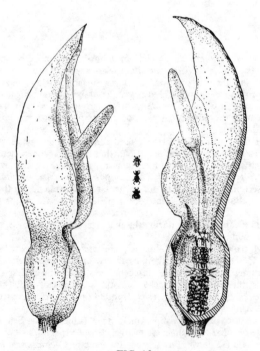

FIG. 13

The next morning anthers open, and the trapped insects are powdered with pollen. The epidermal cells of spadix and obstacles have burnt out their starch and, as they are dying, have lost their turgescence. Their surfaces are caving in and insects' claws can get a foothold on them. So the insects climb up the spadix and alight from there – the spatha being slippery also during the female phase.

Numerous (50–100) small saprophilous dipters and beetles are trapped, the most frequent in Knoll's material being *Sphaerocera subsultans* and *Sphodium tristis*. The relative sizes of blossoms and visitors are indicated in the figure.

Arisarum vulgare Targ. Tozz.

(Araceae. Mediterranean)

Secondary protogynous. Sapromyophilous. Optical trap

The blossom is very like that of *Arum* (see the preceeding), only more primitive, approximately 4 cm long. There are no zones of obstacles, but sliding surfaces have been observed (Vogel in litt.). Approximately ten female flowers are found below about forty males, which open at the same time or a little later. The upper part of the spatha is helm-like and dark, whereas the lower cylindrical part has ten diaphanous vertical zones. Small dipters are attracted by the smell ("impertinently rotten") and generally enter the blossom via the upper part of the spadix. Once inside the blossom, they cannot see the exit, which is also obscured by the dark upper part of the spatha. The diaphanous sides also deceive them and they try to escape through the walls. In the end the insects are said to be exhausted by their vain attempts at getting out this way, so they crawl out via the spadix instead and get thoroughly powdered with pollen on that occasion, if they have not been covered with pollen earlier during their stay.

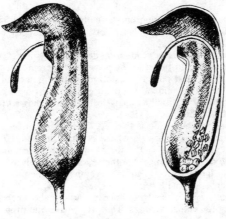

FIG. 14

17.5. BROODPLACE BLOSSOMS

Calluna vulgaris L.

(Ericaceae. Europe)

Protandrous. Entomophilous. Hemitropic, facultatively anemophilous. Bowl- to bell-shaped blossom

The flowers are tetramerous, almost radial, bowl-shaped, *ca.* 4 mm across with pink sepals, more conspicuous than the smaller petals of the same colour. The lower parts of the petals are succulent, their increase in thickness forces the blossom open; the lower petal opens up more than the other ones, making the flower slightly zygomorphic. Also, the lower parts of the filaments are succulent, but their bases are very thin. The nectary forms a continuous, prominent ring under the ovary. Nectar is available to insects

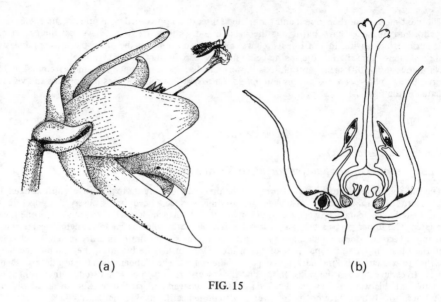

(a)   (b)

FIG. 15

able to insert their probosces between the filaments, forcing the anthers apart. Thecae open laterally, and pollen is liberated when anthers are separated. The horn-like appendages of anthers contribute to the prying-out mechanism. Nectar production is ample, and *Calluna* is visited by a great number of dipters, hymenopters, and lepidopters. The precision and effectivity of these visits for pollination is not very great; but *Calluna* is an important honey plant, both because of its great nectar production and its widespread occurrence.

During the later part of anthesis, nectar production decreases, filaments stretch, and pollen is dispersed by the wind. Wind pollination should therefore be possible. Pollen production is very high. According to Pohl (1937b), *Calluna* has the highest pollen production per unit area of all plants investigated, viz. more than 4×10^9 *tetrads* per m². The number of pollen grains per ovule is comparatively low, viz. of the magnitude of 1000 (Hagerup, 1950a). Pollen analytical evidence shows that great quantities of *Calluna* pollen are actually liberated into the air.

The most interesting type of pollination is that by *Taeniothrips ericae*, discovered by Hagerup in the Faroes, and also demonstrated to take place elsewhere. A small number of *T. ericae* is constantly found in *Calluna* flowers, which are big enough to shelter them completely. The animals are very lively, constantly on the move, and are able to penetrate between anthers and pistil down to the nectary, also reaching parts not immediately available to larger visitors. During this activity, some pollen is set free and adheres to the bodies of the thrips, which are sticky with nectar.

As the males of *T. ericae* are rarer than the females, and are also unable to fly, females fly from one flower to another, searching for sexual partners. They start from the projecting stigmas, and presumably land there, too, thus having the chance to cause both self- and cross-pollination.

After fertilization the *T. ericae* females go down into the flowers again, taking more food in the form of nectar, and also gnawing the succulent parts of filaments and petals – in the latter they deposit their four eggs. The eggs survive the winter in the persistent corollas, and larvae emerge next year. They are found together with imagines in the flowers, and are presumed to pupate in the ground, appearing as imagines only the next season.

T. ericae is thus able to pass its whole active life within the *Calluna* flower. Apparently, it can also live in other plants in the same manner (Hagerup and Hagerup 1953).

(a) *Calluna* flower with departing *T. ericae* in scale. (b) Section through the inner part of a *Calluna* flower with nectary (dotted) and a *T. ericae* egg (black) in the base of one petal.

Trollius europaeus L.

(Ranunculaceae. Europe)

Successive centripetal ripening of stamens and pistils

Entomophilous, eutropic (?) Closed blossom

FIG. 16

Flowers yellow, subglobular, diameter *ca.* 3 cm, with a great number of imbricate, coloured sepals that completely shut off the interior. One row of petals forming narrowly spathulate nectar leaves. The number of observed larger visitors is small. Self-pollination is inevitable, but flowers are apparently self-incompatible. The only regular pollinator observed so far is a small fly, *Chiastochaeta trollii*, which can penetrate between the sepals. The blossom forms a shelter in which the fly spends long periods, feeding on pollen and nectar, causing mess and soil pollination – provided it has visited other plants previously. Males are more often found outside the *Trollius* flowers, females more often inside.

Eggs are deposited at the base of the pistils. The hatching larvae burrow into the pistil and eat some of the developing ovules. After *ca.* 10 days its larval development is completed, and the animal again burrows out of the still young and soft fruit and disappears from the blossom (pupates in the ground?). The larva does not eat all ovules even in the fruitlet it attacks. Presumably other pistils are not attacked.

Left: a young fruit in longitudinal section showing, also, an empty egg-shell (outside) and the damage caused by the larva. Middle: almost ripe fruit with the entrance (small) and exit (larger) holes of the larva. Right: Section of flower with pollinator. (After Hagerup and Peterson 1956.)

Presumably, more regular pollination, especially by bumblebees, may take place in *Trollius* blossoms, certainly in the open flowers of other species.

Yucca species

(Liliaceae, North America)

Herkogamous. Homogamous. Euphilic. Phalenophilous. Broodplace pollination

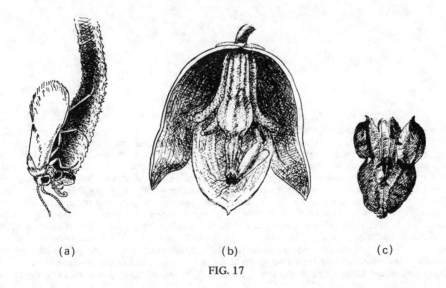

(a)   (b)   (c)

FIG. 17

The flower is hanging, bell-shaped with thick and yellowish-white perianth members (most illustrations show the activities of the pollinator upside down). Most species are completely incapable of autogamy because of the mutual position of anthers and stigma. Intra-ovarian (septal) nectaries are present, but hardly functional in the present pollination syndrome. The pollinator is a moth, *Tegiticula* (*Pronuba*) *yuccasella*. Filaments are stout and stiff, anthers rather small and pollen almost putty-like. The moth apparently does not take any nourishment; the females enter the flower and collect pollen from the anther (a). The presence of a tooth-like appendage permits the transport of comparatively large quantities of pollen. Pollen is transported to the gynoecium and is there deposited in the stigmatic cavity (b). Eggs are at the same time laid in the ovary, and larvae eat some of the developing ovules. (c) Ripe fruit showing exit holes of *Tegiticula* larvae and undeveloped parts of the capsule, where all seeds have been eaten (for a modern discussion cf. Powell and Mackie 1966).

Ficus species

(*Moraceae. Mediterranean*)

Monoecious. Secondary protogynous. Entomophilous. Eutropic. Urnshaped inflorescences

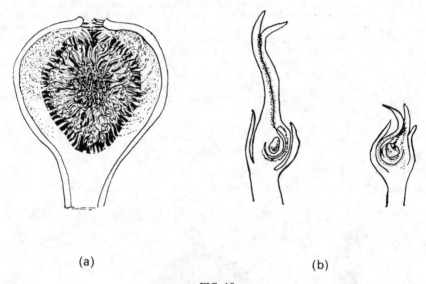

(a)   (b)

FIG. 18

Because of its economic importance, the edible fig, *F. carica*, was the first species to be studied. Its pollination had represented problems to cultivators for millennia, and was taken up for serious study during the latter part of the nineteenth century. In a way the problems, both scientific and practical, were solved. In a broader context, this was unfortunate. *F. carica* has turned out to be an aberrant and, in many ways, very complicated type. Not until Galil and his collaborators, and Ramirez, both in 1969, published the studies on other species was a better understanding of the process possible.

The inflorescences in the genus *Ficus* are urn-shaped so-called syconia, the inner surface of which represents the original surface of a spike turned in, and it carries the flowers. These are extremely reduced: the staminate ones chiefly consist of one to five stamens, whereas the pistillate ones comprise a single pistil, in addition to which three rudimentary perianth members are present. These flowers never emerge from the syconium the only morphological opening of which is a narrow orifice, the ostiolum, with a number of scales representing sterile bracts at the morphologically lower part of the spike.

The syconium exhibits a strong protogyny. Galil divides the normal development into five stages:

A. Immature buds.
B. Pistillate* flowers in anthesis.
C. Intermezzo.
D. Staminate flowers in anthesis.
E. Ripening and seed dispersal.

The general principle of pollination is the following: Female agaonid wasps, only a few millimetres long, loaded with pollen, penetrate into syconia at stage B and oviposit in the pistils of pistillate flowers. The ovules develops into galls. During the intermezzo the larvae develop and at the beginning of phase D the males emerge and impregnate the females still inside the galls. These later get loaded with pollen and fly off to another syconium in stage B, oviposit, etc.

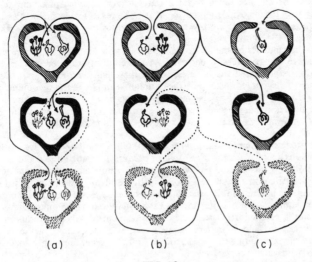

(a)   (b)   (c)

FIG. 19

In the figs studied, pistillate flowers are of two kinds: some short-styled, some long. The ovipositor of the various wasps is sufficiently long to reach the ovule of the short-styled flowers. The egg is deposited in a very precise place in the ovule, and no seed forms. The long-styled are too long for the ovipositor to reach the ovule. After abortive attempts the ovipositor is withdrawn and the ovule, which has been pollinated in the process, can develop into a seed.

The sharp distinction between two classes of pistillate flowers is probably a late phenomenon; the primitive figs presumably contained pistillate flowers with variable style lengths. Also, we must presume that this syndrome developed in an even, tropical climate, where syconia ripened continuously, with the result that the emerging female always found syconia in stage B when they emerged. Figure 19a gives a diagram of pollination and seed setting in three generations of a fig of this type, a so-called monoecious fig.

However, there are other figs, cf. Fig. 19b and c, in which there are two types of syconia, some with staminate and short-styled pistillate flowers, others with long-styled pistillate flowers. These plants (morphologically gynodioecious) are functionally dioecious inasmuch as the one type of syconia only produces pollen but no seeds, the others seed but no pollen. The breeding wasp population is maintained by the galls occurring in the staminate syconia.

Apart from the extreme protogyny, this syndrome is not very far removed from the pollination of *Trollius* and *Yucca*. The main point in Galil's and Ramirez' discoveries was that in the species studied by them pollen transfer in *Ficus* was active, as in *Yucca*. The female wasps possess pockets on both sides of the thorax. They actively load these pockets with pollen before leaving the syconium in which they developed.

To prevent confusion we shall call flowers pistillate–staminate, reserving the terms male–female for the pollinators.

The number and position of staminate flowers in relation to that of the pistillate ones vary between species. In some species staminate flowers are evenly scattered, whereas in others they are concentrated in a separate zone near the ostiolum. In some of the latter type the male wasps cut the filaments before leaving the syconium. The loose anthers fall down and are scattered among the pistillate flowers from which the females are going to emerge. In any case the females open the anthers actively and scoop out the pollen. Galil *et al.* (1973) could observe the process in *F. costaricana* and *F. hemsleyana*. Pollen is taken up by the spoon-like extreme joint of the forelegs, the ariola, which deposits the pollen mainly on the venter of the wasp. From there it is scooped up by combs on the foreleg coxa and deposited in the ventral pockets on both sides of the thorax. Curving the thorax widens the opening of the pockets and facilitates loading. Emptying repeats the process in inverted sequence, pollen being again transferred to the ariola (cf. Galil and Eisikowitch 1973). Pollination is achieved by the active transfer of a small number of pollen grains to the stigmas after each oviposition.

In some syconia the stigmas form a continuous surface, a synstigma, and females bite off the surrounding stigmas after having oviposited in one ovule. The result is that the many useless inquilines and parasites that abound in the syconia are prevented from ovipositing in or near the gall inhabited by the pollinator.

The carbon dioxide contents of the air inside the syconia of some species rises rapidly during phase C. When the males emerge it may reach 10 per cent. At this concentration the males are active, the females inactive. After having impregnated the females, still in their galls, and having cut loose the anthers, the males tunnel through the syconium wall (not the ostiolum!) and leave the syconium. The carbon dioxide is gradually diluted through the exit tunnel, and the females become active, pick up pollen and load their pockets, widen the tunnel if necessary, and leave the syconium for another one in the B phase where they can oviposit. In the thin-walled syconia of other species the atmosphere is "normal" during the ripening process.

This pollination process is what Galil calls ethodynamic; it is based upon active work by the pollinator, as in *Yucca*. The pollinator of *F. carica, Blastophaga psenes*, does not possess thoracic pockets, and the major part of the pollen is transported passively and indiscriminately between the tergites of the wasp; pollen is passively deposited on the stigmas of the pistillate plant in which oviposition has taken place. This is Galil's topocentric pollination. Because of the less precise transfer mechanism, topocentric syconia have a much greater number of staminate flowers than the ethodynamic ones. The surplus pollen on the outside of the females is brushed off before they leave (the outside of) the syconium. *F. carica* is a plant of very long domestication, and there are many different clones, some parthenocarpic. Further, the cultivated fig has an opposite number, the *caprificus*. Both of them are highly specialized domesticated plants that must be propagated vegetatively. Seed propagation gives figs of the "wild" type, the so-called *erinosyce*. Cultivated edible fig and caprificus together function as a dioecious species, whereas *erinosyce* in the main corresponds to the monoecious type. The process of "caprification," i.e. throwing caprificus twigs into the crown of fig-trees, was a well-known technique for safe-guarding the fruit set several thousand years ago. It is understandable that this technique had difficulties in being accepted by practical cultivators in areas where there was no tradition of fig-culture, also after the theoretical basis had been elucidated. According to Condit (1947) one of the pioneers in the field was hooted down when he proposed introduction of *caprificus* and *Blastophaga* to California in order to obtain a crop.

Another complication in *F. carica* pollination is the seasonal character of the Mediterranean climate. Flowering is discontinuous, with three main synchronized seasons. The larvae of the last clutch hibernate in their galls and emerge next spring.

The genus *Ficus* is enormous, more than 1000 species; it is pantropical and subtropical and comprises types ranging from the world's widest trees to creepers, the leaves of which hug the rocks. The syconia vary by a factor of at least 10, and only a few species have been studied so far. The many variations on the general theme discovered by Galil and collaborators indicate that it is too early to generalize. The only generalization that seems safe is that the pollinating syndrome is the same throughout, with species-specific Agaonid wasps doing the work (Wiebes 1963, 1966).

The family Moraceae, to which *Ficus* belongs, is mainly anemophilous. The existence of this extremely complicated entomophilous syndrome is striking, and with possible reservation for the largely unknown pollination of *Dorstenia* (Carauta 1972) it is difficult to trace its evolution.

17.6. CHIROPTEROPHILY

Parkia clappertoniana Keay (Fig. 20)

(Mimosaceae. Central and West Africa)

Epomophorus gambianus lands on the inflorescence and clings to it with its hind feet while licking nectar. Nectar is produced by the upper, sterile flowers (section of inflorescence) and collects in the ring-like depression between sterile and fertile flowers (after Baker and Harris 1957). According to Vogel (1968a) the upper zone in *P. auriculata* inflorescences is male, whereas the (sterile) nectar flowers form an intermediate zone. The lowermost (morphologically top) flowers are female in this species, too.

Agave schottii Engelm. (Fig. 21)
(*Liliceae. Mexico*)

Leptonycteris nivalis lands on the upper part of the inflorescence and crawls downwards, licking nectar from individual flowers. (After Cockrum and Hayward 1962.)

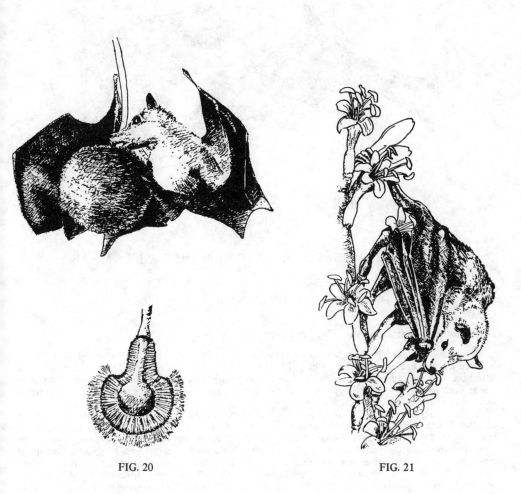

FIG. 20   FIG. 21

Carnegiea gigantea Britton et Rose
(*Cactaceae, Southwestern U.S.A.*)

In this case *Leptonycteris nivalis* stops in the air and licks nectar out of the well-filled cup of the cactus flower. As the bat cannot hover properly, the visits are very short, and it may return to the same flower several times before exhausting the nectar. (After Cockrum and Hayward 1962.)

FIG. 22

17.7. POLLINATION SYNDROMES WITHIN THE PAPILIONACEAE

Astragalus depressus L.

(*Mediterranean*)

Homogamous. Melittophilous. Flag blossom

The genus *Astragalus* is taken here as an example of the simple typical flag blossom. The connection between the alae and carina is effected by projecting tooth-like parts of the edge of the former interacting with pockets in the sides of the carina. Backward projecting spurs on the alae safeguard their return after having been pressed down. The upper part of the carina is open, both anthers and stigma may pass freely out and in when the petals are pressed down.

Visiting insects — mainly bumblebees — will land on the alae and press their heads in under the vexillum in order to reach the nectary at the lower (proximal) end of the ovary. Entry to the nectary can only be gained through the loops formed between the free filament and both sides of the sheath of adnate ones. The flowers are about 1 cm long. Trying to reach the nectar, visitors will by their activity press down the alae on which they are standing, and through the inter-connecting with the carina, also the latter petals. The style and filament sheath, being very stiff, will remain in position and not follow the general downward movement of the corolla and the visitor. The consequence is that the visitor's ventral side makes contact with the emerging sexual organs, and pollen is deposited on and received from the visitor.

In spite of morphological refinements, the mechanism is not infallible: sometimes (in older flowers?) the carina does not return completely to its original position in relation to the sexual organs, some of which may remain outside.

The figures show: Upper left: a complete flower (some anthers projecting). Lower left: flower from which calyx, vexillum, and right-hand ala have been removed, showing the pocket in the carina and the spur on the left-hand ala. Upper right: left-hand ala seen from the inside with spur and marginal tooth. Lower right: filament sheath, free stamen, and style on the background of the left-hand half of the carina.

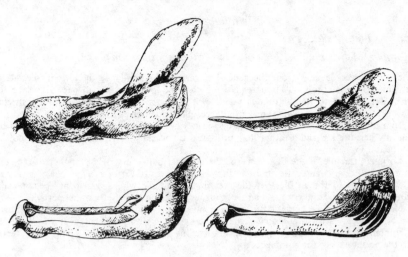

FIG. 23

Coronilla emerus L.

(*Europe*)

Protandrous. Melittophilous. Pollen blossom. Flag blossom

Compared with the more general flag flower in *Astragalus*, the *Coronilla* flower shows two small but interesting modifications.

It forms a classical example of the pump mechanism by which the pollen is forced out of the tip of the carina like a thin sausage. Pumping is effected by a piston consisting of the thickened filament ends and the driving force is derived from the relative pressing down of the flexible petals in relation to the stiff filament tube and ovary. As the flower contains no nectar and pollen is the only known primary attractant, the action of visitors (bumblebees) may be more directly aimed at pressing out the pollen than in a nectariferous flag blossom.

The second noteworthy feature is the absence of a semi-closed tubular proximal part of the corolla. There is no nectar to protect against theft.

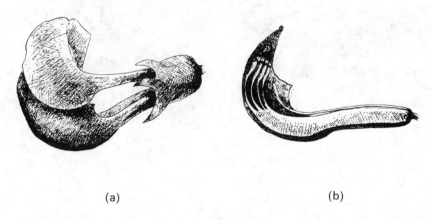

(a)   (b)

FIG. 24

Trifolium medium L.

(Europe, Northern Asia)

Homogamous. Melittophilous. Flag blossom (tube modification)

The clover flower deviates from the general pattern of the family by the extreme connation of the filament tube with the corolla. The consequence is that the lower (proximal) part of the corolla forms a rather heavy tube, where no dilation is possible. The flag mechanism is, therefore, restricted to the outermost free parts (approx. one-third) of the corolla, which gives a completely different mechanical system from that of the ordinary flag flower. Visitors can put only their heads inside the flower: there is no room for their bodies, and a rather long proboscis is necessary for them to reach the bottom of the corolla tube (cf. Fig. 25).

The working principle of the flag flower is the contrast between a rigid system that remains in place (stamen tube and ovary) and a flexible one that bends down. In *Trifolium*, the ovary is very short and does not contribute to the stiffness of the system. Instead the whole system has become stiff by the connation of stamen tube and corolla, and flexibility is restricted to the outermost parts.

The clover head may be classified as a brush blossom, but deviates from the typical form by the inclusion of its sexual part within closed corollas, and by the relatively difficult access to the nectary.

The combination of a "difficult" mechanism, needing great strength, and a long tube effectively bars access to the nectar except for bumblebees.

FIG. 25

Genista tinctoria L.

(Europe, Northern Asia)

Protandrous. Melittophilous. Pollen blossom. Explosive flag blossom.

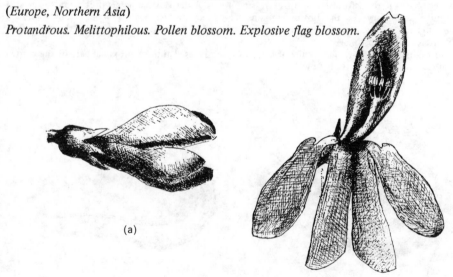

(a)

(b)

FIG. 26

The general shape and function of the *G. tinctoria* blossom is not very different from that of *Astragalus depressus*. The relative size and morphology of the individual components are almost the same. There is no nectar, and all filaments are connate, forming a tube.

The anthers open gradually, four of them before the others. Their filaments shrink, but their pollen is pushed forward into the distal part of the carina by the other stamens which are still actively growing. They open at the beginning of anthesis, and the stigma becomes receptive at the same time.

The upper edges of the carina are connate. As the filament tube and ovary, on the one hand, tend to curve upward, and the alae and carina, on the other hand, tend to curve downward, they keep each other in tension equilibrium. This equilibrium is upset by visitors pressing down the lower part of the corolla. In the flag flower, the alae have a tendency to move down and outwards when depressed, and the outward movement rips open the upper connate edges of the carina, which split apart. The process is almost explosive, and after the first visit the flower is completely devastated; the ovary being pressed against the vexillum, and the four other petals hanging down more or less limply.

Whether the visitors trigger off this mechanism "consciously" or inadvertently when trying to locate non-existing nectar, is not known.

Cytisus scoparius L.

(Europe)

Homogamous. Melittophilous. Pollen blossom. Explosive flag blossom

The general organization and function of the *C. scoparius* flower equals that of *Genista tinctoria* except that the various parts do not after explosion change their position so radically. Neither does the ovary bend up, nor the petals bend down to the same extent. The main movement is here carried out by the free end of the filaments and the style. A second difference is a stronger differentiation between the stamens. The upper five are shorter and form one group, the lower ones have longer free filaments and form another group.

By explosion the short anthers hit the insect ventrally whereas the long ones and the stigma hit it dorsally. The short stamens must be considered feeding stamens. There is no nectar, and there can be no doubt that in this case flowers are visited for their pollen, which the bees brush out of their pubescent coating after the visit. It is easily seen that the visitors receive a little shock at the explosion, but this certainly does not deter them from making these visits. On the other hand, bumblebees are not "interested" in a flower that has already exploded. Such flowers may be visited by other pollen-eating insects, e.g. syrphids.

The figures show (a) a virgin blossom with the stamens lying in tension; (b) after the visit. The style often makes a more or less complete spiral turn after the explosion.

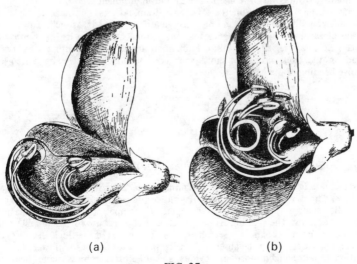

(a)   (b)

FIG. 27

Centrosema virginiana Bentham

(Southwestern U.S.A.)

Homogamous(?). Melittophilous. Inverted flag type

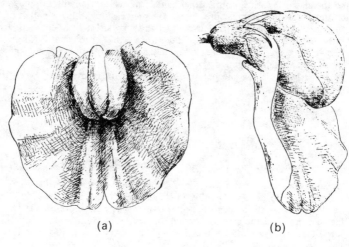

(a)   (b)

FIG. 28

The inverted position of the flowers of *Centrosema* and some related genera (*Clitoria, Canavalia*) transforms the flag-type flower into a gullet blossom. The vexillum still retains its function as advertising part of the blossom, but in addition it also serves as a landing space for visitors – heavy bees of genus *Xylocopa* and others.

Alae and carina are tightly connected – glued together according to Lindman (1902) – and the mechanism requires great strength on the part of the visitors in order for it to be operated. Nectar is as usual found at the base of the ovary. Deposition and removal of pollen is nototribic.

As long as advertising is the only function of the vexillum, no great demands are made on its strength. This changes fundamentally when the landing-place function comes in as well. The force required for operation of the blossom mechanism implies that the animal must stem its legs vigorously against the vexillum, which is, therefore, more stoutly built than usual. It is reinforced by two longitudinal thickenings and on the back also by a short, spur-like appendix which by butting against the calyx prevents excessive bending of the base of the vexillum.

Erythrina crista-galli L.

(Brazil)

Homogamous. Ornithophilous. Gullet (resupinate flag) blossom

At first glance there is not much in the stiff, flaming scarlet *E. crista-galli* flower to suggest its affinity with *Papilionaceae*. The dynamic system of the flag flower, based upon the contrast between the flexible petals and the stiff stamens plus ovary, has completely disappeared, and the carina projects almost knife-like out of the calyx. The vexillum is bent back and is of no use for landing. The alae have been reduced almost out of existence, and are seen hardly projecting out of the thick and stiff calyx. Nectar is produced in great quantities and, in the absence of visits, soon drips out of the flower.

The adaptation to hummingbird visits is obvious, and the fluttering birds cannot help making contact with the projecting anthers and style when they probe for the nectar produced at the base of the ovary.

FIG. 29

Phaseolus multiflorus Lam.

(*South America*)

Homogamous (*V*). *Melittophilous. Modified flag blossom*

The spirally wound carina characterizes the genus *Phaseolus*. In the more primitive representatives, like *P. multiflorus*, the flag blossom pattern is still well developed. The connection between alae and carina is due to adnation exclusively. There are no hooks or spurs translating movements from the one set of petals to the other. The tip of the carina forms about one turn of a spiral. There is secondary pollen presentation of the usual *Lathyrus* type. Hairs on the style brush pollen out of the tip of the carina, which is too narrow to permit the emergence of anthers as well. The style is sufficiently stiff to move through the spiral.

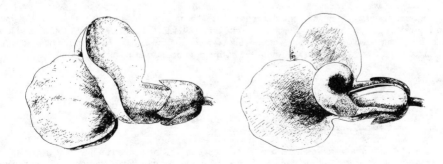

FIG. 30

Phaseolus caracalla L.

(Tropical South America)

In *Ph. caracalla* and other species described by Lindman (1902), the flag blossom pattern has dissolved so much as to be virtually unrecognizable and the flowers form a fantastic, completely asymmetric structure. In the bud stage, the vexillum is spirally wound and encloses the carina spiral (see Troll 1951). This spiral is only partly unwound when the flower opens. The alae are almost asymmetric and may occupy a more or less transversal position in the flower, the left-hand one being then on top. The right-hand ala is resupinate and its morphological outer side turns up. Both outer sides are dark lilac-coloured, their inner sides off-white. The vexillum is white with a yellow nectar guide. The flower is approximately 5 cm across.

Pollinators are large bees, probably *Xylocopas*, which land on the up-turned outer surface of the right-hand ala and proceed towards the nectary which is concealed in a very short corolla tube, and this is again well protected in an exceptionally heavy-walled calyx. Two prominent ridges on the lowermost part of the vexillum enclose the bases of the rest of the flower, and protuberances on the tenth free stamen force the visitor to insert its proboscis in the line of this remarkable structure.

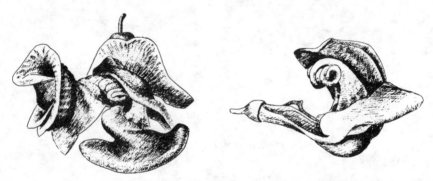

FIG. 31

The carina in this flower forms no less than four full turns. Its tip is situated above the visitor. It is not known if there is a structural connection between the carina and alae. When the carpenter-bee pushes the right-hand ala, the carina, and the sexual parts of the flower downward to reach the nectar the style moves out of the tip of the carina and hits the insect from above. The function of the *Ph. caracalla* blossom is, therefore, no more that of the flag, but of the gullet blossom. The functioning of this remarkable mechanism is safeguarded not only by the stiffness of the style, but also by its being in itself less densely wound than the carina which encloses it. This produces a tension that leads to its emergence. The flower contains much nectar and visits are reported to last for a half minute.

Thus there is no doubt that the remarkable floral structure of *Ph. caracalla* is functional enough, but one may ask if it was really necessary to turn the tip of that rostrum four turns to achieve this effect, or what processes have lead to such an incongruous result. It is noteworthy that cultivated *Phaseolus* species are to a great extent autogamous. Did autogamous biotypes select themselves for European agriculture because European bees on the whole were unable to work the mechanism of these flowers, or was the xenogamic complex breaking down already in the native state because of its inherent complication?

Petalostemon pinnatum Blake

(North America)

Homogamous (?). Hemiphilic. Brush blossom

The genus *Petalostemon* represents a degenerated type in relation to the Papilionaceae in general. The flowers are small, *ca*. 1 cm, and are combined into inflorescences which at first glance look much more like those of Compositae. In other species, e.g. *P. violaceus*, the inflorescence is more spicate. In both kinds they constitute a typical brush blossom.

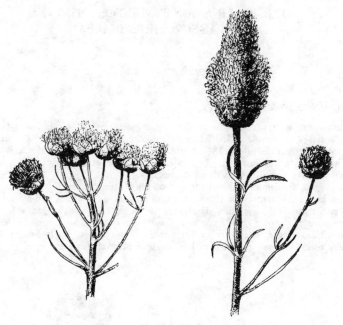

FIG. 32a

The individual flowers are remarkable in their reduction. The ovary is short, and the connate filament sheath very open. Five stamens are fertile, the other four connate stamens form deciduous petaloid staminodes. What has been interpreted as the vexillum, seems to be the last stamen which is also petaloid (a different interpretation of the flower has been given by Wemple and Lersten 1966).

Nectar is available, but the blossoms are also, perhaps chiefly, visited by pollen collectors. Unspecialized visitors, even beetles, play a much greater part than in flag blossoms.

Figure 32b shows from left to right an inflorescence, a flower before and after removal of the calyx.

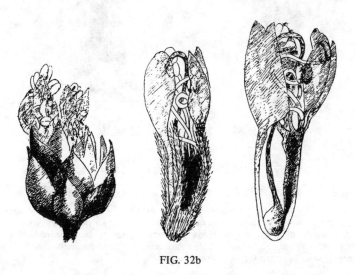

FIG. 32b

17.8. POLLINATION SYNDROMES WITHIN LABIATAE-SCROPHULARIACEAE

Galeopsis speciosa Mill.

(Labiatae. Europe)

Homogamous. Melittophilous. Gullet type

Corolla yellow with a very conspicuous dark violet nectar guide on the lower lip, whose middle constituent petal has a dark violet tip, whereas its proximal part is dark yellow with radiating violet blotches. The total length of the tube is 18–22 mm, of which the upper 6–8 mm is so wide that an insect may insert its head into it. The lower 12–14 mm can only be penetrated by proboscis or tongue. Nectar generally fills the tubes some millimetres above the nectary.

Anthers and stigma are found in the upper part of the blossom, covered by the helm-like upper lip. The fifth stamen being rudimentary or absent, the four remaining ones form two pairs, one of them with somewhat longer filaments than the other. The connective is placed at right angles to the filament and the anthers open centripetally (downward) by valves. These form lids that are hinged at the connective, and the edges of which are fringed. The lower lip is tripartite. At the bases of the two side lobes there are two prominent knobs, which facilitate the work of the bees in the flower (see Fig. 33).

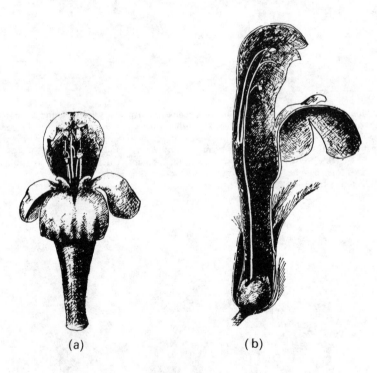

(a)   (b)

FIG. 33

Galeopsis is visited by big bumblebees. A tongue of at least 10 mm is necessary to reach part of the nectar and the visitors must push their heads well into the upper, wide part of the corolla tube in order to reach down. In doing so, their backs will engage the fringed edges of the forward thecae and open them. Similarly, when they withdraw, they open the backward thecae. Similarly for the stigma which is generally projecting from between the two foremost anthers.

Scrophularia nodosa L.

(Scrophulariaceae. Europe, Northern Asia)

Protogynous. Entomophilous, Wasp blossom (?). Bell-shaped

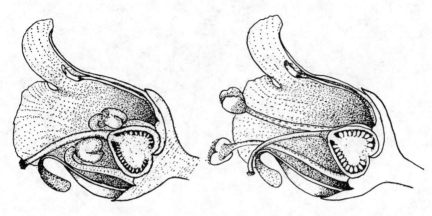

FIG. 34

The zygomorphy of the *S. nodosa* flower is so slight as to be negligible, and it is most naturally considered as a bell-shaped blossom. In the first, or female phase (left) the anthers are bent back, and the receptive stigma occupies a position in the mouth of the flower. The fifth stamen forms a staminode in the upper part of the flower; its function in pollination, if any, remains obscure. The female phase is stated to last for 2 days.

In the male phase (right), the outermost part of the style has bent slightly, and the stigma, which is still available, is more out-of-the-way than before. The filaments have unbent and stretched out, and anthers are exposed at the mouth of the blossom.

Nectar is secreted by a ring-like nectary at the base of the gynoecium. Larger insects lodge their feet on the outside of the corolla and put their heads inside the flower. Smaller insects are able to enter the flower. Their effectivity as pollinators is doubtful. Since Sprengel's days *S. nodosa* has been considered the prototype of a wasp blossom. Whereas there is no doubt that wasps do frequent, perhaps even prefer this blossom, it is also visited by bees.

In comparing the *S. nodosa* flower with flowers of *Salvia, Pedicularis*, etc., the former may be interpreted as a retrograde adaptation to a sternotribic pollinator. Another important point is postfloral nectar production, described by Schremmer (1959) for *S. canina*. Post-floral nectar seems to attract wasps especially.

Linaria vulgaris Mill.

(Scrophulariaceae. Europe, Western Asia)

Homogamous. Melittophilous. Secondary nectar presentation. Closed gullet blossom with spur

The *Linaria* flower represents a refinement in relation to the more central type of the gullet flower represented by *Galeopsis*. Two main features are new. Firstly, the lower lip buckles up so much as to close the entrance to the corolla tube. This means that only insects strong enough to force this obstacle aside can utilize the blossom unless they, like Knoll's *Macroglossa*, are able to insert a thin proboscis through the very small opening between the lips. The base of the lower lip forms a kind of hinge, so that once a sufficient force is applied the whole lip bends down.

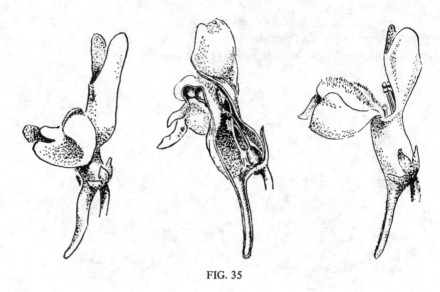

FIG. 35

Secondly, the nectar is not presented on the nectary, but runs down into the corolla tube, in that way being inaccessible to insects with a short proboscis. There are various rows of hairs in the flower; their actual function is not well known. Some of those in the spur have been supposed to lead nectar by capillary action from the nectary to the bottom of the spur. However, it is also possible that some of those hairs are sufficiently unpleasant for bumblebee prosboces to form guiding structures.

Melampyrum pratense L.

(Scrophulariaceae. Europe, Northern Asia)

Homogamous. Melittophilous. Gullet flower (nectar theft)

The *M. pratense* flower is functionally a rather simple gullet type. It is (usually pale) yellow, 15–20 mm long, contrasting against dark, brownish-red bracts. In most other species bracts and calyces are also vividly coloured and form part of the attraction unit.

The axis is horizontal, pedicels rather weak, and flowers bend down under the weight of a pollinator. The landing-place afforded by the lower lip is small, but a very pronounced roughening gives an excellent foothold. By inserting its head into the rather narrow opening of the flower a visitor must dilate the latter, the stigma bends down, and pollen is scattered over the upper side of the insect (head and pronotum). As usual in this type, nectar is found at the base of the ovary.

The illustration shows how thieves gain access to the nectar by biting through the corolla at its base. The hole indicated would correspond to those bitten by, for instance, *Bombus lucorum* (Meidell 1945). A nectar robber of this type may also collect pollen directly on its posterior legs by placing them

FIG. 36

underneath the anthers while sitting on top of the upper lip. By whirring its wings strongly, it shakes pollen out of the anthers and onto the extremities. During the flight to the next blossom the insect moves the pollen into the corbiculae. Whereas a nectar robber does not pollinate, pollination would result from pollen collecting. Visits for the same purpose by *Megachile willoughbyella* differ in that this insect, in a reversed position, places its abdomen under the anthers before whirring the wings. *Megachile* belongs to the abdomen collectors. This visitor also gains access to the nectar by biting through, notwithstanding that its proboscis is sufficiently long to enable it to collect nectar the proper way.

The long-tongued *Bombus* species (*B. jonellus, B. pratorum,* and others), in contrast, enter the *Melampyrum* blossom from the front and carry out pollination in the proper way. However, if holes have been bitten through the corolla previously by other robbers, they will take advantage of the opportunity; but they do not pierce the corolla base themselves. Koeman-Kwak (1973) describes a similar difference between short- and long-tongued bumblebees in *Pedicularis palustris.*

Bartsia alpina L.

(Scrophulariaceae, Europe, North America)

Protogynous. Melittophilous. Late autogamy. Gullet blossom

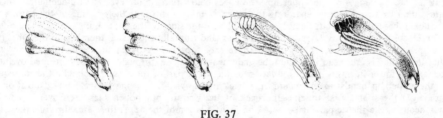

FIG. 37

Bartsia alpina follows the general pattern of the gullet flower with some deviations. Flower size is rather variable, possibly there are some clones (?) with larger, predominantly entomophilous flowers, and others with smaller, predominantly autogamous flowers.

The sombre, dirty violet-coloured spike (bracts and flowers) is not very conspicuous to the human eye. Contingent ultraviolet radiation has not been tested, but the effect in black-and-white photographs may indicate an ability to reflect ultraviolet radiation. However, recorded visits (by bumblebees) are few.

In relation to the general syndrome of melittophily the *B. alpina* flower is anomalous in its lack of a landing-place. The lower lip is very short and its lobes turned in. Bumblebees land frontally and have no difficulty in bending aside the lobes and push their heads into the flower. The longitudinal folds of the corolla permit dilation, especially of the upper part. The dilation of the corolla leads, as usual, to the separation of the anthers and consequent falling out of pollen. One interesting point is the occurrence of spines on the filaments. They guide the visitor and confine it to the lower part of the corolla.

In the smaller flowers the stigmas hardly ever project outside the corolla, and stretching of the latter during anthesis very soon pushes the anthers out to the position of the stigma. The occurrence of matted hairs may to a certain extent counteract autogamy even in such types, but there is hardly any doubt that in the end autogamy will result.

Whether large flowers are ever autogamous and the small flowers always so are questions that still have to be settled.

Figure 37 shows from left to right a large and a small flower in the female stage. Further the same flowers in section (late anthesis for the small one).

Pedicularis oederi Vahl

(Scrophulariaceae. Europe, Asia)

Protogynous (or Homogamous?). Melittophilous. Gullet blossom

P. oederi represents a primitive type within this genus the flowers of which show a series of striking pollination adaptations (cf. Li, 1948—9) and have been studied by Sprague and, more recently and thoroughly, by Macior (cf. references).

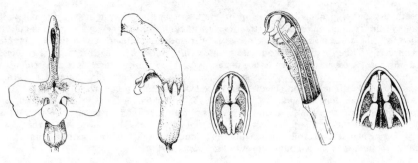

FIG. 38

Flower yellow, top of upper lip dark red inside (staining through), 15–20 mm long, erect, in a narrow spike. Upper lip bent slightly forward, lower one about 45 degrees downward. Upper lip compressed with a strong rib running on each side from the middle of the side of the corolla tube in a bend towards the middle of the free edges of the upper lip. The edges below that rib are everted and echinate. Lower lip lobate with two strong ridges separating the lobes, running in towards the opening of the corolla tube. The lateral lobes are firmly, but flexibly connected with the sides of the upper lip.

The blossom is visited by large bumblebees that land frontally and attach their legs to the lobes of the lower lip. Small wounds show where visitors have got a foothold. Avoiding the echinate edges, the visitor pushes its proboscis and head into the tube underneath them, i.e. through the passage formed by the two ridges on the lower lip. This causes a dilation of the upper part of the corolla tube with the consequence that the upper lip bends down towards the back of the visitors. Owing to its exposed position (and protogyny?) the stigma first touches the back of the visitors, but pollen sifts itself over it practically simultaneously, the thecae parting because of a slight dilation of the upper lip. After the visit, the dilation goes back and the upper lip returns to its former position, ready to perform again at the next visit.

Whereas this mechanism can only be operated by large bumblebees, smaller ones collect pollen by forcing apart the edges of the upper lip, causing the dry pollen to fall down on their abdomen – they work in an inverted position. This behaviour to some extent resembles the work necessary to get at the nectar of *P. palustris* and the rostrate species.

The figures show, from left to right, a flower from the front and the side, longitudinal section of the upper lip, flanked by figures showing (left) the usual, closed position of the pollen containers, and (right) the position when the anthers are forced apart.

Pedicularis silvatica L. (Fig. 39)

(*Scrophulariaceae. Europe*)

The flower in Fig. 39 shows very faintly the tendency towards obliqueness so prominent in the more evolved species. The pink flowers are visited by bumblebees, and the mechanism of pollination – as far as known – is the same as in *P. oederi*.

Pedicularis palustris L. (Fig. 40)

(*Scrophulariaceae. Europe*)

Protogynous (or Homoganous?). Melittophilous. Gullet blossom

Flowers pink, 10–20 mm long. Differs from *P. oederi* in the following: Blossoms ± horizontal, angle between upper and lower lip narrow. Upper lip slightly bent and twisted, the edges near the tip with two filiform teeth supporting the projecting style. The echinate edges on both sides end in a prominent strong tooth. Plane of lower lip twisted about 45 degrees, turning up the morphological right-hand side. Lobes of the lower lip less separate, ridges less prominent.

Owing to the twist of the lower lip, the entrance to the interior of the flower is easier from the left-hand side. This is accentuated by the way in which the lips are connected; on the right-hand side their edges are rolled into each other for some distance, and cannot be separated except by force. The left-hand edges are free, but reinforced and echinate. Also, the nectarium is oblique.

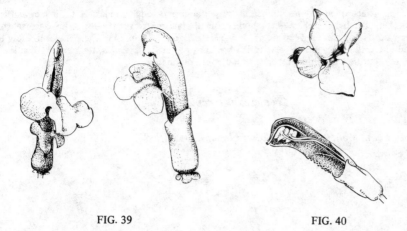

FIG. 39   FIG. 40

Visitors land laterally and approach the blossom from the left-hand side. If they come from the right-hand side (e.g. crawling from another flower) they move into the correct position first. By means of its mandibles, the visitor forcibly opens the upper lip above the echinate edges, just above the projecting strong tooth. It then enters the head there, staying above the echinate parts of the edges. As in *P. oederi*, the opening dilates with the same consequences for pollination, but in this case, the twist and asymmetry of the flower is temporarily suspended because of the dilation. After the visit the flower resumes its old shape.

Pedicularis lapponica L. (Fig. 41)
(*Scrophulariaceae. Circumpolar*)

P. lapponica is similar to *P. palustris*, except that here the lower lip (rather similar to that of *P. oederi*) is twisted 90 degrees.

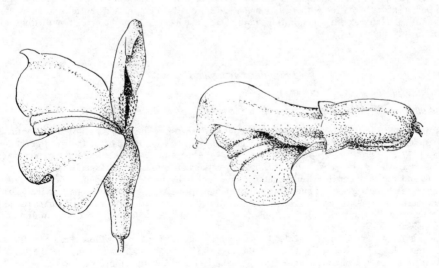

FIG. 41

The treatment accorded to these blossoms by pollinators is rough and leaves its distinct marks on the petals, especially the light ones of *P. lapponica*; small claw marks on the lower lip, against which the visitor stemmed its legs during dilation, and a bigger mark on the upper lip where it was forcibly everted. After some visits, the blossoms, especially the weaker ones of *P. lapponica*, look rather dilapidated and function badly, but by that time they have already been pollinated.

Pedicularis racemosa Dougl.

(Scrophulariaceae. Northwestern U.S.A.)

Protogynous (?). Melittophilous. Gullet blossom, modified

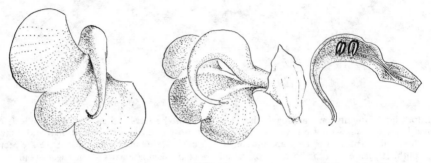

FIG. 42

In its general layout the flower resembles that of *P. palustris* and *P. lapponica*, but differs in the long-rostrate beak of the galea. The rostrum is a channel that supports the style only: the anthers are situated inside (proximal to) the rostrum, and the close juxtaposition of stigma and anthers characteristic of the other species is not found in this one.

The pollination of this species has been described by Macior (1970). According to the latter the flower is nectarless and is visited by pollen collectors only. Bumblebee workers approach the corolla with the antennae extended, grasp the corolla with front and middle legs, and grasp the side of the galea with their mandibles. They rest on their side against the broad lower lip of the corolla and hang in an inverted position while gathering pollen by vibrating the wings. The coiled rostrum of the flower, together with the enclosed style and protruding stigma, contacts the ventral side of the head and anterior thoracic region, while pollen shaken from the anthers within the galea drops to the upturned ventral side of the insects. Foragers occasionally groom pollen into the corbiculae while hanging from the corolla by their mandibles. (From Macior, abbreviated.)

Pedicularis groenlandica Retz.

(Scrophulariaceae. North America)

Homogamous (?). Mellittophilous. Pollen blossom

In this species, the galea is not only rostrate, as in *P. racemosa*, but has a proboscislike continuation which at full anthesis forms one spiral turn, chiefly located in the outer part of the rostrum (cf. Macior 1968a, 1977). As in *P. racemosa*, only the style penetrates the rostrum, the anthers remaining in a bulbous basal part of the galea. Pollinating insects, workers of various *Bombus* species, land on top of the rostrum, and attach themselves to the lateral lower petals by means of their anterior legs. With their mandibles the insects grasp the median ridge of the anterior face of the bulbous part of the galea. The rostrum partially supports the weight of the pollinators (through their middle legs) and curves round their bodies between thorax and abdomen, orientating the stigma "precisely to the center of the anterior abdominal surface".

There is no nectar; pollen is collected by wing vibrations, causing a small yellow cloud of pollen to fall out of the galea and envelop the body of the insect. The wing-beat used in vibrating the blossom is different from that of flight. Pollen adhering to the insect body is then groomed off and transferred to the corbiculae. Residual pollen on the anterior abdominal surface is deposited on the next stigma.

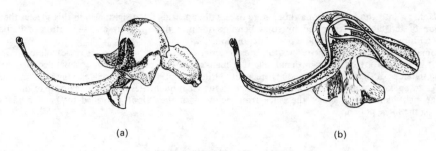

(a)   (b)

FIG. 43

Obviously, this very precise mechanism for its proper working depends on the correct size relations between pollinator and blossom. The head-thorax length of the bumblebee workers in question, 7–8 mm, fit the size of the blossom: 8 mm between the front of the bulbous part of the galea and the loop of the rostrum. The much bigger queens are not adapted to this blossom. Compared with the species mentioned previously, *P. groenlandica* is characterized by (1) nectarless pollen blossoms. In the development of the pollination syndrome in the genus towards greater perfection and exactness, there is reversion to a more original attractant; (2) there is also reversion to sternotribic pollination (if that is more primitive than nototribic), but in contrast to the likewise sternotribic pollination of *P. racemosa*, the stigma in this case passes across the back of the pollinator before it ends on the ventral side, a kind of super-nototriby.

Pedicularis sceptrum-carolinum L.

(Scrophulariaceae. Northern Europe)

Protogynous. Melittophilous. Closed blossom

Blossom erect, yellow with red lobes of the lower lip, about 30 mm long.

The blossoms show the same general features as those of *P. oederi* and *P. palustris*, modified by being permanently closed. *P. capitata* forms an intermediate (Macior 1975), which is still open also to bumblebee workers. The upper lip is wider than in the two other species, and the pollen falls out of the anthers before any visit has taken place. As the edges of the upper lip are bent inwards the pollen is prevented from falling down into the lower lip or the corolla tube. No edges are echinate. The free lobes of the lower lip are very short and bent inwards, embracing the tip of the upper lip and including also the stigma that protrudes slightly from the latter. The twist of the lower lip, easily observed in *P. palustris*, exists also in *P. sceptrum-carolinum*, and is found if one tries to open the blossom from the outside; on the right-hand side, the edges of the two lips are rolled together in such a manner as to be virtually inseparable without destruction of the blossom. The left-hand side can be opened, and such an opened blossom resembles that of *P. palustris*.

Only very strong and "intelligent" bumblebees can operate this blossom. They land on the top of the erect blossom, approach from the left-hand side, and force back the lower lips, stemming their legs against the outside of the flower in a definite manner. Having opened the blossom, the insect enters sideways and

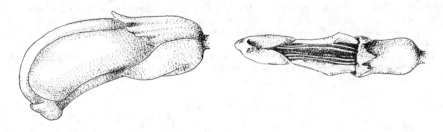

FIG. 44

crawls down into the wide corolla tube, more or less disappearing from sight; also in this species there is a dilation of the blossom followed by some bending down of the upper lip and separation of its bent-in edges with resulting pollination and pollen deposition.

Sometimes, a less "intelligent" visitor tries to enter from the wrong side. It is ridiculous to observe how it is popped out by the right-hand side if it manages to insert its head a very short distance. In spite of the inherent difficulties, the blossoms are avidly visited by bumblebees strong enough and "intelligent" enough to operate them. This is undoubtedly due to the great quantities of nectar available in them.

The right-hand figure shows the galea from below with the "floor", formed by bent-in edges and carrying the loose pollen.

Pedicularis lanceolata Michx.

(Scophulariaceae. North America)

Protandrous. Pollen blossom. Melittophilous. Closed blossom

(a)

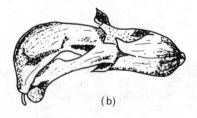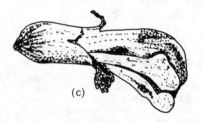

(b)   (c)

FIG. 45

This blossom, the pollination of which has also been described by Macior (1969), is a pollen blossom, like that of *P. groenlandica*, but it is not rostrate, belonging to a different section (cf. Li, 1948–9). Like its closer relative *P. sceptrum-carolinum*, it is closed, but in contrast to the nectar-containing completely closed blossoms of that species, those of *P. lanceolata* are open towards the base. In this respect the blossom forms an interesting parallel to that of *Coronilla emerus* (p. 181), which is distinguished from its nectariferous relatives by the open blossom. However, the sympetalous corolla in *Pedicularis* cannot open up as easily as the choripetalous one of Papilionaceae, and the effect in *P. lanceolata* is mainly achieved by shortening the corolla tube.

Pollinators, bumblebee workers, alight on top of the galea, turn down to the under side of the corolla, and align themselves with the main axis of the flower, head towards base. In this position they grasp the free edge of the galea with their mandibles and its upper side with the middle legs. Pollen is scraped out with the front legs. There is no wing vibration.

Again, there is sternotribic pollination with the stigma making contact with remaining pollen masses in more or less inaccessible parts of the ventral side of the insect.

Salvia patens Cav.

(Labiatae. Mexico)

Protandrous – homogamous. Melittophilous. Gullet blossom

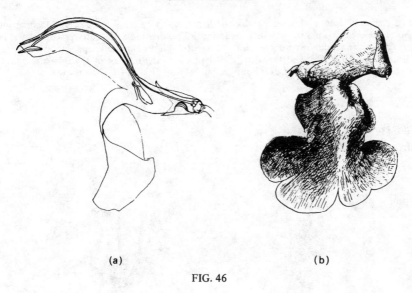

(a)   (b)

FIG. 46

The flowers are dark blue and form an open spike, corolla length 25–30 mm. They are very similar to those of *Galeopsis speciosa*, but differ in the following respects: the position of the flower is more horizontal, the corolla tube is wider and shorter, bell-shaped, and the number of stamens reduced to two. However, the main difference lies in the morphology of the stamens. Filaments are adnate to the corolla, except for a short, stiff end that projects more or less perpendicularly to the general direction of the flower. Anther and filament connected by a link that permits torsions of about 180°. Connective is 15–18 mm long, with very unequal arms. The longer one points upwards and carries a fertile theca, the position of which corresponds to that of the anthers in *Galeopsis*. The other arm is very short and carries a sterile, spoon-like theca at its lower end, position perpendicular to that of the flower. The two connate, sterile lower thecae between them close, more or less completely, the entrance to the corolla tube.

The nectary is situated at the base of the ovary. Immediately above it, the corolla is constricted by a bulge in the lower part of the tube.

Bumblebees land on the very large lower lip and push towards the nectary in the bottom of the corolla tube. To reach it, they must push back the valve formed by the two connate sterile thecae obstructing the path. This activates a lever mechanism, causing the fertile thecae to dip down and hit the back of the visitor, leaving the pollen there. The stigma is independent of this mechanism; it reaches out of the flower and at any rate during the later part of anthesis dips so far down as to brush the back of the visitor even before the anther dips down.

This general type of pollination mechanism is found in many *Salvia* species, mostly with blue flowers. Generally there are reduced flowers as well as the normal ones; smaller and/or unisexual (female with more or less reduced androecium).

Salvia splendens Sellow

(*Labiatae.Brazil*)

Protandrous – homogamous. Ornithophilous. Tube flower

FIG. 47

With its fiery red colour of both corolla and calyx, and with its absence of any landing space, this flower shows the ornithophilous, or rather hummingbird pollination syndrome very distinctly. For a full evaluation this flower should be compared with the melittophilous *S. patens* type.

The lever mechanism is out of function, and the path to the interior of the flower is unobstructed. Pollen deposition is inevitable when a bird puts its bill into the flower. Cf. Trelease 1881.

Mentha species

(*Labiatae. Europe*)

Protandrous. Myophilous. Brush blossom

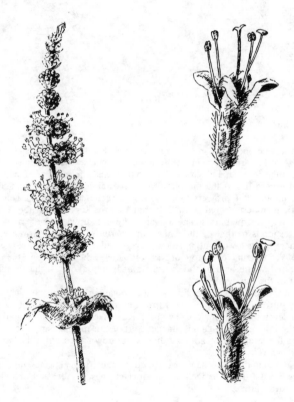

FIG. 48

Together with related genera, *Mentha* represents the brush blossom type within the Labiatae. The simpler flowers of these genera must have developed from the type as represented in *Galeopsis* by reduction. They still show traces of zygomorphy, and the reduction of the fifth stamen, functional in the gullet flower, is maintained in this pseudo-actinomorphic one. The long-exserted stamens and stigma belong to the syndrome of the brush blossom.

In addition to hermaphroditic flowers there are also more or less exclusively female ones, sometimes on separate stocks. They are smaller than the hermaphroditic type (2–3 against 3–5 mm corolla tube length) and are said to be visited after them.

Visitors are chiefly coleopters and, especially, dipters, even if the ample nectar secretion at the base of the ovary may occasionally also attract higher grade pollinators. In *Thymus*, with a similar flower, bees are very active.

Coleus frederici Taylor

(Labiatae. Angola)

Flag (reversed gullet) blossom

Little is known ..oout the natural pollinators of *Coleus* species. The very intricate blossom mechanism certainly suggests visits by eutropic insects or in some instances by birds. However, the flower depicted here is hardly ornithophilous. No bird could bend its beak through the tube of this flower. The blue colour would also suggest other classes of pollinators.

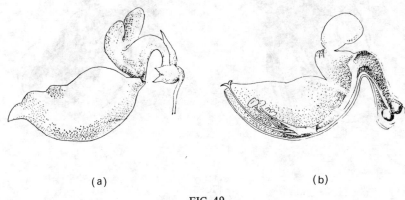

(a)   (b)

FIG. 49

C. frederici is quoted here as an example of a reversed gullet (functional flag) blossom. Whereas reversion in *Centrosema* is due to resupination, in *Coleus* it is a result of transference of stamens and style from the upper to the lower part of the flower.

The general pattern of the *Labiatae-Scophulariaceae* flower is greatly changed. The two lobes usually forming the galea are small and separated. They project flatly above the rest of the blossom and form a kind of visual background. The lateral lobes of the lower lip are almost rudimentary, whereas the central one forms a great, laterally compressed pouch down into which the sexual organs project. This lobe is hinged to the rest of the flower. The corolla tube is comparatively stiff and sharply bent.

The stamens are adnate to the corolla up to the level of the hinge of the lower lip. From there and half-way further up they are connate and form a stiff channel enclosing the style. This latter structure, which is unlike anything formed in an ordinary gullet blossom, gives the necessary stiffness to the sexual parts. Therefore, anthers and style remain in place when the pouch-like median lobe bends down under the visitor. The bending takes place in the hinge at the base of the median lobe.

In (b) corolla and filament tube have been sectioned almost to the top.

17.9. POLLINATION SYNDROMES IN ORCHIDS

The study of pollination of orchids is almost a science of its own, and it is impossible here to give any idea of the infinite variation present. We have, instead, presented the three main types to show the basic pattern of functions. For details we must refer to van der Pijl and Dodson (1966).

Cypripedium calceolus L.

(Northern hemisphere)

Homogamous. Semi-trap blossom. Attraction by deceit

In this species, five tepals are so small and dull-coloured as almost to be of no visual importance in the pollination syndrome. In contrast, the labellum is large, shining, yellow, and sack-like. In other species, the labellum is often less obtrusive and the other tepals more conspicuous. Odour production is localized in lateral tepals (Stoutamire 1967).

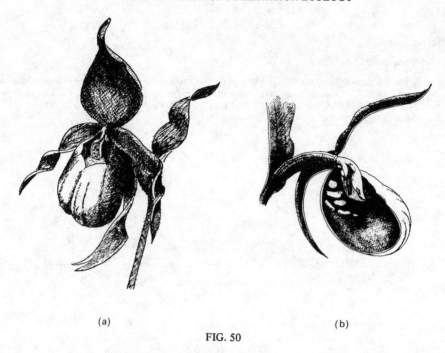

(a)   (b)

FIG. 50

The centripetal part of the labellum narrows to a channel over which the gynostemium forms a lid, topped by a shield-like staminode that hides most of the operative parts. The stigma forms another shield-like structure with the receptive surface turned downwards. The two functional anthers open one on each side of the basal part of the stigma. The outer part of the channel is covered, but the gynostemium contracts centripetally, leaving an exit hole on each side at the base.

The attractant of the *Cypripedium* flower has been under discussion for a long time. No food seems to be present. The gnawing of hairs reported may perhaps — if observations are correct — be interpreted as "emergency reactions" of trapped insects. According to Daumann (1968), the attraction is by deceit (odour), and the visitors, small solitary bees of the genus *Andrena*, slide into the semi-trap labellum, from which there is only an exit along the gynostemium. The pollination mechanism in other species seems to be similar, cf. the fungus mimesis in *C. debile* (Stoutamire *loc. cit.*).

The small space in the labellum prevents flying, and according to Daumann (*loc. cit.*) the inner surfaces are so slippery that they cannot be negotiated. So, at any rate in this species, there would be no necessity for the turned-in rim of the labellum to function as a barrier, making it impossible for any insect to crawl out of the flower that way. Accordingly, the visitor must squeeze out through the channels underneath the gynostemium. In so doing, it cannot help making dorsal contact first with the stigma and then with one of the anthers. There is thus a functional protogyny. The pollen is loose, but viscid, and smears over the back of the visitor.

The *C. calceolus* labellum is so strongly bent that an insect can hardly, from a position in the bottom of the flower, see the exits. But in the bend there are several light windows towards which it will be attracted, so discovering the actual exits later. In Cypripediinae with a straight labellum (*Paphiopedilum*) the exits are directly visible from the bottom of the labellum, and there are no windows. Within this group, the edges of the labellum are straight or even flanged outwards.

Orchis maculata and related species

(Europe)

Homogamous. Melittophilous. Open gullet flower. No known attractant

The operative parts of the flower are the labellum and the gynostemium. The five other tepals form part of the attractive apparatus only. Visitors — generally bees, sometimes flies — land on the labellum, their axis parallel to that of the flower, and orient themselves towards the centre of the flower, the stigmatic cavity, and the entrance to the spur at the base of the labellum.

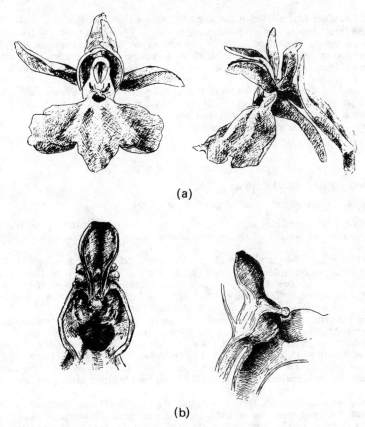

(a)

(b)

FIG. 51

The one remaining anther forms the top of the gynostemium. It has a comparatively broad connective; the two thecae open by slits, their contents forming two coherent masses, pollinia. The pollinium continues in a thin stalk, the caudicle, leading down toward the rostellum.

Of all organs in the orchid flower, the rostellum is the most controversial one, and no opinion is postulated about its morphological origin. It has been interpreted both as a transformed stigmatic branch, as transformed stamens, and as an organ that has arisen *de novo*. As the stigma is in many orchids tripartite this suggests that there all three stigmatic branches still have retained their function.

Operationally, the rostellum is a projecting knob, the interior cells of which have partly degenerated and form two viscid balls, surrounded by a viscid fluid. The caudicles adhere to the epidermis of the upper surface of the rostellum. Rupturing very easily it leaves two small discs of epidermis attached to the caudicles. The viscid balls are attached to the under side of the discs.

On being touched by an insect penetrating into the flower (this can be simulated by a sharp pencil or any other suitable object) the epidermis of the rostellum ruptures as described above, and the remainder bends back exposing the viscid balls which immediately adhere to the surface of the intruder. The adhesive qualities of the viscid substances are remarkable; they cement the discs to the animal which, when withdrawing, pulls the whole pollinarium (viscidium, caudicle, and pollinium) out of the anther.

Originally, the pollinaria project straight up and would in this position hit the anther of the next flower visited. By asymmetric drying out of its cells the caudicle bends forwards in half a minute's time and place the pollinia in such a position that they hit the stigmatic cavity of the next flower. The surface of the stigma is very sticky, and a major part, sometimes all, of the pollinium adheres, leaving only caudicles and discs as witnesses of a successful pollination activity on the part of the insect.

The pouch-like lower part of the rostellum switches back immediately there is no more contact with the intruding object, thus protecting the viscidium of the remaining pollinarium should – as often happens – only one of them be removed by the first visitor.

Within the whole of the Basitonae there is little variation on this pattern, chiefly dealing with the mutual distance between the viscidia and their fusion into one – corresponding to the lateral distances between the two sides of the insect organ to which pollinaria adhere, from thin probosces to fat body parts, and also from a top to a lateral position and further to a position on the underside of the proboscis as in the non-resupinate flowers of *Nigritella*. Attraction is by nectar in most genera, sexual in *Ophrys*. Nectar is collected in a spur, the length of which varies according to the preferred pollinator. In *Ophrys*, there is no spur. *Orchis* is another exception; its attraction is unknown; there is no nectar in the spur of many species (Daumann 1941, 1971).

In many ways the genus *Orchis* and its closest relatives are unstable. It is doubtful if bumblebees are the principal pollinators within this group. Butterflies and moths are known from various species, even within the genus *Orchis* itself (*O. ustulata*), and in the Faeroes, where the genus *Bombus* plays no rôle, Hagerup has shown that flies (*Eristalis*) pollinate orchids.

Cattleya species

(*Tropical South and Central America*)
Homogamous. Melittophilous. Nectar attraction

Cattleya is a good representative of the group of acrotonic orchids which differ from the basitonic ones (e.g. *Orchis*) in the position of the anther and in details of the mechanism of the withdrawal of the pollinium.

As in *Orchis* the operative parts of the flower are labellum and gynostemium, which in this case form a gullet-like blossom inside the flower. The other tepals, some of them very conspicuous, form attraction organs only. Large, semi-social bees are the adapted pollinators for the blossom type illustrated here (some *Cattleya* spp. deviate). They land on the labellum and orient themselves, again, towards the centre of the flower. Nectar is present in a nectary embedded in the ovary.

Instead of the orthotropic position occupied by the anther in *Orchis*, it is bent in this case, with its main axis perpendicular to the axis of the gynostemium. Its dorsal half is more or less immersed in the tissue of the latter (the morphological value of which we leave aside). Caudicles are present and project slightly out of the anther, but are not attached to anything. There are no slits in the front of the thecae.

The rostellum forms a broad lip, the upper part of which is formed by a firm tissue with intact epidermis, whereas the lower part has degenerated into a viscid matter hanging freely exposed under the rostellum.

On entering, an insect will not come into contact with this viscid mass, as the upper, firm part of the rostellum bends back and covers up, protecting the viscid substance. However, on withdrawing, the insect will first bend this firm part of the rostellum the other way, so exposing its lower part "and a surprising quantity of viscid matter is forced over the edges and sides, and at the same time into the tip of the anther" (Darwin). The effect is to glue the caudicles to the insect. When the pollinia are then pulled out of the anther, the upper part of it must be lifted, like a lid, and in many species it breaks off.

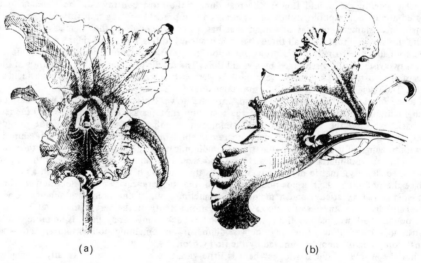

(a)   (b)

FIG. 52

The further steps, drying-out of the caudicles, the position of pollinia in relation to the stigmatic cavity, etc., are in principle like the corresponding features in *Orchis* pollination.

The most important operational differences in comparison with *Orchis* are: (1) the lid effect of the anther, and (2) the fact that the mechanism is operated on the return journey out of the flower. If the entering insect already had pollinia glued to its back, these would have been deposited in the stigmatic cavity before the anther was removed. The total effect must be one of functional protogyny, which tends to safeguard allogamy.

Within the acrotonic orchids the variations chiefly concern the number, form, and cohesiveness of pollinia, and the form and attachment of caudiculae.

In *Listera ovata* there are no caudiculae, and pollinia are cemented directly to the insect. The viscid matter is, in this blossom, sealed up, and ejected from the rostellum when touched. It then dries in a few seconds. Because of the lack of caudiculae, pollinia cannot carry through the movements necessary to come into position for hitting the stigmatic cavity of the next flower. Instead, the rostellum moves out of the way in a day's time, leaving free the passage to the stigma.

More important is the introduction of an intermediate link, inasmuch as the caudicles in many genera attach themselves permanently to the epidermis of the upper side of the rostellum. Together with the underlying tissue this then forms a so-called pedicel or stipe, which separates from the rest of the rostellum when the mechanism is activated by a withdrawing pollinator. The viscid substance present under or at the tip of the rostellum cements the pedicel to the insect, which then pulls the pollinia out of the thecae. The – highly variable – pedicel assumes the same function as the caudicles in *Orchis*; among orchids with pedicels the caudicles are very short or rudimentary. Tensions in the pedicel form the basis of the explosive mechanism in Catasetinae by which, on being triggered, the pollinarium is flung through the air and lands with great precision on the visitor. These species are, partly, diclinous.

The specimen used for the illustration is a cultivar (cv. Tityus). In the longitudinal section the pollinium is drawn black. There is a distinct stylar canal leading from the stigmatic cavity.

Most interesting is the observation of Dodson's that because of lack of balance between odour, dimension, and stickiness the flowers of some hybrids in nature become veritable death-traps, killing the visiting bees.

Nigritella nigra L.

(Europe)

Psychophilous. Brush blossom

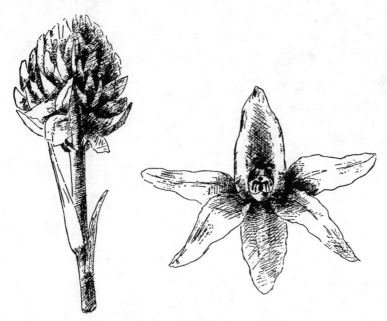

FIG. 53

Nigritella represents the brush blossom type among orchids. Most of the usual differentiation between the members of the orchid perianth have been lost, and the labellum is not appreciably different from the other tepals. The resupination of the flower — meaningless in a brush blossom — has also disappeared.

There is still a short, nectariferous spur on the labellum. Although it is but *ca*. 2 mm long, the nectar can only be utilized by butterflies because the opening is too narrow for the coarser probosces of other insects. Many butterflies have been recorded as visitors in spite of the dark, purple-brownish colour which is not one typical for the syndrome of psychophily.

The pollination principle is similar to the classical one in *Orchis* except that because of the position of the flower, pollinia attach themselves to the lower side of the proboscis of the visitor, and the movements of the caudiculae — to bring pollinia in position to hit the stigma — differ accordingly.

CHAPTER 18

EPILOGUE

Pollination ecology provides examples of some of the most precise, most intricate, and most amazing adaptations in nature. They are there, whatever the explanation may be for their coming into existence. Sprengel ascribed it to the Lord's wisdom; today's scientists tend to prefer other explanations. But the facts are the same, both those we know and the immense number yet to be discovered from observations to be made in those regions of the world where pollination has not been adequately studied.

Nature has not always used the same means to reach the same goal. Each taxon is, at any one time, limited by a certain morphogenetic potentiality, outside of which it cannot go without losing its identity (Lamprecht 1959). The self-incompatibility or strong inbreeding depression of old crop plants like rye and maize shows how strong these limits can be. Selection pressure will have to work within the limit of this morphogenetic potentiality in order to produce a certain morphological and functional result.

In pollination ecology, as in all other phylogenetic evaluations one can distinguish two tendencies which require explanation — an explanation that, so far, has not been given in adequate terms anywhere. The first is what we may call the orthogenetic tendency; the tendency for a given development once started to carry through to the end, towards an over-specialization so delicately balanced that very small maladjustments will be sufficient to throw it out of balance, and place these populations in a very unfortunate position with regard to selection. No doubt, pollination has produced many "sabretoothed tigers", which in their over-perfection have suffered the fate of that notorious animal. And one gets the impression that similar cases are on the point of occurring under our eyes today: *Angraecum, Yucca, Ophrys.*

The other tendency or problem is but another aspect of the same thing: the relative merits of the specialized versus the generalized, primitive types. This problem exists in pollination ecology as everywhere: for example why is *Lepidodendron* extinct while *Lycopodium* is still extant? We cannot answer or discuss this problem here. Our reason for mentioning it is that many of the objections that have been and are still being raised against the interpretation of adaptive mechanisms in pollination ecology fail to recognize the general validity of such problems. In terms of number of visits the specialized blossoms are definitely not favoured. One may wait for hours or days to see a visitor to an orchid while neighbouring allotropous blossoms may teem with insects. When there is, nevertheless, a tendency, almost an orthogenesis, towards greater specialization, the only reason must be that the few visits which do take place in the specialized blossoms, are in some way much more valuable. We may form an idea of some reasons for that, especially as a result of speciation processes under the influence of pollination ecology, but we may still be far from realizing all aspects of the problem.

No pollinator is exempt from errors; even animals that, like bumblebees, are during their whole life dependent on blossoms and their products, may die from poisonous nectar. No pollination mechanism is infallible; nectar thieves can deprive the most refined blossoms of their attractants, or external circumstances, ranging from bad weather to man exterminating the pollinator, may deprive them of all their pollinating agents. These, and a hundred other things, may happen. The "wrong" pollinator may visit a blossom, sometimes causing pollination, like the syrphids in *Lonicera periclymenum*, sometimes not, like moths introducing their probosces in *Linaria vulgaris* without working the mechanism. What Baker *et al.* (1971) call minor pollinators, who come in after the main party to partake of the left-overs, may be important pollinators (cf. Beattie 1969). In pollination ecology too many interpretations seem to imply that visitors are machines, always reacting in the same manner on the same impulse. Even if insects are chiefly led by their instincts, one can easily generalize too much about their behaviour. What they do today and in a particular place is not necessarily the same as what they will do tomorrow and in another place. But this fact does not invalidate the observations of pollination ecologists through almost 200 years, and it does not prevent the adaptations observable in pollination ecology from being some of the most precise and marvellous external adaptations in the living world. Nor do they detract from the fascination of studying them, a fascination which led professional writers like Hyman, or even poets like Maeterlinck, to write about pollination with a poetic vigour which provides, perhaps, compensation for their lack of factual information.

REFERENCES

Titles of Russian-language papers are given in translation. Authors' names and titles of journals are transliterated.

Adlerz, W. C. 1966 Honeybee visits and watermelon pollination. – J. econ. entomol. 50: 28–30.

Agthe, C. 1951 Über die physiologische Herkunft des Pflanzennektars. – Ber. schweiz. bot. Ges. 61: 240–247.

Alcorn, S. M., S. E. McGregor, G. D. Butler jr., E. B. Kurtz jr. 1959 Pollination requirements of the saguaro cactus (*Carnegiea gigantea*). – Cact. succ. J. Am. 31, 2: 134–138.

Alcorn, S. M., S. E. McGregor, G. Olin 1961 Pollination of saguaro cactus by doves, nectar-feeding bats, and honeybees. – Science 133: 1594–1595.

Alpatov, V. V. 1946 The mutual help of insects and entomophilous plants as an example of interspecific symbiosis. – Zool. zh. 25: 325–328.

Amadon, D. 1947 Evolution in Hawaiian birds. – Evolution 1: 63–68.

Amoore, J. E., M. Rubin, D. Apotheker, R. Lutmer, J. Johnston, A. Sandval, H. Saunders 1962 The stereochemical theory of olfaction. – Proc. sci. sect. Spec. suppl. 37. Toilet goods assos. 12 pp.

Anderson, E. 1939 Recombination in species crosses. – Genetics 24: 668–698.

Arnell, H. W. 1903 Om dominerande blomningsföreteelser i Södra Sverige. – Ark. bot. 1: 287–376.

Arroyo, M. T. and P. H. Raven 1975 The evolution of subdioecy in morphologically gynodioecious species of Fuchsia sect. Eucliandra (Onagraceae). – Evolution 29: 500–511.

Ascher, P. D. and J. S. Peloquin 1966 Effect of floral ageing on the growth of compatible and incompatible pollen tubes in *Lilium longiflorum*. – Am. J. bot. 53: 99–102.

Ash, J. S., P. Hope Jones, R. Melville 1961 The contamination of birds with pollen and other substances. – Br. birds 54: 93–100.

Ashton, P. S. 1969 Speciation among tropical forest trees: some deductions in the light of recent evidence. – Biol. J. Linn. soc. 1: 155–196.

Atsatt, P. R. and D. Strong 1970 The population biology of annual grassland hemiparasites. I: The host environment. II: Reproductive patterns in *Orthocarpus*. – Evolution 24: 278–291, 598–612.

Aufsess, A. von 1960 Geruchliche Nahorientierung der Biene bei entomophilen und ornithophilen Blüten. – Z. vergl. Physiol. 43: 469–498.

Ayensu, E. A. 1974 Plant and bat interaction in West Africa. – Ann. Mo. bot. gard. 61: 702–727.

Baird, A. M. 1938 A contribution to the life history of *Macrozamia reidlei*. – J. R. soc. W. Aust. 25: 153–175.

Baker, H. G. 1955 Self-compatibility and establishment after 'long-distance' dispersal. – Evolution 9: 347–348.

Baker, H. G. 1960 Reproductive methods as factors in speciation in flowering plants. – Cold Spring Harbor symp. quant. biol. 24: 177–191.

Baker, H. G. 1961a The adaptations of flowering plants to nocturnal and crepuscular pollinators. – Quart. rev. biol. 36: 64–73.

Baker, H. G. 1961b *Ficus* and *Blastophaga*. – Evolution. 15: 375–379.

Baker, H. G. 1962 Heterostyly in Connaraceae with special reference to *Byrsocarpus coccineus*. – Bot. gaz. 123: 206–221.

Baker, H. G. 1966 The evolution, functioning and breakdown of heteromorphic incompatibility systems. I: The Plumbaginaceae. – Evolution 20: 349–368.

Baker, H. G. 1975 Sugar concentrations in nectars from hummingbird flowers. – Biotropica 7: 37–41.

Baker, H. G. and I. Baker 1973 Amino acids in nectar and their evolutionary significance. – Nature 241: 543–545.

Baker, H. G. and I. Baker 1975 Studies of nectar-constitution and pollinator–plant coevolution. – In Gilbert and Raven 1975: 100–140.

Baker, H. G., I. Baker, P. A. Opler 1973 Stigmatic exudates and pollination. – In Brantjes and Linskens 1973: 47–60.

Baker, H. G., R. W. Cruden, I. Baker 1971 Minor parasitism in pollination biology and its community function: the case of Ceiba acuminata. – Bioscience 21: 1127–1129.

Baker, H. G. and B. J. Harris 1957 The pollination of Parkia by bats and its attendant evolutionary problems. – Evolution 11: 449–460.

Baker, H. G. and B. J. Harris 1959 Bat-pollination of the silk-cotton tree, Ceiba pentandra (L.) Gaertn. (sensu lato) in Ghana. – J. W. Afr. sci. ass. 5: 1–9.

Baker, H. G. and P. H. Hurd jr. 1968 Intrafloral ecology. – Ann. rev. entomol. 13: 385–414.

Barrett, S. A. H. 1977 Breeding systems in Eichhornea and Pontederia, tristylous genera of the Pontederiaceae. – Diss. abstr. 38 B: 3526.

Bateman, A. J. 1947 Contamination of seed crops. I: Insect pollination. – J. genet. 48: 257–276.

Bateman, A. J. 1951 The taxonomic discrimination of bees. – Heredity 5: 271–280.

Bateman, A. J. 1968 The role of heterostyly in Narcissus and Mirabilis – Evolution 22: 645–646.

Bawa, K. S. 1974 Breeding systems of tree species of a lowland tropical community. – Evolution 28: 85–92.

Beattie, A. J. 1969 The floral biology of three species of Viola. – New phytol. 68: 1187–1201.

Beattie, A. J. 1971 Itinerant pollination in a forest. – Madroño 21: 120–124. Quoted from Biol. Abstr. 56: 65670.

Beccari, O. 1877–8 Piante nuove o rare dell'arcipelago malese della Nuova Guinea. – In Beccari: Malesia 1 (2–3).

Beck von Managetta, G. 1914 Die Pollennachahmung in den Blüten der Orchideengattung Eria. – Sitzungsber. Akad. Wiss. Wien. Mat.–natw. Kl. 123: 1033–1046.

Benl, G. 1938 Die genetischen Grundlagen der Blütenfarben (Sammelreferat). – Z. indukt. Abstamm. Vererbungsl. 74: 242–329.

Bentley, B. 1976 Plants bearing extrafloral nectaries and the associated ant community: interhabitat difference in the reduction of herbivore damage. – Ecology 57: 815–820.

Bentley, B. L. 1977 The protective function of ants visiting the extrafloral nectaries of Bixa orellana (Bixaceae). – J. ecol. 65: 27–38.

Berg, R. Y. 1959 Seed dispersal, morphology and taxonomic position of Scoliopus, Liliaceae. – Skr. nor. vidensk.-akad. Oslo. I. Mat.-natw. Kl. 1959, 4: 56 pp.

Bhaskar, V. and B. A. Razi 1974 Nocturnal germination in Impatiens – Curr. sci. 43: 626–628.

Blagoveshchenskaya, N. N. 1970: The evolution of pollination in angiosperms in the light of the theory of mutually adaptive parallel evolution. – Bot. zh.: 1301–1303.

Blake, S. F. and A. Roff 1953–6 The honey flora of SE Queensland. – Queensl. agric. J. 76–82.

Bohart, G. E. 1971 Management of habitats for wild bees. – Proc. tall timber conf. ecol. anim. contr. habit. man. February 25–27, 1971: 253–266.

Bohart, G. E. and W. P. Nye 1960 Insect pollination of carrots in Utah. – Utah Agric. exp. stn. Utah state univ. Bull. 419. 16 pp.

Bohart, G. E., W. P. Nye, L. R. Hawthorn 1970 Onion pollination as affected by different levels of pollinator activity. – Utah agric. exp. stn. Utah state univ. Bull. 482: 57 pp.

Bohart, G. E. and Youssuf, N. N. 1972 Notes on the biology of Megachile (Megachiloides) umatillensis Mitchell (Hymenoptera: Megachilidae) and its parasites. – Trans. R. entomol. soc. London 124: 1–19.

Bohn, C. W. and G. N. Davis 1964 Insect pollination is necessary for the production of muskmelons (Cucumis melo v. reticulatus). – J. apicult. res. 3: 61–63.

Bonnier, G. 1879 Les nectaires. – Ann. sci. nat. Bot. 8: 1–212.

Bornmann, C. H. 1972 Welwitschia mirabilis: paradox of the Namib desert. – Endeavour 31: 95–98.

Brantjes, N. B. M. 1973 Sphingophilous flowers, function of their scent. – In Brantjes and Linskens 1973: 27–46.

Brantjes, N. B. M. 1976 Riddles around the pollination of Melandrium album (Mill.) Garcke (Caryophyllaceae) during the oviposition by Hadena bicruris Hufn. (Noctuidea, Lepidoptera). I–II. – Proc. K. ned. acad. wet. Ser. C. 79: 1–12, 127–141.

Brantjes, N. B. M. and J. A. A. M. Leeman 1976 Silene otites (Caryophyllaceae) pollinated by nocturnal Lepidoptera and mosquitoes. – Acta bot. neerl. 25: 281–295.

Brantjes, N. B. M. and H. F. Linskens (eds.) 1973 Pollination and dispersal. – Symp. Nijmegen univ. 14 December 1973 offered to L. v. d. Pijl. 125 pp.

Brewbaker, J. L. 1959 Biology of the angiosperm pollen grain. – Indian J. genet. plant breed. 19: 121–133.

Brian, A. D. 1957 Differences in the flowers visited by four species of bumblebees and their causes. – J. anim. ecol. 26: 71–98.

Brodie, J. L. 1951 The splash-cup mechanism in plants. – Can. J. bot. 29: 224–234.

Bruch, C. 1923 Coleopteros fertilizadores de "Prosopanche Burmeisteri" de Bary. − Physis, Rev. soc. argent. cienc. nat. 7: 82−88.

Bryhn, A. 1897 Beobachtungen über das Ausstreuen der Sporen bei den Splachnaceen. − Biol. Zentralbl. 17: 48−55.

Büdel, A. 1959 Das Mikroklima der Blüte in Bodennähe. − Z. Bienenforsch. 4: 131−140.

Burkhardt, J. 1964 Color discrimination in insects. − Adv. insect physiol. 2: 131−173.

Cady, L. R. and E. R. Rotherham 1970 Australian native orchids in colour. − Reed (Melbourne). 111 pp.

Cammerloher, H. 1923 Zur Biologie der Blüte von Aristolochia grandiflora Swartz. − Österr. bot. Z. 72: 180−198.

Cammerloher, H. 1929 Zur Kenntnis von Bau and Funktion extrafloraler Nektarien. − Biol. gen. 5: 281−302.

Cammerloher, H. 1931 Blütenbiologie. − Berlin (Borntraeger).

Campbell, C. G. and S. H. Nelson 1973 Staining embedded pollen tubes. − Can. J. plant sci. 53: 415.

Campbell, W. D. 1963 Warblers as pollinators in Britain. − Br. birds 56: 111−112.

Carauta, J. P. P. 1972 Dorstenia hirta Desvaux (Moraceae)-figuerilha; estudo de sua biologia floral. − At. soc. biol. Rio Janeiro 16: 7−11.

Carey, F. M., J. J. Lewis, J. L. MacGregor, M. Martin-Smith 1959 Pharmacological and chemical observations on some toxic nectars. − J. pharm. pharmacol. Suppl. II: 269T−274T.

Cautwell, G. E., H. Shimanuki, H. J. Retzer 1970 Containerized bees airdropped into cranberry bogs. − Am. bee J. 111: 272−273.

Cazier, M. A. and E. C. Linsley 1974 Foraging behavior of some bees and wasps at Kallstroemia grandiflora flowers in Southern Arizona and New Mexico−Am. mus. novit. 2546. 20 pp.

Chapman, P. D. 1966 Floral biology and the fruitfulness of Jamaica allspice (Pimenta dioica) − Bee world 47 Suppl: 125−130.

Chase, V. C. and P. H. Raven 1975 Evolutionary and ecological relationships between Aquilegia formosa and A. pubescens (Ranunculaceae), two perennial plants. − Evolution 29: 474−486.

Chatin, A. 1855 Mémoire sur le Vallisneria spiralis L. − Paris.

Cholodny, H. G. 1944 Pollination in Salvia glutinosa. − Bot. zh. 29: 108−113.

Churchill, D. M. and P. Christensen 1970 Observations on pollen harvesting by brush-tongued lorikeets. − Aust. J. zool. 18: 427−437.

Clausen, J. and W. M. Hisey 1960 The balance between coherence and variation in evolution. − Proc. nat. acad. sci. 46: 494−506.

Clements, F. E. and F. L. Long 1923 Experimental pollination. − Carnegie inst. Wash. Publ. 336.

Clifford, H. T. 1962 Insect pollinators of Plantago lanceolata L. − Nature 193: 196.

Cockrum, E. L. and B. J. Hayward 1962 Hummingbird bats. − Nat. hist. 71, 8: 38−43.

Coe, H. J. and F. M. Isaac 1965 Pollination of the Baobab (Adansonia digitata L.) by the lesser bush baby (Galego crassicaudatum E. Geoffroy). − E. Afr. Wildl. J. 3: 123−124.

Collins, F. C., V. Lenmargkol, J. P. Jones 1973 Pollen storage of certain agronomic species in liquid air. − Crop sci. 13: 493−494. Quoted from Biol. abstr. 57: 5892.

Colwell, R. K., B. J. Betts, P. Bunuel, F. L. Carpenter, P. Feinsinger 1974 Competition for the nectar of Centropogon valerii by the hummingbird Colibri thalassinus and the flower-piercer Diglossa plumbea, and its evolutionary implications. − Condor 76: 447−452.

Condit, I. J. 1947 The fig. − Chronica botanica (Waltham, Mass.). 222 pp.

Corner, E. J. H. 1964 The life of plants. − Weidenfeld and Nichols. 315 pp.

Correns, C. 1891 Zur Biologie und Anatomie der Salvienblüte. − Jahrb. wiss. Bot. 22: 190−237.

Correvon, H. and M. Pouyanne 1916 Un curieux cas de mimétisme chez les Ophrydées. − J. soc. nat. d'hortic. France 29: 23−84.

Cramer, J. M., A. D. J. Meeuse, P. A. Teunissen 1975 A note on the pollination of nocturnally flowering species of Nymphaea. − Acta bot. neerl. 24: 489−490.

Crepet, W. L. 1972 Investigation of North American cycadeoids: Pollination mechanisms in Cycadeoidea. − Am. J. bot. 59: 1048−1056.

Crepet, W. L., D. L. Dilcher, T. W. Potter 1974 Eocene angiosperm flowers. − Science 781−782.

Cruden, R. W. 1972a Pollination biology of Nemophila menziesii with comments on the evolution of oligolectic bees. − Evolution 26: 373−389.

Cruden, R. W. 1972b Pollination in high-elevation ecosystems: relative effectiveness of birds and bees. − Science 176: 1438−1440.

Cruden, R. W. 1976 Intraspecific variation in pollen−ovule ratios and nectar secretion. − Ann. Mo. bot. gard. 68: 247−289.

Cumber, R. A. 1953 Some aspects of the biology and ecology of bumblebees bearing upon the yields of red-clover seed in New Zealand. − N.Z. J. sci. technol. Ser. B 34: 217−240.

Dahl, E. and E. Hadać 1940 Maur som blomsterbestøvere. − Nytt mag. naturv. 81: 46−48.

Darwin, C. 1876 The effects of cross- and self-fertilization in the vegetable kingdom. − London. 482 pp.

Darwin, C. 1877 The different forms of flowers on plants of the same species. – London. 352 pp.
Darwin, C. 1890 The various contrivances by which orchids are fertilized by Insects. 2nd edn. – London. 300 pp. (1st edn. 1862).
Darwin, F. 1877 On the glandular bodies on *Acacia sphaerocephala* and *Cecropia peltata* serving as food for ants. With an appendix on the nectar-glands of the common brake fern, *Pteris aquilina*. – J. Linn. soc. Botany 15: 398–408.
Daumann, E. 1930a Nectarien und Bienenbesuch bei *Opuntia monacantha* Haw. – Biol. gen. 6: 353–375.
Daumann, E. 1930b Nektarabscheidung in der Blütenregion einiger Araceen. – Planta 12: 38–48.
Daumann, E. 1932 Über postflorale Nektarabscheidung. – Beih. bot. Zentralbl. 49, 1: 720–734.
Daumann, E. 1935 Die systematische Bedeutung des Blütennektariums der Gattung *Iris*. – Beih. bot. Zentralbl. 53B: 525–625.
Daumann, E. 1941 Die anbohrbaren Gewebe und rudimentären Nektarien in der Blütenregion. – Beih. bot. Zentralbl. 61A (1942): 12–82.
Daumann, E. 1960 On the pollination ecology of *Parnassia* flowers. A new contribution to the experimental flower ecology. – Biol. plant. 2: 113–125.
Daumann, E. 1963. Zur Frage nach dem Ursprung der Hydrogamie, zugleich ein Beitrag zue Blütenökologie von *Potamogeton*. – Preslia 35: 23–30.
Daumann, E. 1966 Pollenkitt, Bestäubungsart und Phylogenie. – Novit. bot. univ. carol. pragensis 1966: 19–28.
Daumann, E. 1967 Zur Blütenmorphologie und Bestäubungsökologie von *Veratrum album* ssp. *lobelianum* (Bernh.) Rchb. –Österr. bot. Z. 114: 134–148.
Daumann, E. 1968 Zur Bestäubungsökologie von *Cypripedium calceolus* L. – Österr. bot. Z. 115: 434–446.
Daumann, E. 1970a Zur Frage nach der Bestäubung durch Regen (Ombrogamie). – Preslia 42: 220–224.
Daumann, E. 1970b Zur Blütenökologie von *Digitalis*. – Preslia 42: 317–329.
Daumann, E. 1971 Zum Problem der Täuschblumen. Preslia 43: 304–317.
Daumer, K. 1958 Blumenfarben, wie sie die Bienen sehen. – Z. vergl. physiol. 41: 49–110.
Davis, P. H. and V. H. Heywood 1963 Principles of angiosperm taxonomy. – Oliver & Boyd (London). 556 pp.
Degener, O. 1945 Plants of Hawaii national park. – Degener (Oahu). 314 pp.
Delevoryas, T. 1968 Investigations of North American cycadeoids: structure, ontogeny, and phylogenetic consideration of cones of *Cycadeoidea*. – Palaeontographica 121 B: 122–133.
Delpino, F. 1868–75 Ulteriori osservazione sulla dicogamia nel regno vegetale. I–II. – Att. soc. ital. sci. nat. Milano 11–12.
Dement, W. A. and P. H. Raven 1974 Pigments responsible for ultraviolet patterns in flowers of *Oenothera* (Onagraceae). – Nature 225: 705–706.
Demianowicz, Z. 1964 Charackteristik der Einartenhonige. – Ann. Abeille 7: 273–288.
Dessart, P. 1961 Contribution a l'étude des Ceratopogonoidae (Diptera). – Bull. agric. Congo belge 52: 525–540.
Devreux, M. and C. Malingraux 1960 Pollen de *Elaeis guineesis*. – Bull. agric. Congo belge 51: 543–566.
Diels, L. 1916 Käferblumen bei den Ranales und ihre Bedeutung für die Phylogenese der Angiospermen. – Ber. dtsch. bot. Ges. 34: 758–774.
Docters van Leeuwen, W. M. 1929 Mierenepiphyten II. – De trop. nat. 18: 131–135; Ber. dtsch. bot. Ges. 47: 50–99.
Docters van Leeuwen, W. M. 1931 Vogelbesuch an den Blüten von einigen *Erythrina*-Arten auf Java. – Ann. jard. bot. Buitenz. 42: 57–95.
Docters van Leeuwen, W. M. 1933 Biology of plants and animals occurring in the higher parts of Mount Pangrango-Gedeh in West Java. – Verh. Ak. wet. Amst. afd. natuurk. 31: 1–278.
Docters van Leeuwen, W. M. 1954 On the biology of some Javanese Loranthaceae and the role birds play in their life-history. Beaufortia 4: 102–207.
Dodson, C. H. 1975 Coevolution of orchids and bees. – In Gilbert and Raven: 91–97.
Dodson, C. H., R. L. Dressler, H. G. Hills, R. M. Adams, N. H. Williams 1964 Biologically active components in orchid fragrances. – Science 164: 1243–1249.
Dodson, C. H. and G. P. Frymire 1961a Preliminary studies in the genus *Stanhopea*. – Ann. Mo. bot. gard. 48: 137–171.
Dodson, C. H. and G. P. Frymire 1961b Natural pollination of orchids. – Ann. Mo. bot. gard. 49: 133–139.
Dormer, K. J. 1960 The truth about pollination in *Arum*. – New phytol. 59: 209–281.
Dorr, J. and E. C. Martin 1966 Pollination studies on highbush blueberry *Vaccinium corymbosum*. – Mich. agr. exp. sta. quart. bull. 48: 437–448.
Douglas, G. E. 1957 The inferior ovary. – Bot. rev. 23: 1–46.

Doull, K. M. 1973 Relationships between pollen, broodrearing and consumption of pollen supplements by honeybees. – Apidologie 4: 285–293.

Downes, J. A. 1971 The evolution of blood-sucking diptera: an evolutionary perspective. – In A. M. Tallin (ed.): Ecology and physiology of diptera. A symposium. Toronto Univ. Press.

Downes, J. A. 1974 The feeding habits of adult chironomidae. – Ent. tidskr. 95 Suppl.: 84–90.

Doyle, J. D. 1945. Developmental lines in pollination mechanisms in the Coniferales. – Sci. proc. R. Dublin soc. N.S. 24: 43–62.

Doyle, J. D. and A. Kam 1944 Pollination in *Tsuga pattoniana* and in species of *Abies* and *Picea*. – Sci. proc. R. Dublin soc. N.S. 23: 57–70.

Dressler, R. L. 1957 The genus *Pedilanthus* (Euphorbiaceae). – Contr. Gray herb. Harvard univ. 182 pp.

Dressler, R. L. 1970 How to study orchid pollination without any orchids. – Am. J. bot. 57: 749.

Dressler, R. L. 1971 Dark pollinia in hummingbird-pollinated orchids, or do hummingbirds suffer from strabismus. – Am. nat. 105: 80–82.

Dufay, C. 1961 Étude de phototropisme des Noctuides (Lepidopt.) Réactions de *Noctua pronuba* L. en presence de deux sources lumineuses de même intensité. – C. r. Paris 252: 1866–1869.

Dvorak, F. 1968 A contribution to the study of the variability of the nectaries. – Preslia 40: 13–17.

Dyakowska, J. and J. Zurzycki 1959 Gravimetric studies on pollen. – Bull. acad. pol. sci. Cl. II, Sér. sci. biol. 7: 11–16.

Ebbels, L. 1969 Pollination of *Puya chilensis* by *Turdus merula* in the Isles of Scilly. – Ibis III: 615.

Ehrlich, P. R. and L. E. Gilbert 1973 Population structure and dynamics of the tropical butterfly *Heliconus ethilla*. – Biotropica 5: 69–82.

Eisikowitch, D. and J. Galil 1971 Effect of wind on the pollination of *Pancratium maritimum* L. (Amaryllidaceae) by hawkmoths (Lepidoptera: Sphingidae). – J. anim. ecol. 40: 673–678.

Eisikowitch, D. and S. R. J. Woodell 1975 The effect of water on pollen germination in two species of Primula. – Evolution 28: 692–694.

Eisner, T., R. E. Silbergeld, D. Aueshansky, J. E. Carrel, H. C. Howland 1969 Ultra-violet video-viewing: the television camera as an insect eye. – Science 166: 1172–1174.

Elias, T. S. 1972 Morphology and anatomy of foliar nectaries of *Pithecellobium macradenium* (Leguminosae). – Bot. gaz. 133: 38–42.

Elias, T. S. and H. Gelband 1975 Nectar: its production and function in Trumpet Creeper. – Science 189: 289–291.

Elias, T. S., W. R. Rozich, L. Newcombe 1975 The foliar and floral nectaries of *Turnera ulmifolia*. – Am. J. bot. 62: 570–576.

Elliot, G. F. Scott 1890 Ornithophilous flowers in South Africa. – Ann. bot. 4: 265–280.

Ernst, A. 1953 Primärer und sekundärer Blütenmonomorphismus bei Primeln. – Österr. bot. Z. 100: 235–255.

Ernst, A. 1955 Self-fertility in monomorphic primulas. – Genetica 27: 391–448.

Ernst-Schwarzenbach, M. 1945 Zur Blütenbiologie einiger Hydrocharitaceen. – Ber. schweiz. bot. Ges. 55: 33–69.

Essig, B. F. 1971 Observations on pollination in *Bactris*. – Principes 15: 20–24.

Essig, B. F. 1973 Pollination in some New Guinea palms. – Principes 17: 75–83.

Estes, J. A. and R. W. Thorp 1975 Pollination ecology of *Pyrrhopappus carolinianus* (Compositae). – Am. J. bot. 62: 148–159.

Fægri, K. 1962 Palynological studies in a bumblebee nest. – Veröff. geobot. Inst. Rübel Zürich 37: 60–67.

Fægri, K. 1965 Reflections on the development of pollination systems in African Proteaceae. – J. S. Afr. bot. 31: 133–136.

Fægri, K. 1974 Major ecological steps in the development of plant life. – First silver jubilee lecture. Birbal Sahni inst. palaeobot. Lucknow (1973). 10 pp.

Fægri, K. and J. Iversen 1975 Textbook of pollen analysis. 3rd edn. – Munksgaard (København). 296 pp.

Fechner, G. H. and R. W. Funsch 1966 Germination of blue spruce and ponderosa pine pollen after 11 years storage at 0° to 4°C. – Silvae genet. 15: 164–166.

Ford, H. A. and N. Forde 1976 Birds as possible pollinators of Acacia pycnantha. – Aust. J. bot. 24: 793–795.

Fordham, F. 1946 Pollination of *Calochilus campestris*. – Vict. nat. 62: 199–201.

François, J. 1964 Observations sur l'héterostylie chez *Eichhornia crassipes*. – Bull. K. acad. overz. wet. 3: 501–519.

Frankie, G. W. 1975 Tropical forest phenology and pollinator plant evolution. – In Gilbert and Raven 1975: 192–209.

Frankie, G. W. 1976 Pollination of widely dispersed tropical trees by animals in Central America, with an emphasis on bee pollination. – In Bailey, J. and B. T. Stiles (Eds.) Varieties, breeding and conservation of tropical forest trees. Academic Press (New York).

Free, J. B. 1962a Studies on the pollination of fruit trees by honeybees. – J. R. hort. soc. 87: 302–309.

Free, J. B. 1962b The behaviour of honeybees visiting field beans (*Vicia faba*). – J. anim. ecol. 31: 497–502.

Free, J. B. 1966 The foraging behaviour of bees and its effects on the isolation and speciation of vascular plants. – Bot. soc. Brit. Isles conf. rep. 1965: 76–91.

Free, J. B. 1968 Dandelion as a competitor to fruit trees for bee visits. – J. appl. ecol. 5: 169–178.

Free, J. B. 1970 Insect pollination of crops. – Academic Press (New York). 544 pp.

Free, J. B. and I. H. Williams 1972 The transport of pollen on the body of honeybees (*Apis mellifera* L.) and bumblebees (*Bombus* spp. L.). – J. appl. ecol. 9: 609–615.

Free, J. B. and I. H. Williams 1973 Genetic determination of honeybee (*Apis mellifica*) foraging preferences. – Ann. appl. biol. 73: 137–141.

Frey-Wyssling, A. and E. Häusermann 1960 Deutung der gestaltlosen Nektarien. – Ber. Schweiz. bot. Ges. 70: 150–162.

Fries, R. E. 1903–4 Beiträge zur Kenntnis der Ornithophilie in der südamerikanischen Flora. – Ark. bot. 1: 384–440.

Frisch, K. von 1914 Der Farbensinn und Formensinn der Biene. – Zool. Jahrb. Abt. allg. Physiol. Tiere 35: 1–182.

Frisch, K. von 1923, 1924 Über die "Sprache" der Bienen. – Zool. Jahrb. Abt. allg. Zool. 40: 1–186.

Frisch, K. von 1950 Bees. – Cornell Univ. Press.

Fröman, I. 1944 De senkvartära strandförskjutningarna som växtgeografisk faktor til belysning av murgrönans geografi i Skandinavien–Baltikum. – Geol. för. Stockh. förh. 66: 655–681.

Fryxell, P. G. 1957 Mode of reproduction of higher plants. – Bot. rev. 23: 135–233.

Fuchs, G. V. 1975 Die Gewinnung von Pollen und Nektar bei Käfern. – Nat. Mus. 104: 45–54.

Gabriel, W. J. 1968 Dichogamy in *Acer saccharum*. – Bot. gaz. 129: 334–338.

Galil, J. 1973a Topocentric and ethodynamic pollination. – In Brantjes and Linskens 1973: 85–100.

Galil, J. 1973b Pollination in dioecious figs. Pollination of *Ficus fistulosa* by *Ceratosolen hewitti*. – Gard. bull. 26: 303–311.

Galil, J. and D. Eisikowitch 1969 Further study in the pollination ecology of *Ficus sycomorus* L. – Tijds. entomol. 112: 1–13.

Galil, J. and D. Eisikowitsch 1973 Further studies on pollination ecology in *Ficus sycomorus*. II: Pocket filling and emptying by *Ceratosolen arabicus* Mayr. – New phytol. 73: 515–520.

Galil, J., W. Ramirez, D. Eisikowitch 1973 Pollination of *Ficus costaricana* and *F. hemsleyana* by *Blastophaga esterae* and *B. Tonduzi* in Costa Rica (Hymenoptera, Chalcidoidea, Agaonidae). – Tijds. entomol. 116: 175–183.

Galil, J. and M. Zeroni 1967 On the pollination of *Zizyphus spina-christi* in Israel. – Israel J. bot. 16: 71–77.

Galil, J., M. Zeroni, D. Bar Shalom (Bogoslavsky) 1973 Carbon dioxide and ethylene effects in the coordination between the pollinator *Blastophaga quadraticeps* and the syconium in *Ficus religiosa*. – New phytol. 72: 1113–1127.

Ganders, F. R. 1975 Heterostyly, homostyly, and fecundity in *Amsinckia spectabilis* (Boraginaceae). – Madroño 23: 56–62.

Geisler, G. and W. Steche 1962 Natürliche Trachten als Ursache für Vergiftungserscheinungen bei Bienen und Hummeln. – Z. Bienenforsch. 6: 72–99.

Gilbert, L. E. 1972 Pollen feeding and reproductive biology of *Heliconius* butterflies. – Proc. nat. acad. sci. 69: 1403–1407.

Gilbert, L. E. and P. Raven 1975 Coevolution of animals and plants. – Univ. Texas Press. (Austin). 246 pp.

Glauert, L. 1958 The honey mouse. – Aust. mus. mag. 12: 284–286.

Goebel, K. 1924 Die Entfaltungsbewegungen der Pflanzen. 2nd edn. – Fischer (Jena).

Good, R. K. and S. S. Saini 1971 Pollination studies in *Lycopersicum esculentum* Mill. – Himachal. J. agric. res. 1: 65–70. Quoted from Biol. abstr. 54: 8113.

Goot, V. S. van der and R. A. J. Grabandt 1970 Some species of the generae *Malastomum, Platycheirus* and *Pryophaene* (Diptera, Syrphidae) and their relation to flowers. – Entomol. Ber. 30: 135–143.

Gottsberger, G. 1967, 1972 Blütenbiologische Beobachtungen an brasilianischen Malvaceen. (I)–II. – Österr. bot. Z. 114: 349–378; 120: 439–509.

Gottsberger, G. 1970 Beiträge zur Biologie von Annonaceen-Blüten. – Österr. bot. Z. 118: 237–279.

Gottsberger, G. 1971 Colour change of petals in Malva-viscus arboreus flowers. – Acta bot. neerl. 20: 381–388.

Gottsberger, G. 1974 The structure and function of the primitive Angiosperm flower – Acta bot. neerl. 23: 461–471.

Gottsberger, G., J. Schrauwen, H. F. Linskens 1973 Die Zucker-Bestandteile des Nektars einiger tropischen Blüten. – Port. act. biol. Ser. A 13: 1–8.

Gouriay, W. G. 1950 *Puya alpestris* in its native land. – J. R. hort. soc. 75: 399–402.

Graenicher, S. 1909 Wisconsin flowers and their pollination. – Bull. Wis. nat. hist. soc. 7: 19–77.
Grant, K. 1966 A hypothesis concerning the prevalence of red coloration in California hummingbird flowers. – Am. nat. 100: 85–98.
Grant, K. and V. Grant 1964 Mechanical isolation of *Salvia apiana* and *Salvia mellifera*. – Evolution 18: 197–212.
Grant, K. and V. Grant 1965 Flower pollination in the Phlox family. – Columbia Univ. Press (New York). 180 pp.
Grant, K. A. and V. Grant 1968 Hummingbirds and their flowers. – Columbia Univ. Press (New York). 115 pp.
Grant, V. 1949a Arthur Dobbs (1750) and the discovery of the pollination of flowers by insects. – Bull. Torr. bot. 76: 217–219.
Grant, V. 1949b Pollination systems as isolating mechanisms in angiosperms. – Evolution 3: 82–97.
Grant, V. 1950a The protection of the ovules in flowering plants. – Evolution 4: 179–201.
Grant, V. 1950b The flower constancy of bees. – Bot. rev. 16: 379–398.
Grant, V. 1950c The pollination of *Calycanthus occidentalis*. – Am. J. bot. 37: 291–297.
Grant, V. 1952 Isolation and hybridization between *Aquilegia formosa* and *A. pubescens* – Aliso 2: 341–359.
Grant, V. 1955 Cross-fertilization. – Encycl. am. 8: 230–234.
Grant, V. 1963 The origin of adaptations. – Columbia Univ. Press (New York). 606 pp.
Grant, V. 1976 Isolation between *Aquilegia formosa* and *A. pubescens*: a reply and reconsideration. – Evolution 30: 625–628.
Greenewalt, C. H. 1963 Photographing hummingbirds in Brazil. – Nat. geogr. mag. 123: 100–116.
Gregory, D. P. 1963 Hawkmoth pollination in the genus *Oenothera*. – Aliso 5: 357–419.
Gregory, P. H. 1957 Deposition of airborne *Lycopodium* spores on cylinders. – Ann. appl. biol. 38: 357–376.
Grinfeld, E. K. 1959 The importance of nocturnal insects in sunflower pollination. – Agrobiologiya 4: 634–635.
Grinfeld, E. K. 1973 Anthophilous evolution of the Apoidea. – Vestn. Leningr. univ. Ser. biol. 28: 7–15.
Grinfeld, E. K. and I. V. Issi 1958 The role of beetles in plant pollination. – Uch. zap. Leningr. gos. univ. 240: 148–150.
Grout, A. J. 1926 Spore dispersal in *Sphagnum*. – Bryologist 29: 55.
Gubin, A. 1936 Bestäubung und Erhöhung der Samenernte bei Rotklee (*Trifolium pratense*) mit Hilfe der Bienen. – Arch. Bienen. 17: 209–269.
Haas, A. 1952 Die Mandibeldrüse als Duftorgan bei einigen Hymenopteren. – Naturwissenschaften 39: 20.
Haeger, J. S. 1955 The non-blood feeding habits of *Aedes taeniorhynchus* (Diptera: Culicidae) on Sanibel Island, Florida. – Mosq. news 15: 21–26.
Hagerup, E. and O. Hagerup 1953 Thrips pollination of *Erica tetralix*. – New phytol. 52: 1–7.
Hagerup, O. 1932 On pollination in the extremely hot air at Timbuctu. – Dan. bot. ark. 8, 1. 20 pp.
Hagerup, O. 1943 Myre-bestøvning. – Bot. tidsskr. 46: 116–123.
Hagerup, O. 1950a Thrips pollination in *Calluna*. – K. dan. vidensk. selsk. Biol. medd. 18, 4.
Hagerup, O. 1950b Rain-pollination – K. dan. vidensk. selsk. Biol. medd. 18, 5. 19 pp.
Hagerup, O. 1951 Pollination in the Faroes – in spite of rain and poverty in insects. – K. dan. vidensk. selsk. Biol. medd. 18, 15, 78 pp.
Hagerup, O. 1952 The morphology and biology of some primitive orchid flowers. – Phytomorphology 2: 134–138.
Hagerup, O. 1954 Autogamy in some drooping Bicornes flowers. – Bot. tidsskr. 51: 103–106.
Hagerup, O. and V. Peterson 1956 Botanisk atlas. – Munksgaard (København).
Halket, A. C. 1936 Bud pollination and self-fertilization in *Melandrium noctiflorum* Fries. – J. bot. 74: 310–316.
Hallermeier, M. 1922 Ist das Hangen der Blüten eine Schutzeinrichtung? – Flora 115: 75–101.
Hanson, C. H. and T. A. Campbell 1972 Vacuum-dried pollen of alfalfa (*Medicago sativa* L.) – viable after eleven years. – Crop. sci. 12: 874. Quoted from Biol. abstr. 55: 40892.
Harris, J. A. 1914 On a chemical peculiarity of the dimorphic anthers of *Lagerstroemia indica*. – Ann. bot. 28: 499–506.
Harris, T. M. 1956 The investigation of a fossil plant. – Proc. R. instn. 36 (136): 1–11.
Harrison, C. R. and J. Arditti 1976 Post-pollination phenomena in orchid flowers. VII: Phosphate movement among floral segments. – Am. J. bot. 63: 911–918.
Hartog, C. den 1964 Over de oecologie van bloeiende *Lemna trisulca*. – Gorteria 2: 68–72.
Haskell, G. 1943 Spatial isolation of seed crops. – Nature 152: 591–592.
Haskell, G. 1954 Correlated responses to phylogenetic selection in animals and plants. – Am. nat. 88: 5–20.

Haslerud, H.-D. 1974 Pollination of some Ericaceae in Norway. – Nor. J. bot. 21: 211–216.
Haumann-Merck, L. 1912 Observations éthologiques et systématiques sur deux espèces argentines du genre *Elodea*. – Rec. inst. bot. Leo Errera 9: 33–39.
Hawkins, R. P. 1961 Observations on the pollination of red clover by bees. I: The yield of seeds in relation to the number and kinds of pollinators. – Ann. appl. biol. 49: 55–65.
Hayase, H. 1963 *Cucurbita* crosses. XV: Flower pollination at 4 a.m. in the production of *C. pepo* × *C. moschata* hybrids. – Jap. J. breed. 3: 76–89.
Hedberg, O. 1964 Features of afro-alpine plant ecology. – Acta phytogeogr. sue. 49.
Hedge, I. 1960 Two remarkable new Salvias from Afganistan. – Notes R. bot. gard. 23: 163–166.
Heide, F. F. R. 1923 Biologische underzoekingen bij landbouwgewassen. I: – Meded. alg. proefstat. landb. Buitenzorg 14. 46 pp.
Heide, F. F. R. 1927 Observation on the pollination of some flowers in the Dutch East Indies. – Dan. bot. ark. 5: 1–24.
Heinrich, B. 1972 Energetics of temperature regulation and foraging in a bumblebee, *Bombus terricola* Kirby. – J. comp. physiol. 77: 49–64.
Heinrich, B. 1973 The energetics of the bumblebee. – Sci. Am. 228: 96–102.
Heinrich, B. 1975 Energetics of pollination. – Ann. rev. ecol. syst. 6: 139–170.
Heinrich, B. 1976 The foraging specializations of individual bumblebees. – Ecol. monogr. 46: 105–128.
Heinrich, B. and P. H. Raven 1972 Energetics and pollination ecology. – Science 176: 597–602.
Heiser, C. B. 1962 Some observations on pollination and compatibility in *Magnolia*. – Proc. Indiana acad. sci. 72: 259–266.
Heithaus, E. R., P. A. Opler, H. G. Baker 1974 Bat activity and pollination of *Bauhinia pauletii*: Plant-pollinator coevolution. – Ecology 55: 412–419.
Hendryck, R. 1953 Zur Frage der phylogenetischen Bedeutung der Gnetinae. – Preslia 25: 111–134.
Heslop-Harrison, J. 1958 Ecological variation and ethological isolation. – Uppsala univ. årsskr. 6: 150–158.
Heslop-Harrison, J. 1959 Apomixis, environment, and adaptation. – Presented biosyst. symp. IX bot. congr. Montreal 1959. Mimeographed.
Heslop-Harrison, J. 1963 Sex expression in flowering plants. – In Meristems and differentiation. Brookhaven symp. biol. 16: 109–124.
Heslop-Harrison, J. and Y. Heslop-Harrison 1975 Enzymic removal of the proteinaceous pellicle of the stigma prevents pollen tube entry in the Caryophyllaceae. – Ann. bot. 39: 163–165.
Hesselmann, H. 1919 Iakttagelser över skogsträdpollens spridnings-förmåga. – Medd. statehs skogförsöksanst. 16.
Heusser, C. E. 1912 Kunstmatige bestuiving bij den oliepalm (*Elaeis guineensis* Jacq.). – Teysmannia 32: 434–445.
Heywood, V. H. (ed.) 1973 Taxonomy and ecology. – Academic Press (London). 370 pp.
Hickman, J. C. 1974 Pollination by ants: a low-energy system. – Science 184: 1290–1292.
Hiepko, P. 1966 Zur Morphologie, Anatomie und Funktion des Discus bei Paeoniaceae. – Ber. dtsch. bot. Ges. 79: 233–245.
Hills, H. G. 1972 Floral fragrances and isolating mechanisms in the genus *Catasetum* (Orchidaceae). – Biotropica 4: 61–76. Quoted from Biol. abstr. 56: 64898.
Hocking, B. 1968 Insect-flower associations in the high Arctic with special reference to nectar. – Oikos 19: 359–387.
Hocking, B. and C. D. Sharplin 1965 Flower basking by arctic insects. – Nature 206: 215.
Horovitz, A. 1976 Edaphic factors and flower colour distribution in the Anemoneae (Ranunculaceae). – Plant. syst. evol. 126: 239–242.
Horovitz, A., J. Galil, D. Zohary 1975 Biological flora of Israel. 6. *Anemone coronaria* L. – Israel J. bot. 24: 26–41.
Howell, D. J. 1974 Bats and pollen: physiological aspects of the syndrome of chiropterophily. – Comp. bioch. physiol. A. Comp. physiol. 48: 263–276.
Hubbard, H. G. 1895 Insect fertilization of an aroid plant. – Insect life 7: 340–345.
Huber, H. 1956 Die Abhängigkeit der Nektarsekretion von Temperatur, Luft- und Bodenfeuchtigkeit. – Planta 48: 47–98.
Hurd, P. D. jr. and E. A. Linsley 1963 Pollination of the unicorn plant (Martyniaceae) by an oligolectic, corolla-cutting bee (Hymenoptera: Apoidea). – J. Kans. entomol. soc. 36: 248–252.
Hurd, P. D. jr., E. A. Linsley, A. E. Michelbacher 1974 Ecology of the squash and gourd bee, *Peponapis pruinosa*, on cultivated cucurbits in California (Hymenoptera: Apoidea). – Smiths. contrib. zool. 168. 17 pp.
Hurd, P. D., E. A. Linsley, T. W. Whitaker 1971 Squash and gourd bees (*Peponapis, Xenoglossa*) and the origin of the cultivated cucurbits. – Evolution 25: 218–234.
Huth, H.-H. and D. Burkhardt 1972 Der spektrale Sehbereich eines Violettohr-Kolibris. – Naturwissenschaften 59: 650.

Hyman, S. E. 1960 Some surrealist botany. – Texas quart. 3: 113–127.

Ilse, D. 1928 Über den Farbensinn der Tagfalter. – Z. vergl. Physiol. 8: 658–692.

Ingold, C. T. 1971 Fungal spores, their liberation and dispersal. – Oxford Univ. Press. 302 pp.

Ingram, C. 1967 The phenomenal behaviour of a South African Gladiolus. – J. R. hort. soc. 92: 396–398.

Iversen, J. 1940 Blütenbiologische Studien. I: Dimorphie und Monomorphie bei *Armeria*. – K. dan. vidensk. selsk. Biol. medd. 15, 8.

Iwanamy, Y. 1973 Accelerating of the growth of *Camellia sasanqua* pollen by soaking in organic solvents. – Plant physiol. 52: 508–509.

Iyengar, M. O. P. 1923 Two instances of short-cuts by animals to nectaries. – J. Ind. bot. soc. 2: 85–88.

Jaeger, P. 1954 Les aspects actuels de problème de la cheiropterogamie. – Bull. inst. fr. Afr. noire 16: 796–821.

Janzen, D. H. 1966 Coevolution of mutualism between ants and acacias in Central America. – Evolution 20: 249–275.

Janzen, D. H. 1971 Euglossine bees as long-distance pollinators of tropical plants. – Science 171: 203–205.

Johow, F. 1900–1 Zur Bestäubungsbiologie chilenischer Blüten. – Verh. dtsch. wiss. Ver. Santiago 3: 1–22, 4: 345–424.

Johri, B. M. 1971 Differentiation in plant tissue cultures. – Proc. 58 Ind. sci. congr. II. Pres. addr. Botany.

Jones, C. E. jr. and P. V. Rich 1972 Ornithophily and extrafloral color patterns in *Columnea florida* Morton (Gesneriaceae). – Bull. S. Calif. acad. sci. 71: 113–116.

Jong, P. C. de 1976 Flowering and sex expression in *Acer* L. – Meded. landbouwhoogesch. Wageningen 76, 2. 201 pp.

Jost, L. 1907 Über die Selbststerilität einiger Blüten. – Bot. Ztg. 65: 77.

Judd, W. W. 1971 Wasps (Vespidae) pollinating helleborine, *Epipactis helleborine* (L.) Crantz, at Owen Sound, Ontario. – Proc. entomol. soc. Ont. 102: 115–118.

Kaplan, S. M. and D. L. Mulcahy 1971 Mode of pollination and floral sexuality in *Thalictrum*. – Evolution 25: 659–668.

Karr, J. R. 1976 An association between a grass (*Paspalum virgatum*) and moths. – Biotropica 8: 284–285.

Kaufmann, T. 1973 Biology and ecology of *Tysora tessmannii* (Homoptera, Psyllidae) with special reference to its role as pollinator in cocoa in Ghana, W. Africa. – J. Kans. entomol. soc. 46: 285–293.

Kaul, R. B. 1970 Evolution and adaptation of inflorescences in the Hydrocharitaceae. – Am. J. bot. 57: 708–716.

Kauta, K. 1960 Intra-ovarian pollination in *Papaver rhoeas* L. – Nature 188: 683–684.

Kendall, D. A. and M. E. Solomon 1973 Quantities of pollen on the bodies of insects visiting apple blossoms. – J. appl. ecol. 10: 627–634.

Kennedy, H. 1973 Notes on Central American Maranthaceae. I: New Species and records from Costa Rica. – Ann. Mo. bot. gard. 60: 413–426.

Kerner, A. 1873 Die Schutzmittel des Pollens gegen die Nachteile vorzeitiger Dislocation und gegen die Nachteile vorzeitiger Befeuchtung. – Ber. Naturwiss. – med. Ver. Innsbruck. 2–3.

Kerner von Marilaun, A. 1898 Pflanzenleben II. 2nd edn. Berlin–Wien.

Kerr, K. E. 1960 Evolution of communication in bees and its role in speciation. – Evolution 14: 386–387.

Kevan, P. G. 1972. Insect pollination of high arctic flowers. – J. ecol. 60: 831–847.

Kevan, P. 1975 Sun-tracking solar furnaces in High Arctic flowers: significance for pollen and insects. – Science 189: 723–726.

Khan, S. A. R. 1929 Pollination and fruit formation in lichi (*Nephelium litchi*). – Agr. J. India 24: 183–187.

Kho, Y. A. and J. Baër 1970 Die Fluorescensmikroskopie in der botanischen Forschung. – Zeiss. inf. 18: 54–57.

Kikuchi, T. 1962, 64 Studies on the coaction among insects visiting flowers. I. VI. – Sci. rep. Tohôku univ. 4 Ser. (Biol.) 28: 17–22; 30: 143–150.

Kinkaid, T. 1963 The ant plant, *Orthocarpus pusillus* – Trans. Am. micr. soc. 82: 101–105.

Kirchner, O. von 1925 Über die sogenannten Pollenblumen und die Ausbeutestoffe der Blüten. – Flora 118–119: 312–330.

Knoch, E. 1899 Untersuchungen über die Morphologie, Biologie und Physiologie der Blüte von *Victoria regia*. – Biol. bot. 47. 60 pp.

Knoll, E. 1921–6 Insekten und Blumen. – Abh. zool.–bot. Ges. Wien. 12: 1–645.

Knoll, F. 1923 Über die Lückenepidermis der *Arum*–Spatha. – Österr. bot. Z. 72: 246–254.

Knoll, F. 1930a Über die Laubblattnektarien von Catalpa bignonbides und ihre Insektenbesuche. – Biol. gen. 4: 541–570.
Knoll, F. 1930b Über Pollenkitt und Bestäubungsart. – Z. Bot. 23: 609–675.
Knoll, F. 1956 Die Biologie der Blüte. – Springer (Berlin). 164 pp.
Knox, R. B. and J. Heslop-Harrison 1971 Pollen-wall proteins: Localization of antigenic and allergenic proteins in the pollen-grain walls of Ambrosia spp. (ragweeds). – Cytobios 4: 49–54.
Knox, R. B., R. R. Willems, A. E. Ashford 1972 Role of pollen-wall proteins as recognition substances in interspecific incompatibility in poplars. – Nature 237: 381–383.
Knox, R. B., R. R. Willing, L. D. Pryor 1972 Interspecific hybridization in poplars using recognition pollen. – Silvae genet. 21: 65–69.
Knuth, P. (and E. Loew) 1895–1905 Handbuch der Blütenbiologie. I–III, 2. Engelmann (Leipzig). 2973 pp.
Koeman-Kwak, M. 1973 The pollination of Pedicularis palustris by nectar thieves (short-tongued bumble-bees). – Acta bot. neerl. 22: 608–615.
Koski, V. 1970 A study of pollen dispersal as a mechanism of gene flow in conifers. – Metsäntutkimuslaitoksen julkaisuja (Comm. inst. forest. fenn.) 70, 4. 78 pp.
Kraai, A. 1962 How long do honeybees carry germinable pollen on them? – Euphytica 11: 53–56.
Kubitzki, K. 1969 Pollendimorphie und Androdiözie bei Tetracera (Dilleniaceae). – Naturwissenschaften 56: 219–220.
Kugler, H. 1930–42. Blütenökologische Untersunchungen mit Hummeln. I. – Planta 10: 288–290; II. – Ber. dtsch. bot. Ges. 49: III–X. – Planta 16, 19, 23, 25, 29, 32.
Kugler, H. 1936 Die Ausnutzung der Saftmalsumfärbung bei den Rosskastanienblüten durch Bienen und Hummeln. – Ber. dtsch. bot. Ges. 54: 394–400.
Kugler, H. 1942 Hummelblumen. Ein Beitrag zum Problem der "Blumenklassen" auf experimenteller Grundlage. – Ber. dtsch. bot. Ges. 60: 128, 134.
Kugler, H. 1943 Hummeln als Blütenbesucher. – Ergebn. biol. 19: 143–323.
Kugler, H. 1955a Zum Problem der Dipterenblumen. – Österr. bot. Z. 102: 529–541.
Kugler, H. 1955b, 1970 Einführung in die Blütenökologie. 1st edn., 2nd edn. – Fsicher (Stuttgart). 345 pp.
Kugler, H. 1956 Über die optische Wirkung von Fliegenblumen auf Fliegen. – Ber. dtsch. bot. Ges. 69: 387–398.
Kugler, H. 1962 UV-Musterung auf Blüten und ihr Zustandekommen. – Ber. dtsch. bot. Ges. 75: 49–54.
Kugler, H. 1973 Zur Bestäubung von Scaevola plumieri (L.) Vahl und Ipomoea pes-caprae Sweet, zwei tropischer Strandpflanzen. – Flora 162: 381–391.
Kugler, H. 1975 Die Verbreitung anemogamer Arten in Europa. – Ber. dtsch. bot. Ges. 88: 441–450.
Kühn, A. and F. Pohl 1924 Zum Nachweis des Farbenuntersceidungsvermögens der Bienen. – Naturwissenschaften 12: 116–118.
Kullenberg, B. 1950 Bidrag til kännedomen om Ophrys- arternas blombiologi. – Sven. bot. tidskr. 44: 446–464.
Kullenberg, B. 1956a On the scents and colours of Ophrys flowers and their specific pollinators among aculeate hymenoptera. – Sven. bot. tidskr. 50: 25–46.
Kullenberg, B. 1956b Field experiments with chemical sexual attractants on aculeate hymenoptera males. – Zool. bidr. Uppsala 31: 253–254.
Kullenberg, B. 1961 Studies in Ophrys pollination. – Zool. bidr. Uppsala 34: 1–340.
Kullenberg, B. 1973a Field experiments with chemical sexual attractants on aculeate hymenoptera males. II. – In Kullenberg and Stenhagen 1973: 31–42.
Kullenberg, B. 1973b New observations on the pollination of Ophrys (Orchidaceae). – In Kullenberg and Stenhagen 1973: 9–13.
Kullenberg, B. and G. Bergström 1976 The pollination of Ophrys orchids. – Bot. not. 129: 11–19.
Kullenberg, B., G. Bergström, B. Bringer, B. Carlberg, B. Cederberg 1973 Observations on the scent marking by Bombus Latr. and Psithyrus Lep. males and localization of site of production of the secretion. – In Kullenberg and Stenhagen 1973: 23–30.
Kullenberg, B. and E. Stenhagen 1973 The ecological station of Uppsala University on Öland 1963–73. – Zoon. Suppl. 1. 151 pp.
Laczynska-Hulewiczawa, T. 1958 Badania nad samplodnoscia koniczny czervonez di-i-tretaploidalny. – Ros. nauk. roln. Ser. A. 79: 157–160.
Lagerberg, T., J. Holmboe, R. Nordhagen 1957 Våre ville planter. VI, I.–Tanum (Oslo). 269 pp.
Lam, H. J. 1961 Reflections on angiosperm phylogeny. K. ned. akad. wet. Proc. Ser. C. 54: 251–276.
Lamprecht, H. 1959 Der Artbegriff, seine Entwicklung und experimentelle Klarlegung. – Agri hortique genet. 17: 105–264.
Lange, J. H. de, A. P. Vincent, J. H. de Leeuw 1974 Pollination studies in Mimelo tangelo. – Agroplantae 5: 40–54. Quoted from Biol. abstr. 58: 49701.

Leach, D. G. 1972 The ancient curse of the Rhododendron. – Am. hortic. 51, 3: 20–29.
Leppik, E. E. 1955 *Dichromena ciliata*, a noteworthy entomophilous plant among Cyperaceae. – Am. J. bot. 42: 455–458.
Leppik, E. E. 1957 A new system for classification of flower types. – Taxon 6: 64–67.
Leppik, E. E. 1961 Phyllotaxis, anthotaxis and semataxis. – Acta biotheor. 14: 1–28.
Lester, E. W. and C. Chadby 1965 Use of exogenous growth substances in promoting pollen tube growth and fertilization in barley-rye crosses. – Can. j. genet. expt. 7: 511–518.
Levin, D. A. 1968. The breeding system of *Lithospermum carolinense*: adaptation and counteradaptation. – Am. nat. 102: 42–443.
Levin, D. A. 1970 Reinforcement or reproductive isolation: plants versus animals. – Am. nat. 104: 571–581.
Levin, D. A. 1972a The adaptedness of corolla-color variants in experimental and natural populations of *Phlox drummondii*. – Am. nat. 106: 56–70.
Levin, D. A. 1972b Competition for pollinator service: a stimulus for the development of autogamy. Evolution 26: 668–674.
Levin, D. A. and W. W. Anderson 1970 Competition for pollinators between simultaneously flowering species. – Am. nat. 104: 455–467.
Levin, D. A. and D. E. Berube 1972 *Phlox* and *Colias*: the efficiency of the pollinator system. – Evolution 26: 242–250.
Levin, D. A. and H. W. Kersten 1973 Assortative pollination for stature in *Lythrum salicaria*. – Evolution 27: 144–152.
Levin, D. A., H. W. Kersten, M. Niedzlek 1971 Pollination flight directionality and its effect on pollen flow. – Evolution 25: 113–118.
Lewis, D. 1955 Sexual incompatibility. – Sci prog. 172: 593–603.
Lex, R. E. 1961 Pollen dimorphism in *Tripogandra grandiflora* – Baileya 9: 53–56.
Lex, T. 1954 Duftmale an Blüten. – Z. vergl. Physiol. 36: 212–234.
Li, H.-L. 1948–49 A revision of the genus *Pedicularis* in China. I–II. – Proc. acad. nat. sci. Phil. 100: 205–378: 101: 1–214.
Lindauer, M. 1971 The functional significance of the honeybee waggle dance. – Am. nat. 105: 89–96.
Lindman, C. A. M. 1900 Die Blüteneinrichtungen einiger südamerikanischer Pflanzen. I. Leguminosae. – Bih. K. sven. vetensk. akad. förhandl. 27, III (1901–2): 14.
Linhart, Y. B. 1973 Ecological and behavioural determinants of pollen dispersal in hummingbird-pollinated *Heliconia*. – Am. nat. 107: 511–523.
Linskens, H. F. 1967 Inkompatibilität der Phanerogamen. – Handb. Pflanzenphysiol. 18. Springer (Heidelberg).
Linskens, H. F. 1975 The physiological basis of incompatibility in angiosperms. – Biol. J. Linn. soc. 7. Suppl. 1: 143–152.
Linskens, H. F. and M. L. Suren 1969 Die Entwicklung des Polliniums von *Asclepias curassavica*. – Ber. dtsch. bot. Ges. 82: 527–534.
Linsley, E. G. 1958 The ecology of solitary bees. – Hilgardia 27: 543–599.
Linsley, E. G. 1962 Ethological adaptations of solitary bees for the pollination of desert plants. – Meddel. 7 Sver. Fröodlareförb. 189–197.
Linsley, E. G. and J. M. MacSwain 1959 Ethology of some *Ranunculus* insects with emphasis on competition for pollen. – Univ. Calif. publ. entomol. 16, 1.
Linsley, E. G., J. W. MacSwain, P. H. Raven, R. W. Thorp 1963–73 Comparative behavior of bees and Onagraceae. I–V. – Univ. Calif. publ. entomol. 33: 1–98, 70: 1–84, 71: 1–67.
Linsley, E. G., C. M. Rich, S. G. Stephens 1966 Observations on the floral relationships of the Galapagos carpenter bee. – Pan-pac. entomol. 42: 1–18.
Loew, E. 1884 Beobachtungen über den Besuch von Insekten an Freilandpflanzen der botanischen Gärten zu Berlin. – Jahrb. bot. Gart. Berlin 3: 69–118, 253–296.
Loew, E. 1895 Das Leben der Blüten. – Dümmler (Berlin).
Lokoschus, F. S. and J. L. W. Keulart 1968 Eine weitere Funktion der Mandibeldrüse der Arbeiterinnen. Produktion eines Pollenkeimungsstoffes. – Z. Bienenforsch. 9: 333–334.
Loper, G. M. and G. D. Waller 1970 Alfalfa flower aroma and flower selection among honeybees. – Crop sci. 10: 66–68. Quoted from Shuel 1970 in Bee world 51: 63–69.
Lorch, J. 1966 The discovery of sexuality and fertilization in higher plants. – Janus 53: 212–235.
Lorougnon, G. 1973 Le vecteur pollinique chez les *Mapania* et les *Hypolytrum*, cyperacées du sous-bois des forêts tropicales ombrophiles. – Bull. jard. bot. nat. Belg. 43: 33–36.
Louveaux, J. 1960 Recherches sur la récolte du pollen par les abeilles. – Ann. l'abeille 3.
Lüttge, U. 1960 Über die Zusammensetzung des Nektars und den Mechanismus seiner Sekretion. I. – Planta 56: 189–212.
Lüttge, U. 1966 Funktion und Struktur der pflanzlichen Drüsen. – Naturwissenschaften 53: 96–103.

Løken, A. 1950 Bumblebees in relation to *Aconitum septentrionale* in western Norway (Eidfjord). – Nor. entomol. tidsskr. 8: 1–15.

Løken, A. 1958 Pollination studies in apple orchards of western Norway. – Proc. Xth int. congr. entomol. 4: 961–966.

McCann, C. 1952 The tui and its food plants. – Notornis 1952: 6–14.

McGregor, S. E. 1976 Insect pollination of cultivated crop plants. U.S. Dept. agricult. Agricult. handb. 496. 411 pp.

McGregor, S. E., S. K. Alcorn, E. B. Kurtz jr., G. D. Butler jr. 1959 Bee visitors to saguaro flowers. – J. econ. entomol. 52: 1002–1004.

Macior, L. W. 1966 Foraging behavior of Bombus (Hymenoptera: Apidae) in relation to Aquilegia pollination. – Am. j. bot. 53: 302–309.

Macior, L. W. 1968a Pollination adaptation in *Pedicularis groenlandica*. – Am. J. bot. 55: 927–932.

Macior, L. W. 1968b Pollination adaptation in *Pedicularis canadense* – Am. J. bot. 55: 1031–1035.

Macior, L. W. 1969 Pollination adaptation in *Pedicularis lanceolata*. – Am. J. bot. 56: 853–859.

Macior, L. W. 1970 The pollination ecology of *Pedicularis* in Colorado. – Am. J. bot. 57: 716–728.

Macior, L. W. 1971 Co-evolution of plants and animals–systematic insight from plant-insect interaction. – Taxon 20: 17–28.

Macior, L. W. 1974 Behavioral aspects of coadaptation between flower and insect pollinator. – Ann. Mo. bot. gard. 61: 760–769.

Macior, L. W. 1975 The pollination ecology of *Pedicularis* in the Yukon territory. – Am. J. bot. 62: 1065–1072.

Macior, L. W. 1977 The pollination ecology of *Pedicularis* (Scrophulariaceae) in the Sierra Nevada of California. – Bull. Torr. bot. cl. 104: 149–154.

Mackensen, O. and W. P. Nye 1969 Selective breeding of honeybees for alfalfa pollen collection: sixth generation and outcrosses. – J. apicult. res. 8: 9–12.

McNaughton, J. H. and J. G. Harper 1960 External breeding barriers between *Papaver* species – New phytol. 59: 15–26.

McWilliams, J. R. 1958 The role of the micropyle in the pollination of *Pinus*. – Bot. gaz. 120: 109–117.

McWilliams, J. R. 1959 Interspecific incompatibility in *Pinus* – Am. J. bot. 46: 452–453.

Maeterlinck, M. 1907 L'intelligence des fleurs.

Mahabale, T. S. 1968 Spores and pollen of water plants and their dispersal. – Rev. palaeobot. palynol. 7: 285–296.

Maheshwari, P. 1960 Evolution of the ovule. – Birbal Sahni inst. paleobot. Lucknow. 13 pp.

Mandl, K. 1924 Über *Cypripedilum macranthos* Swartz, useine Varietäten und sein natürlicher Bastard mit *C. calceolus* L. – Österr. bot. Z. 73: 267–271.

Manning, A. 1956 The effects of honey-guides. – Behaviour 9: 114–139.

Manning, A. 1957 Some evolutionary aspects of the flower constancy of bees. – Proc. R. phys. soc. 25: 67–71.

March, G. L. and R. M. F. S. Sadleir 1972 Studies on the band-tailed pigeon (*Columba fasciata*) in British Columbia. II: Food resources and mineral-getting activity. – Syesis 5: 279–285. Quoted from Biol. abstr. 55: 48159.

Marden, L. 1963 The man who talks to hummingbirds. – Nat. geogr. mag. 123: 80–99.

Martin, F. W. 1967 Distyly, self-incompatibility, and evolution in *Melochia*. – Evolution 21: 493–499.

Martin, F. W. 1969 Compounds from the stigmas of ten species. – Am. J. bot. 56: 1023–1027.

Martin, F. W. and R. Ruberte 1972 Inhibition of pollen germination and tube-growth by stigmatic substances – Phyton Rev. int. bot. exp. 30: 119–125.

Matthes, D. and E. Scicha 1966 Funktionsmorphologische Studien an Mundwerkzeugen pollenfressender Coleopteren (Malachiidae). – Naturwissenschaften 53: 364.

Maurizio, A. 1942 Pollenanalytische Beobachtungen 10–12. Schweiz. Bienenztg. II: 524–534.

Maurizio, A. 1950a Bienenvergiftungen mit pflanzlichen Wirkstoffen. – Proc. VIII int. bot. congr. Stockholm 1950: 190–192.

Maurizio, A. 1950b The influence of pollen feeding and brood rearing on the length of life and physiological condition of the honeybee. – Bee world 31: 9–12.

Maurizio, A. 1953 Weitere Untersuchungen an Pollenhöschen. – Beih. schweiz. Bienenztg. 2: 486–556.

Maurizio, A. 1959 Pollenkeimung hemmende Stoffe im Korper der Honigbiene. – XVIIth int. Bienenzücht. Kongr. Bologna- Roma 1958.

Maurizio, A. 1975 Der Honig. – Ulmer (Stuttgart). 212 pp.

Mayr, E. 1947 Ecological factors in evolution. – Evolution 1: 263–288.

Mazokin-Poshniakov, G. A. and E. E. Grasvekaya 1966 Recognition of colour combinations and the comparison of this ability in bees and wasps. – Zh. obshch. biol. 27: 112–116.

Medler, J. F. 1962. Effectiveness of domiciles for bumblebees. – Proc. 1st int. congr. pollination. Medd. 7 Sver. fröodlareförb. 8 pp.

Meeuse, A. D. J. 1972 Palm and pandan pollination: primary anemophily or primary entomophily. – Botanique 3: 1–6.

Meeuse, A. D. J. 1974 The different origins of petaloid semaphylls. – Phytomorphology 23: 88–99.

Meeuse, A. D. J. 1975 Changing floral concepts: Anthocorms, flowers and anthoids. – Acta bot. neerl. 24: 23–36.

Meeuse, B. J. D. 1959 Beetles as pollinators. – Biologist 42: 22–32.

Meeuse, B. J. D. 1961 The story of pollination. – Ronald (New York) 243 pp.

Meeuse, B. J. D. 1966 Production of volatile amines and skatoles at anthesis in some arum lily species. – Plant. physiol. 41: 343–347.

Meeuse, B. J. D. 1968 Enige factoren die van invloed zijn op het openen en dichtgaan van bloemen. – Vakbl. biol. 1968: 155–165.

Meeuse, B. J. D. 1975 Thermogenic respiration in aroids. – Ann. rev. plant physiol. 26: 117–126.

Meeuse, B. J. D. and R. G. Buggeln 1969 Time, space, light, and darkness in the metabolic flare-up of the *Sauromatum* appendix. – Acta bot. neerl. 18: 159–172.

Mehra, P. N. 1950 Occurrence of hermaphrodite flowers and the development of female gametophytes in *Ephedra intermedia* Schrenk. et Mey. – Ann. bot. N.S. 14: 164–180.

Meidell, O. 1945 Notes on the pollination of *Melampyrum pratense* and the "honeystealing" of bumblebees and bees. – Bergens mus. årb. 1944 Natv. rk. 11.

Mell, R. 1922 Beiträge zur Fauna sinica. I: Die Vertebraten Südchinas. – Arch. Naturgesch. Ser. A 88: 1–134.

Michener, C. D. 1965 A classification of the bees of the Australian and South Pacific region. – Bull. Am. mus. nat. hist. 139: 1–132.

Miller, R. B. 1973 Maintaining species integrity in Aquilegia: are pollinators relevant? – Abstr. lect. 1st ICESB. 1 p.

Moll, J. W. 1934 Phytography as a fine art. – Brill (Leiden). 543 pp.

Morcombe, M. K. 1968 Australia's western wildflowers. – Landfall Press (Perth). 112 pp.

Morley, B. D. 1971 A hybrid swarm between two hummingbird-pollinated species of *Columnea* (Gesneriaceae) in Jamaica. – J. Linn. soc. Botany 64: 81–96.

Mosebach, G. 1932 Über die Schleuderbewegungen der explodierenden Staubgefässe und Staminodien bei einigen Urticaceen. – Planta 16: 70–115.

Mulcahy, D. L. and D. Caporello 1970 Pollen flow within a tristylous species: *Lythrum salicaria* – Am. J. bot. 57: 1027–1030.

Müller, H. 1873 Die Befruchtung der Blumen durch Insekten. – Engelmann (Leipzig). 479 pp.

Müller, H. 1881 Alpemblumen, ihre Befruchtung durch Insekten. Engelmann (Leipzig), 611 pp.

Müller, L. 1926 Zur biologischen Anatomie der Blüten von *Ceropegia woodii* Schlechter – Biol. gen. 2: 799–814.

Müller, L. 1929 Anatomisch-biomechanische Studien an maskierten Scrophulariaceenblüten. – Österr. bot. Z. 79: 193–214.

Mulligan, G. A. and P. G. Kevan 1973 Color, brightness, and other floral characteristics attracting insects to the blossoms of some Canadian weeds. – Can. J. bot. 51: 1939–1952.

Neal, M. C. 1929 In Honolulu gardens. – Bernice C. Bishop mus. spec. publ. 13. 336 pp.

Nielsen, E. Tetens 1963 Om myg og deres levevis. – Nat. verden 46: 129–160.

Niknejad, M. and M. Khosh-khuni 1972 Natural cross-pollination in gram (*Cicer arietinum* L.). – Ind. J. agric. sci. 42: 273–274. Quoted from Biol. abstr. 55: 65966.

Nilsson, H. 1947 Totale Inventierung der Mikrotypen eines Minimiareals von *Taraxacum officinale*. – Hereditas 33: 119–142.

Nordhagen, R. 1932 Zur Morphologie und Verbreitungsbiologie der Gattung *Roscoea* Sm. – Bergens mus. årb. 1932. Natv. rk. 4, 57 pp.

Nordhagen, R. 1938 Studien über die monotypische Gattung *Calluna* Salisb. I–II. – Bergens mus. årb. 1937 Natv. rk. 4.

Nunez, J. A. 1967 Sammelbienen markieren versiegte Futterquellen durch Duft. – Naturwissenschaften 54: 322–323.

Ockendon, D. J. and L. Currah 1977 Self-pollen reduces the number of cross-pollen tubes in the styles of Brassica oleracea. – New phytol. 78: 675–680.

Olberg, G. 1951 Blüte und Insekt. – Akadem. Verlagsges. (Leipzig).

Omarov, D. S. 1973 Free wind pollination of winter barley. – S.–KH. biol. 8: 374–377. Quoted from Biol. abstr. 57: 5918.

Ordetz, G. S. Ris 1952 Flora Apicola de la America tropical. – Habana.

Ornduff, R. 1966 The origin of dioecism from heterostyly in *Nymphoides* (Menyanthaceae). – Evolution 20: 309–314.

Ornduff, R. 1970 Incompatibility and the pollen economy of *Jepsonia parryi*. – Am. J. bot. 57: 1036–1041.

Ornduff, R. 1972 The breakdown of trimorphic incompatibility in *Oxalis corniculatus*. – Evolution 26: 52–65.

Oster, G. and B. Heinrich 1976 Why do bumblebees major? A mathematical model. – Ecol. monogr. 46: 129–133.

Overland, L. 1960 Endogenous rhythm in opening and odor of flowers of *Cestrum nocturnum*. – Am. J. bot. 47: 378–382.

Palmer-Jones, T. and I. W. Forster 1972 Measures to increase the pollination of lucerne (*Medicago sativa* Linn.). – N.Z. J. agric. res. 15: 186–193.

Pande, G. K., R. Pakrash, M. A. Hassam 1972 Floral biology of barley (*Hordeum vulgare* L.). – Ind. J. agric. sci. 48: 697–703. Quoted from Biol. abstr. 56: 29310.

Pandey, K. K. 1960 Evolution of gametophytic and sporophytic systems of self-incompatibility in angiosperms. – Evolution 14: 98–115.

Pascher, A. 1959 Zur Blütenbiologie einer aasblumigen Liliacee und zur Verbreitungsbiologie abfallender geflügelter Kapseln. – Flora 148: 153–178.

Pedersen, P. H., H. B. Johansen, J. Jørgensen 1961 Pollen spreading in diploid and tetraploid rye. Distance of pollen spreading and risk of intercrossing. – K. vet. landbohøjsk. 1961: 68–86.

Pennell, F. W. 1948 The taxonomic significance of an understanding of floral biology. – Brittonia 6: 301–308.

Percival, M. S. 1947 Pollen collecting by *Apis mellifera*. – New phytol. 46: 142–173.

Percival, M. S. 1955 The presentation of pollen in certain angiosperms and its collection by *Apis mellifera*. – New phytol. 54: 353–368.

Percival, M. S. 1961 Types of nectar in angiosperms. – New phytol. 60: 235–281.

Percival, M. S. 1965 Floral biology. – Pergamon (Oxford). 239 pp.

Petter, J. J. 1962 Recherches sur l'ecologie et l'ethologie des lemuriens malgaches. – Mém. mus. hist. nat.

Pijl, L. van der 1933 Welriekende vliegenbloemen bij *Alocasia pubera*. – De trop. nat. 22: 210–214.

Pijl, L. van der 1936 Fledermäuse und Blumen. – Flora 31: 1–40.

Pijl, L. van der 1937a Disharmony between Asiatic flower-birds and American bird-flowers. – Ann. jard. bot. Buitenz. 48: 17–26.

Pijl, L. van der 1937b Biological and physiological observations on the inflorescence of *Amorphophallus*. – Rec. trav. bot. neerl. 34: 157–167.

Pijl, L. van der 1939 Over de meeldraden van enkele Melastomataceae. – De trop. nat. 28: 169–172.

Pijl, L. van der 1951 On the morphology of some tropical plants: *Gloriosa*, *Bougainvillea*, *Honckenya* and *Rottboellia*. – Phytomorphology 1: 185–188.

Pijl, L. van der 1953 On the flower biology of some plants from Java. – Ann. bogor. 1: 77–99.

Pijl, L. van der 1954 *Xylocopa* and flowers in the Tropics. I–III. – Proc. K. ned. ak. wet. Ser. C. 57: 413–23, 541–562.

Pijl, L. van der 1955 Some remarks on myrmecophytes. – Phytomorphology 5: 190–200.

Pijl, L. van der 1956 Remarks on pollination by bats in the genera *Freycinetia*, *Duabanga* and *Haplophragma* and on chiropterophily in general. – Acta bot. neerl. 5: 135–144.

Pijl, L. van der 1957 Dispersal of plants by bats (chiropterochory). – Acta bot. neerl. 6: 291–315.

Pijl, L. van der 1960–61 Ecological aspects of flower evolution. I–II. – Evolution 14: 403–416; 15: 44–59.

Pijl, L. van der 1969 Evolutionary action of tropical animals on the reproduction of plants. – Biol. J. Linn. soc. 1: 85–92.

Pijl, L. van der 1972 Functional considerations and observations on the flowers of some Labiatae. – Blumea 20: 93–103.

Pijl, L. van der and C. Dodson 1966 Orchid flowers. Their pollination and evolution. – Univ. Miami Press (Coral Gables). 214 pp.

Pohl, F. 1929 Beziehung zwischen Pollenbeschaffenheit, Bestäubungsart und Fruchtknotenbau. – Beih. bot. Zentralbl. 46 I: 247–285.

Pohl, F. 1930 Kittstoffreste auf den Pollenoberflüchen windblütiger Pflanzen. – Beih. bot. Zentralbl. 46 I: 286–305.

Pohl, F. 1935 Ein Fall von zweckloser Proterandrie (*Butomus umbellatus*). – Ber. dtsch. bot. Ges. 53: 779–782.

Pohl, F. 1937a Die Pollenkorngewichte einiger Pflanzen und ihre ökologische Bedeutung. – Beih. bot. Zentralbl. 57: 112–172.

Pohl, F. 1937b Die Pollenerzeugung der Windblütler. – Beih. bot. Zentralbl. 56: 365–470.

Pojar, J. 1973 Pollination of typically anemophilous salt marsh plants by bumblebees, *Bombus terricola occidentalis* Gren. – Am. midl. nat. 89: 448–451.

Pojar, J. 1974 Reproductive dynamics of four plant communities of southwestern British Columbia. – Can. J. bot. 52: 1819–1834.

Ponomarev, A. N. 1966 Some adaptations to anemophily in grasses. – Bot. zh. 51: 28–39.
Ponomarev, A. N. and Yu. N. Prokrudin 1975 Dynamic anemophily and its significance for grass taxonomy and speciation. XIIth int. bot. congr. Sect. 6 Struct. bot. Symp. 3 Flower struct. pollin. ecol. Leningr. 7 pp.
Ponomarev, A. N. and M. B. Rusakova 1968 The daily rhythm of pollination and its role in the speciation in grasses. – Bot. zh. 53: 1371–1383.
Porsch, O. 1905 Beiträge zur histologischen Blütenbiologie. I: Über zwei neue Insektenanlockungsmittel der Orchideenblüten. – Österr. bot. Z. 55: 165–173, 227–235, 253–260.
Porsch, O. 1922 Methoden der Blütenbiologie. – Abderhaldens Handb. biol. Arbeitsmeth. 11, 1: 395–514.
Porsch, O. 1924, 1929 Vogelblumenstudien. I–II. – Jahrb. wiss. Bot. 63: 553–706; 70: 181–277.
Porsch, O. 1926a – 30 Kritische Quellenstudien über Blumenbesuch durch Vögel. I – V. – Biol. gen. 2: 217–240; 3: 171–206, 475–548; 5: 157–210; 6: 133–246.
Porsch, O. 1926b Vogelblütige Orchideen. – Biol. gen. 2: 107–136.
Porsch, O. 1931a Grellrot als Vogelblumenfarbe. – Biol. gen. 9: 647–674.
Porsch, O. 1931b Crescentia, eine Fledermausblume. – Österr. bot. Z. 80: 32–44.
Porsch, O. 1934–36 Säugetiere als Blumenbesucher und die Frage der Säugetierblume. – Biol. gen. 10: 657–685; 11: 171–185; 12: 1–8.
Porsch, O. 1936b Säugetierblumen. – Forsch. Fortschr. 12: 207.
Porsch, O. 1937a Die Bestäubungseinrichtungen der Loxanthocerei. – Cactaceae (jahrb. dtsch. Kakt. – Ges.) 1937: 15–19.
Porsch, O. 1938–39 Das Bestäubungsleben der Kakteenblüte. – Cactaceae (Jahrb. dtsch. Kakt. – Ges.) 1938 – 39.
Porsch, O. 1941 Ein neuer Types Fledermausblumen. – Biol. gen. 15: 283–294.
Porsch, O. 1950 Geschichtliche Lebenswertung der Kastanienblüte. – Österr. bot. Z. 97: 269–321.
Porsch, O. 1956 Windpollen und Blumeninsekt. – Österr. bot. Z. 103: 1–19.
Porsch, O. 1958 Alte Insektentypen als Blumenausbeuter. – Österr. bot. Z. 104: 115–164.
Poulton, E. B. 1938 The conception of the species as interbreeding communities. – Proc. Linn. soc. London 150: 225–226.
Powell, J. A. and R. A. Mackie 1966 Biological interrelationships of moths and Yucca whipplei. – Univ. Cal. publ. entomol. 42.
Prance, G. T. 1976 The pollination and androphore structure of some Amazonian Lecythidaceae. – Biotropica 8: 235–241.
Prance, G. T. and A. B. Anderson 1976 Studies of the floral biology of neotropical Nymphaeaceae 3. – Acta amazonica 6: 163–170.
Prazmo, 1960 Genetic studies in the genus Aquilegia L. I: Crosses between Aquilegia vulgaris L. and Aquilegia ecalcarata Maxim. – Acta soc. bot. pol. 29: 57–77.
Priesner, E. 1973 Reaktionen von Riechrezeptoren männlicher Solitair-bienen auf Inhaltsstoffe von Ophrys – Blüten. – In Kullenberg and Stenhagen 1973: 43–51.
Pruzcinsky, S. 1960 Über Trocken- und Feuchtluftresistenz des Pollens. – Sitzungsber. österr. Akad. Wiss. Mat. – natw. Kl Abt. 1. 169: 43–100.
Pryce-Jones, J. 1944 Some problems associated with nectar, pollen and honey. – Proc. Linn. soc. London 153: 129–174.
Pulliam, H. R. 1974 On the theory of optimal diet. – Am. nat. 108: 59–74.
Ramirez, W. 1969 Fig wasps: mechanism of pollen transfer. – Science 163: 580–581.
Rattray, G. 1913 Notes on the pollination of some South African cycads. – Trans. R. soc. S. Afr. 3: 259–270.
Raven, P. H. 1972. Why are bird-pollinated flowers predominantly red? – Evolution 26: 674 –.
Raw, A. 1974 Pollen preference of three Osmia species (Hymenoptera). – Oikos 25: 54–60.
Rempe, H. 1937 Untersuchungen über die Verbreitung des Blütenstaubes durch die Luftströmungen. – Planta 27: 93–147.
Resvoll, T. R. 1918 En utpreget selvbestøver. – Nyt mag. naturvid. 56: 131–135.
Rhein, W. von 1952–3 Über die Duftlenkung der Bienen beim Raps im Jahre 1952 und ihre Ergebnisse. – Hess. Biene 88: 192–194.
Ribbands, C. R. 1955 The scent perception of the honeybee. – Proc. R. soc. B 143: 367–379.
Roberts, H. F. 1929 Plant hybridization before Mendel. – Facsimile reprint 1965. Hafner (New York). 374 pp.
Roggen, H. P. J. R. 1972 Scanning electron microscopical observation on compatible and incompatible pollen-stigma interaction in Brassica. – Euphytica 21: 1–10.
Roggen, H. P. J. R. and A. J. van Dijk 1974 Electric aided and bud pollination: which method to use for seed production in cole crops (Brassica oleracea)? – Euphytica 22: 260–263. Quoted from Biol. abstr. 57: 8229.
Sachs, J. 1875 Geschichte der Botanik vom 16. Jahrhundert bis 1860. – (München).

Sazima, M. and I. Sazima 1975 Quiropterofilia en *Lafoensia pacari* St. Hil. (Lythraceae), na Seira do Cipo, Minas Gerais. – Cienc. cult. 27: 405–416.

Schlising, R., F. R. Hainsworth, F. G. Stiles 1972 Energetics of foraging rate and efficiency of nectar extraction by humming birds. – Science 176: 1351–1352.

Schmid, R. 1975 Two hundred years of pollination biology: an overview. – Biologist 57: 26–35.

Schmucker, T. 1932 Physiologische und ökologische Untersuchungen an Blüten tropischer *Nymphaea* Arten. – Planta 16: 376–412.

Schoenichen, W. 1902 Achtzig Schemabilder aus der Lebensgeschichte der Blüten. – Braunschweig (Goeritz). 156 pp.

Scholze, E., H. Pichler, H. Herau 1964 Zur Entfernugsschätzung der Bienen nach dem Kraftaufwand. – Naturwissenschaften 51: 69–70.

Schremmer, F. 1941 Sinnesphysiologie und Blumenbesuch des Falters von *Plusia gamma* L. – Zool. Jahrb. Abt. Syst. Ökol. 74: 375–435.

Schremmer, F. 1953 Blütenbiologische Beobachtungen an Labiaten. – Österr. bot. Z. 100: 8–24.

Schremmer, F. 1955 Über anormalen Blütenbesuch und das Lernvermögen besuchender Insekten. – Österr. bot. Z. 102: 551–571.

Schremmer, F. 1959 Blütenbiologische Beobachtungen in Istrien. – Österr. bot. Z. 106: 177–202.

Schremmer, F. 1960 *Acanthus mollis*, eine europäische Holzbienenblüte. – Österr. bot. Z. 107: 84–105.

Schremmer, F. 1963 Wechselbeziehungen zwischen Pilzen und Insekten. Beobachtungen an der Stinkmorchel, *Phallus impudicus* L. ex Pers. – Österr. bot. Z. 110: 380–400.

Schremmer, F. 1969 Extranuptiale Nektarien. Beobachtungen an *Salix elaeagnos* Scop. und *Pteridium aquilinum* (L.) Kuhn. – Österr. bot. Z. 117: 205–222.

Schrottky, C. 1908 Blumen und Insekten in Paraguay. – Z. wiss. Insekten.-Biol. 4: 22–26, 47–52, 73–78.

Schuster, J. 1932 Cycadaceae. – Pflanzenreich IV, 1. 168 pp.

Semerikov, L. F. and N. V. Glotov 1971 Evolution of the isolation in populations of durmast oak (*Quercus petraea* Liebl.). – Genetika 7: 65–71.

Sernander, R. 1906 Über postflorale Nektarien. – Bot. stud. F. B. Kjellmann: 276–287.

Sharma, M. 1970 An analysis of pollen loads of honeybees from Kangra, India. – Grana 10: 35–42.

Sharp, M. A., D. R. Parker, P. R. Ehrlich 1974 Plant resources and butterfly selection. – Ecology 55: 870–875.

Shaw, R. C. 1962 The biosystematics of *Scrophularia* in western North America. – Aliso 5: 147–148.

Silberbauer-Gottsberger, I. 1973 Blüten- und Fruchtbildung von *Butia Leiospatha* (Arecaceae). – Österr. bot. Z. 121: 171–185.

Simpson, B. B., J. L. Neff, D. Siegler 1977 Krameria, free fatty acids and oil collecting bees. – Nature 267: 150–151.

Simpson, J. 1966 Repellency of mandibular gland scent of worker honeybees. – Nature 209: 531–532.

Skovgaard, O. S. 1970 Den Kaukaisiske honningbi som rødkløverbestøver. – Tidsskr. planteavl 59: 877–878.

Smith, A. C. 1949 Additional notes on *Degeneria vitiensis*. – J. Arnold arbor. 30: 1–9.

Smith, E. B. 1970 Pollen competition and relatedness in *Haplopappus* section *Isopappus* (Compositae). II. – Am. J. bot. 57: 874–881.

Soderstrom, T. R. and Calderon, C. E. 1971 Insect pollination in tropical rain forest grasses. – Biotropica 3: 1–16.

Sols, A., E. Cadenas, F. Alvaredo 1960 Enzymatic basis of mannose toxicity in honeybees. – Science 131: 297–298.

Soria Vasco, S. de J. 1970 Studies on *Forcipomyia* spp. midges (Diptera, Ceratopogonidae) related to the pollination of *Theobroma cacao* L. – Diss. abstr. 31: 2744 B.

Springensguth, W. 1935 Physiologische und ökologische Untersuchungen über extraflorale Nektarien und die sie besuchenden Insekten. – Sitzungsber. Abh. naturf. Ges. Rostock 3. F. 5: 31–111.

Stäger, R. 1902 Chemischer Nachweis von Nektarien bei Pollenblumen und Anemophilen. – Beih. bot. Zentralbl. 12: 34–43.

Stanley, R. G. and H. F. Linskens 1974 Pollen. Biology, biochemistry, management. – Springer (New York). 307 pp.

Stebbins, G. L. 1957 Self-fertilization and population variability in higher plants. – Am. nat. 91: 337–354.

Stebbins, G. L. 1958 Longevity, habitat, and release of genetic variability in higher plants. – Cold Spring Harbor symp. quant. biol. 23: 365–378.

Stephen, W. P. 1961 Artificial nesting sites for the propagation of the leaf-cutter bee *Megachile* (*Eutricheraea*) *rotundata* for alfalfa pollination. – J. econ. entomol. 54: 989–993.

Stevens, P. F. 1976 The altitudinal and geographical distributions of flower types in *Rhododendron* section *Vireya*, especially in the Papuasian species, and their significance. – Biol. J. Linn. soc. 72: 1–34.

Stiles, F. G. 1971 Time, energy, and teritoriality of the Anna humming bird (*Calypta anna*). – Science 173: 818–821.

Stirton, C. H. 1976 Thuranthos: notes on generic status, morphology, phenology and pollination biology. – Bothalia 12: 161–165.

Stone, D. 1957 Studies in population differentiation and variation in *Myosurus* of the Ranunculaceae. – Thesis Ph.D. Univ. Cal. Berkeley.

Stout, A. B. 1926 The flower behaviour of avocados. – Mem. N.Y. bot. gard. 7: 145–203.

Stout, A. B. 1933 Dichogamy in flowering plants. – Bull. Torr bot. club 55: 141–153.

Stoutamire, W. P. 1967 Flower biology of the Lady's slipper (Orchidaceae: *Cypripedium*). – Mich. bot. 6: 159–175.

Stoutamire, W. P. 1968 Mosquito pollination of *Habenaria obtusata* (Orchidaceae). – Mich. bot. 7: 203–212.

Stoutamire, W. P. 1974 Australian terrestrial orchids, thynnid wasps, and pseudocopulation. – Am. orch. soc. bull. 1974: 13–18.

Straw, R. M. 1955 Hybridization, homogany, and sympatric speciation. – Evolution 9: 441–444.

Straw, R. M. 1972 A Markov model for pollination constancy and competition. – Am. nat. 106: 597–620.

Straw, R. U. 1956 Adaptive morphology of the *Pentstemon* flower. – Phytomorphology 6: 112–119.

Stuessy, T. F. 1972 Revision of the genus *Melampodium*. – Rhodora 74: 1–70, 161–222.

Süssenguth, K. 1936 Über den Farbwechsel von Blüten. – Ber. dtsch. bot. Ges. 54: 409–417.

Swan, K. 1961 The ecology of the high Himalayas. – Sc. Am. 205: 68–78.

Swynnerton, C. F. M. 1915 Short cuts by birds to nectaries. Short cuts to nectaries by blue-tits. – J. Linn. soc. London 43: 381–422.

Sørensen, T. 1970 Beretning om Botaniskhaves virksomhed for året 1969. København 1970.

Takhtajan, A. 1973 Evolution und Ausbreitung der Blütenpflanzen. – Fischer (Jena). 189 pp.

Tanaka, H. 1972 Pollination of *Plectranthus inflexus*. – J. Jap. bot. 47: 249–254.

Taylor, G. R. 1954 Sex in history. – Ballantine (New York). 320 pp.

Thien, L. B. 1974 Floral biology of *Magnolia* – Am. J. bot. 61: 1037–1045.

Thien, L. B. and F. Utech 1970 The mode of pollination in *Habenaria obtusata* (Orchidaceae). – Am. J. bot. 57: 1031–1035.

Thijsse, J. P. 1934 De bloemen en haar vrienden. – Verkade (Zaandam). 93 pp.

Thomson, G. M. 1927 The pollination of New Zealand flowers by birds and insects. – Trans. proc. N.Z. inst. 57: 106–125.

Tikhmenev, E. A. 1974 On the pollen viability of arctic grasses. – Bot. zh. 59: 1520–1524.

Todd, F. E. and S. E. McGregor 1960 The use of honeybees in the production of crops. – Ann. rev. entomol. 5: 265–278.

Toledo, V. H. 1974 Observations on the relationships between hummingbirds and *Erythrina* species. – Lloydia 37: 482–487.

Trelease, W. 1881 The fertilization of *Salvia splendens* by birds. – Am. nat. 15: 265–269.

Troll, C. 1943 Thermische Klimatypen der Erde. – Petermanns Mitt. 89: 81–89.

Troll, W. 1922 Über Staubblatt- und Griffelbewegungen und ihre teleologische Deutung. – Flora N.F. 15: 191–250.

Troll, W. 1929 *Roscoea purpurea* Sm., eine Zingiberacee mit Hebelme-chanismus in den Blüten. – Planta 7: 1–29.

Troll, W. 1951 Botanische Notizen. II–III. – Akad. Wiss. Lit. (Mainz) Abh. mat. -natw. Kl. 1951: 81–117.

Tschermak-Seysenegg, E. 1957 Blütenbiologische Beobachtungen an *Co-diaeum variegatum* am Fensterbrett. – Ber. dtsch. bot. Ges. 70: 449–452.

Uphof, J. C. T. 1938 Cleistogamic flowers. – Bot. rev. 4: 21–50.

Valle, J. J. and D. R. Cirino 1972 Biologia floral del irupe, *Victoria cruziana* D'Orb. – Darwiniana 17: 477–497.

Vello, F. and W. S. Magalhaes 1971 Estudos sobre a participacao de formiga cacarema (*Azteca chartifex spiriti* Forel) na polinizacao do cacaueiro na Bahia. – Rev. theobroma 1: 29–42. Quoted from Biol. abstr. 55: 18933.

Vereshchagina, V. A. 1965 The ecology of flowering and pollination of *Oxalis acetosella*. – Bot. zh. 50: 1078–1091.

Visser, T. 1955 Germination and storage of pollen. – Meded. landbouwhogesch. Wageningen 55: 1–68.

Vogel, S. 1950 Farbwechsel und Zeichnungsmuster bei Blüten. – Österr. bot. Z. 97: 44–100.

Vogel, S. 1954 Blütenbiologische Typen als Elemente der Sippengliederung. – Bot. Stud. 1: 1–338.

Vogel, S. 1955 Über den Blütendimorphismums einiger südafrikanischer Pflanzen. – Österr. bot. Z. 102: 486–500.

Vogel, S. 1958 Fledermausblumen in Südamerika. – Österr. bot. Z. 105: 491–530.
Vogel, S. 1959 Organographie der Blüten kapländischer Ophrydeen. 1–11. – Akad Wiss. Litt. (Mainz) Abh. Mat.-natw. Kl. 1959: 265–532.
Vogel, S. 1961 Die Bestäubung der Kesselfallenblüten von Ceropegia – Beitr. Biol. pfl. 36: 159–237.
Vogel, S. 1963 Duftdrüsen im Dienste der Bestäubung – Akad. Wiss. Litt. (Mainz) Abh. Mat.-natw. Kl. 1962: 605–763.
Vogel, S. 1965a Kesselfallenblumen. – Umschau 65: 12–16.
Vogel, S. 1965b Mutualismus und Parasitismus in der Nutzung von Pollenträgern. – Verh. dtsch. zool. Ges. 1975: 102–110.
Vogel, S. 1966 Parfümsammelnde Bienen als Bestäuber von Orchideen und Gloxinia. – Österr. bot. Z. 113: 302–361.
Vogel, S. 1967 Iris fulva Ker.-Gawl., eine Kolibri-Blume.–Jahrb. dtsch. Lilienges. 1967: 3–11.
Vogel, S. 1968a, 1969a Chiropterophilie in der neotropischen Flora. I – III. – Flora 157: 562–602; 158: 185–222, 269–323.
Vogel 1968b Scent organs of orchid flowers and their relation to insect pollination. – Proc. 5th world orch. conf. Long Beach (Cal.) 1966: 253–259.
Vogel, S. 1969 Flowers offering fatty oil instead of nectar. – Abstr. Xlth int. bot. congr. : 229.
Vogel, S. 1972 Pollination von Orchis papilionacea L. in den Schwarmbahnen von Eucera tuberculata F. – Jahresber. naturw. Ver. Wuppertal 25: 67–74.
Vogel, S. 1973 Fungus mimesis of fungusgnat flowers. – In Brantjes and Linskens 1973: 13–18.
Vogel, S. 1974 Ölblumen und ölsammelnde Bienen. – Trop. subtrop. Pflanzenw. (7): 283–547.
Vogel, S. 1975 V. Blütenökologie. – Fortschr. bot. 37: 379–392.
Vogel, S. 1975a Oelblumen der Holarktis. – Naturwissenschaften 63: 44–45.
Vogel, S. 1975b Campanula rotundifolia (Campanulaceae) Pollination durch Apis mellifica (Hymenoptera) – Melittophilie. – Encyclopaedia cinematographica E 2049. 7 pp.
Vogel, S. 1976a Lysimachia: Oelblumen der Holarktis. – Naturwissenschaften 63: 44–45.
Vogt, P. 1966 Bienen unterscheiden Flimmerfrequenzen im Verhaltensexperiment. – Naturwissenschaften 53: 536.
Vogt, P. 1966 Bienen unterscheiden Flimmerfrequenzen im Verhaltensexperiment. – Naturwissenschaften 53: 536.
Vuillemier, B. S. 1967 The origin and evolutionary development of heterostyly in plants. – Evolution 21: 210–226.
Wafa, A. K., S. H. Ibrahim, M. A. Ewers 1973: Activity of honeybees on alfalfa, Medicago sativa, L., and its competitor plants (Hymenoptera, Apoidea). – Bull. soc. entomol. Egypte 56 (1972) 227–233. Quoted from Biol. abstr. 57: 7384.
Wahl, O. 1966 Besitzen höselnde Bienen einen Spursinn für den Nährwert des Sammelgutes? – Z. Bienenforsch. 8: 229–235.
Waller, G. D. 1970 Attracting honeybees to alfalfa with citral, geraniol and anise. – J. apicult. res. 9: 9–12.
Waller, G. D., G. M. Loper, R. L. Berdel 1974 Olfactory discrimination by honeybees of terpenes identified from volatile oil of alfalfa flowers. – J. apicult. res. 13: 191–197.
Wanndorp, H. E. 1974 Calotropis gigantea (Asclepidaceae) and Xylocopa tenuiscapa (Hymenoptera, Apoidea). – Sven. bot. tidskr. 68: 25–32.
Watt, W. B., P. C. Hoch, S. G. Mills 1974: Nectar resource use by Colias butterflies: chemical and visual experiments. – Oecologia (Berlin) 14: 353–374.
Webb, D. A. 1957 The vasculum and the microtome. – Adv. sci. 55: 183–190.
Wemple, D. K. and N. E. Leersten 1966 An interpretation of the flower of Petalostemon. – Brittonia 18: 117–125.
Wendelbo, P. 1965 The genus Pedicularis in Afghanistan with notes on the floral morphology of P. bicornuta. – Nyt. mag. bet. 12: 123–134.
Werth, E. 1915 Kurzer Überblick über die Gesamtfrage der Ornithophilie. – Bot. Jahrb. 53: 313–378.
Werth, E. 1956a Bau und Leben der Blumen. – Enke (Stuttgart). 204 pp.
Werth, E. 1956b Zur Kenntnis des Androeceums der Gattung Salvia und seiner stammesgeschichtlichen Wandlung. – Ber. dtsch. bot. Ges. 69: 381–386.
Wester, J. 1910 Pollination experiments with Annona. – Bull. Torr. bot. cl. 37: 522–534.
Wettstein, R. von 1888 Über die Kompositen der österreichisch-ungarischen Flora mit zuckerausscheidenden Hüllschuppen. – Sitzungsber. Akad Wiss. Wien. Mat. -natw. Kl. 97: 570–589.
Whigham, D. 1974 An ecological life history study of Uvularia perfoliata L. – Am. midl. nat. 91: 343–359.
Whitehouse, H. L. K. 1950 Multiple-allelomourph incompatibility of pollen and style in the evolution of the angiosperms. – Ann. bot. N.S. 14: 199–215.

Wiebes, J. T. 1963 Taxonomy and host preferences of Indo-Australian fig wasps of the genus *Ceratosolen* (Agaonidae). – Tijdshr. entomol. 106: 1–112.

Wiebes, J. T. 1966 Provisional host catalogue of fig wasps (Hymenoptera, Chalcidoidea). – Zool. Verh. Leiden 83. 44 pp.

Wiesmann, R. 1962 Geruchsorientierung der Stubenfliege, *Musca domestica*. – Z. angew. entomol. 50: 74–81.

Williams, N. H. and C. H. Dodson 1972 Selective attraction of male euglossine bees to orchid floral fragrances and its importance in long distance pollen flow. – Evolution 26: 84–95.

Willis, J. C. 1895 Contribution to the natural history of the flower. I–II. – J. Linn. soc. London 30: 51–63, 284–298.

Wilson, B. H. and M. Lieux 1972 Pollen grains in the guts of field collected tabanids in Louisiana. – Ann. entomol. soc. Am. 65: 1264–1266. Quoted from Biol. abstr. 55: 31711.

Winder, J. A. and P. Silva 1972a Cacao pollination: microdiptera of cacao plantations and some of their breeding places. – Bull. entomol. res. 61: 651–655. Quoted from Biol. abstr. 55: 54301.

Winder, J. A. and P. Silva 1972b Pesquisa sobre a polinizacao do cacaueiro por insectos na Bihia. – Rev. theobroma 2: 36–46. Quoted from Biol. abstr. 56: 42390.

Witherell, P. C. 1972 Can hairless honeybees collect pollen? – Am. bee. J. 112: 119–131.

Wodehouse, R. P. 1935 Pollen grains – McGraw-Hill (New York). 574 pp.

Wodehouse, R. P. 1945 Hayfever plants. – Chronica botanica (Waltham, Mass.). 245 pp.

Wolf, E. 1933 Das Verhalten der Bienen gegenüber flimmerndenn Feldern und bewegten Objekten. – Z. vergl. Physiol. 20: 151–161.

Wolf, L. L. 1975 Energy intake and expenditure in a nectar-feeding sunbird. – Ecology 56: 92–104.

Wolf, L. L. and F. R. Hainsworth 1971 Time and energy budgets of territorial hummingbirds. – Ecology 52: 980–988.

Wolff, T. 1950 Pollination and fertilization of the fly ophrys, *Ophrys insectifera* L. in Allindelille fredskov, Denmark. – Oikos 2: 20–59.

Wolff, T. 1951 Ecological investigations on the fly ophrys, *Ophrys insectifera* L. in Allindelille fredskov, Denmark. – Oikos 3: 71–97.

Wright, W. 1953 Some practical applications of pollen-dispersion studies. – J. for. 51.

Yasuda, K. K. 1939 Parthenocarpy induced by stimulation of pollination in some higher plants. – Mem. fac. sci. agric. Taihoku imp. univ. 27: 1–50.

Zander, E. 1935–49. Beträge zur Herkunftbestimmung bei Honig. I–IV. – Liedloff *et al.* (Berlin, Leipzig), Ehrenwirth (München).

Zawortnik, T. J. 1972 A new subgenus and species of *Megandrena* from Nevada, with notes on its foraging and mating behaviour (Hymenoptera; Andrenidae). – Proc. entomol. soc. Wash. 74: 61–75. Quoted from Biol. abstr. 54: 31877.

Zeisler, M. 1938 Über die Abgrenzung der eigentlichen Narbenfläche mit Hilfe von Reaktionen. – Beih. bot. Zentralbl. A 58: 308–318.

Zimmermann, J. G. 1932 Uber die extrafloralen Nektarien der Angiospermen. – Beih. bot. Zentralbl. Abt. 1, 49: 99–196.

INDEX OF PLANT NAMES

If only one species of a genus is mentioned, its name is indexed together with that of the genus, in parentheses. Similarly for genus vs. family etc. The bold page numbers refer to Case histories or definitions/discussions.

Abies (homolepis) 12
Abutilon 123
Acacia pycnantha 64
Acanthus 91–2, 95
Acer (pseudoplatanus) 26, 35–6
Achillea millefolium 50
Aconitum (septentrionale) 47, 90, 92, 113
Acorus calamus 141
Acrotonae 202
  (subfam. Orchid.)
Adansonia 60, 132
  digitata 122
  gregorii 130
Aeolanthus 150
Aeschynomene 41
Aesculus (hippocastanum) 36, 83
Agave (schottii) 130, 132, **179**
Akebia 21
Alchemilla 140
Algae
Alfalfa see *Medicago sativa* 9
Alisma (plantago) 38, 40
Allium 42, 162
Alocasia pubera 74
Aloë (ferox) 127–8
Ambrosia 34, 38
Amentiferae 34, 61, 134
Amorphophallus 77
  titanum 101, 105
  variabilis 71, 94
Amphitechna 133
Anacamptis pyramidalis 118
Anapalina 98
Andropogon 138
Anemone (coronaria) 60, 72, 74, 82
Angraecum (sesquipedale) 45, 118, 205
Annonaceae 27, 73–5, 77, 94, 100, 104
Antennaria alpina 141
Anthirrinum (-eae) 149
Antholyza ringens 127
Anthurium 70
Anthyllis 145
Aquilegia 68, 92, 154
  formosa, pubescens 153
Araceae 65, 70–1, 82, 84, 86, 94, 103
Arachis hypogaea 145

Arachnis flos-aeris 85
Araujoa 115
Archangelica 113
Arctostaphylos otayensis 56
Arisaema (laminatum) 86, 104, 156
Arisarum proboscideum 105
  vulgare 86, **173**
Aristolochia (-ceae) 57, 103–5
  clematitis 106
  grandiflora 94, 104
  tricaudata 104
Armeria 32, 137
Arnebia echioides 83
Artemisia 38
Artocarpus heterophyllus 74
Arum 42, 52, 85, 94
  conophalloides 79, 106
  maculatum 96, 104
  nigrum 71, **172**
Aruncus 103
Asarum 105
Asclepias (-adaceae) 18, 21–22, 52, 103, 115, 120
Aseroë 104
Aster 97, 115
Astragalus (depressus) 142, 144, **180**, 183
Athyrium filix-femina 1
Avena sativa 137

Bacteria 9
Balsa see *Ochroma*
Bambusoideae 58
Banana (*Musa sapientium*) 132
Banksia attenuata 122
Baobab (*Adansonia*) 60, 132
Barley see *Hordeum*
Barnadesia 82
Bartsia (alpina) 140, 191
Basitonae 200
Bassia 130
Bauhinia 53, 132, 143
  megalandra 129
  pauletii 132
Beans (*Phaseolus*) 137
Begonia 58
Berberidaceae 17, 27

Betula (-ceae)   37–9, 89
  verrucosa   36
Bignoniaceae   132–3
Bikkia comptonii   90
Bixa orellana   110
Blastophaga psenes   178
Bocconia   40
Boerlagiodendron   123
Bombax (-caceae)   123–4, 128, 132–3
  malabaricum   133
  valetonii   133
Boraginaceae   32, 89
Bougainvillea   117
Bovistaceae   9
Brassia   75
Brassica   29
Bromeliaceae   74, 128
Buddleia   117
Bulbophyllum   23
Burmanniaceae   103
Butomus   24, 29

Cactaceae   64, 104, 117, 128, 130, 132–3
Cadaba   128
Caesalpinia pulcherrima   119
Caesalpiniaceae   32, 143
Caladenia   76
Calceolaria (*uniflora*)   17, 46, 69, 123
Calliandra   52
Callitriche autumnalis, hamulata   41
Calluna vulgaris   16, 35, 68, 72, 120, 135, 160,
  173
Calochilus   75
Calonyction bona-nox   118
Caltha   90, 115
Calycanthus (-aceae)   23, 28, 70, 94, 100–1
  occidentalis   170
Calypso   51
Calystegia   90
Camellia sasanqua   30
Camoensia   145
Campanea   132
Campanula (-ceae)   17, 27, 86, 90, **166**
Camptosema nobile   143–6
Canavallia   110, 184
Canna   16, 17
Cannabis   34
Capparidaceae   32, 117, 133
Caprifoliaceae   27
Cardamine chenopodifolia   139
Carica papaya   162
Carnegiea (*gigantea*)   47, 56, 132, 140, **179**
Caryophyllaceae   27, 30, 72
Cassia   17, 53, 62, 123, 143
Castanea (*vesca*)   37, 91, 134
Castilleja (*coccinea*)   21, 91–2, 121, 147–8
Casuarina   121
Catasetum (-inae)   22, 26, 70, 80, 203
Catopheria   147
Cattleya (*aurantiaca*)   129, **202**
Caytoniales   20

Ceiba (*pentandra*)   40, 43, 131–3
Celosia argentea   109
Centaurea   17–8, 22
  cyanus   136
Centrosema (*virginiana*)   91, 113, 145–6, **184**,
  199
Ceratophyllum   41
Ceropegia   86, 104, 106
Cestrum nocturnum   117
Chamaenerium angustifolium   27, 56
Characeae   77
Chestnut see *Castanea vesca*
Chincona   162
Chrysothrix capensis   135
Chrysanthemum   90
  carinatum   85
  leucanthemum   20
Cicendia filiformis   115
Cicer arietinum   49
Cichorium (*intybus*)   89, 152
Circaea   102
Cirrhopetalum   104
Citrullus   113
Citrus   126
Clathrus   104
Claviceps   9, 159
Clianthus puniceus   146
Clitoria   145, 184
Cocoa see *Theobroma*
Coffea   58
Coleus frederici   199
Columnea florida   128
Commelina coelestis   31, 62, 137
Commelinaceae   32
Compositae   16, 17, 27–9, 38, 82, 89, 117, 126
Coniferae   11, 26, 29, 38, 61
Connaraceae   33
Conopodium (*majus*)   81, 90
Convolvulus arvensis   85
Cornflower (*Centaurea cyanus*)   136
Corniculatae (sect. *Oxalis*)   33
Cornus (*mas*)   IX, 16, 20, 82, 101
Coronilla emerus   144–5, **181**, 196, 198
Coryanthes   171
Corydalis   91, 151
Corylus (*avellana*)   18, 36–8
Corytholoma   128
Cosmos   82
Cotton (*Gossypium*)   61, 161
Cranberry (*Oxycoccus*)   160
Crescentia (*cujete*)   129, 133
Crocus   124
Cruciferae   27
Cryptanthemis   139
Cryptocoryne griffithii   101
Cucumber (*Cucumis*)   162–3
Cuphea   128
Curcurbita (-ceae)   47, 71, 163
Cycadeoidea   71, 72
Cycads   11, 58, 73, 77, 81, 140
Cycas (*circinalis*)   18, 86
Cyclanthaceae   77

Cyperaceae   34, 37, 134—5
Cypripedilinae   79, 104
Cypripedium calceolus   86, 94, 154, **199**
Cypripedium debile   105, 200
Cytisus   22, 52
   racemosus   144
   scoparius   62, 91, 144, **183**

Dalechampia   111
Dama di noche see *Costrum*
Dandelion see *Taraxacum*
Darlingtonia   42, 104
Daucus carota   162
Degeneria   16, 101
Delphinium   92, 113
Dendrobium   129
Desmodium   144
Dicentra   91
Dichromena   135
Dictamnus   91, 95
Digitalis   18, 85
   lanata, purpurea   147
Dillenia   123
Disa   129
Dodecatheon (pauciflorus)   136, 157
Dorstenia   178
Drakea   76
Drimys (brasiliensis)   21, 81
Drymonia   132
Dryopteris filix-mas   1
Dumoria heckelii   130
Durio   129, 133

Eichhornea crassipes   33
Elaeis guineensis   73, 163
Elaeocarpus ganitrus   124
Elleanthus capitatus   129
Elodea   41
Encephalartos   11, 74
Eperua (falcata)   129, 132
Ephedra *(campylopoda)*   11, 12, 63
Epidendrum o'brienianum   94
Epipactis (palustris)   23, 109
Epiphyllum   126
Eragrostis   135
Eria (vulpina)   71, 111
Erica (-ceae)   17, 52, 120
Eriope   150
Erysimum amoenum, nivale   157
Erythrina   123, 125—7, 146
   caffra   146
   crista-galli   146, **184**
   variegata var.   146
Eucalyptus (diversifolia)   121, 125—7, 129
Eucomis   104
Eugenia cauliflora   132
Eupatorium (cannabinum)   90, 94
Euphorbiaceae   65, 110, 128
Euphrasia   150
Eusphace (sect. *Salvia*)   148
Exacum   17, 58, 62, 83

Fagaceae   101, 134
Fagus (silvatica)   15, 36
Ferns (Filices)   10
Ficus   15, 22, 29, 71—2, 90, 109, 135, **176**
   carica   176, 179
   costaricana, hemsleyana   178
Fraxinus   34
Freycinetia arborea   122—3
Freycinetia funicularis   123
Freycinetia insignis   129—30
Fuchsia (fulgens)   126—8
Fumaria (-ceae)   29, 91
Fungi   9, 63

Galeopsis (speciosa)   22, 146—8, **188—9**, 197—8
Galium (hercynicum)   81, 87, 140
Garcinia   17
Gardenia   117
Genista tinctoria   144, 151, **182**
Gentiana acaulis   86, 90
Gentianaceae   32
Geranium   28, 85, 89
Gesneriaceae   132
Ginkgo   11—12, 140
Gladiolus grandis   83
Glaux maritima   110
Glechoma   147
Gloriosa rothschildiana   21
Gloxinia speciosa   70
Gnetum   12
Goodeniaceae   17
Gossampinus (heptaphyllus)   123, 132—3
Gossypium   61, 161
Gramineae   34, 36—9, 60, 103, 134
Grasses *see* Gramineae
Guava *(Psidium guajava)*   132
Gymnadenia (conopea)   92, 118
Gymnospermae   10, 63

Habenaria obtusata   106
Halophila   41
Hardwickia   145
Hedera   102
   colchica   109
   helix   103, 109, 141
Helianthus   28
Heliconia (rostrata)   56, 127
Helleborus   60
Heracleum   113, 160
Hesperis tristis   90, 118
Heteropogon contortus   135
Hibiscus henningsianus   110
Hieracium   140
Hordeum sativum   37, 137, 159
Hottonia   40
Houttynia   82
Hydnoraceae   72, 103
Hydrilla   41
Hydrochariataceae   41
Hyptis   150

Ilex   102, 162

Impatiens 27, 92, 117
Ipomoea (albivena) 13, 110, 130
Ipomopsis aggregata 152
Ipomopsis congesta 101
Iris (-idaceae) 19, 21, 32, 69, 91, 95, 110
   florentina 78
    fulva 128
    pseudacorus 22, 95, 106
    sibirica 85
Ischaemum muticum 135

Jatropha curcas 109
Juncaceae 27, 34, 36, 134
Junci genuini 37
Jungermanniaceae 79

Kallstroemia grandiflora 61, 69
Kapok *v. Ceiba*
Kentrosiphon 92
Kigelia 129, 131–3
Krameriaceae 69

Labiatae 27, 91, 95, 113–14, 137, **146–50,**
   188–9, 197–99
Laburnum 29
Lagerstroemia indica 62, **167**
Laminaria 19
Lamium amplexicaule 138
Lantana 115, 117
Lathyrus (rotundifolius) 145
Lavandula 147
Lecytidaceae 101
Leea robusta 109
Leguminosae 27, **142–6**
Lemnaceae 141
Lemna trisulca 41
Leonotis (leonurus) 127, 148
Lepidodendron 205
Lespedeza 145
Leucadendron discolor 101
Leucospermum 76
Liliaceae 72
Lilium, lilies 117
Limosella 137
Linaceae 33
Linaria 66, 92, 95
   minor 150
   vulgaris 21, 85, 94–5, 149, **189,** 206
Listera ovata 99, 120, 203
Lithocarpus (densiflorus) 91, 101
Lithospermum caroliniense 31, 139
Lobelia 121, 124, 140
   dortmanna 40, 138
   fulgens 126
Lonicera (periclymenum) 94, 96, 121, 206
Loranthus (-aceae) 62, 123, 125
Lotus 144
Loxanthocerei 128
Lupinus pilosus 83
Lycopersicum 137
Lycopodium 163, 205
Lythraceae 33

Lythrum (salicaria) 31–2, 50

Macrozamia (tridentata) 11
Madhuca 130
Magnolia (-aceae) 52, 77, 90, 100–1
Maize see *Zea*
Malpighiacea 65, 69
Malvaceae 27
Malvaviscus 83, 92, 128
Manilhot glaziovii 109
Mapania 135
Maranthaceae 112
Margravia 129, 132
Markhamia 132
Masdevallia (rosea) 105, 129
Maxillaria 71
Medicago (sativa) 80, 111, 120–1, 144, 160–2
Melampyrum (pratense) 53, 68, 149–50, **190**
Melandrium album 72–3
Melastoma (-taceae) 17, 53
Melilotus 143
Melochia 136
Mentha (-eae) 62, 145, 147, **198**
Mentzelia tricuspis 75
Mesadenia 21
Mesocereus marginatus 120
Mimetes 95
   hartogii 22, 91
   hirta 92
Mimosoideae 17, 52, 92, 143
Mimulus (cardinalis) 22, 31, 91, 126
Mirabilis froebelii 45
Monarda 126
Monotropa 52
Moraceae 178
Mosses (Bryophyta) 9
Mucor 9
Mucuna 126, 132–3
Musa 131–3
   fehii 130
   paradisiaca 131
   sapientium 132
Muscari comosum 81, 83
Mussaenda 33
Mutisia (-e) 82, 126
Myosotis (discolor) 83, 85, 92
Myosurus 138
Myriophyllum 40
Myristica 77
Myrmecodia 66
Myrtaceae 61, 91
Myxomycetes · 9

Nactosphace (sect. *Salvia*) 148
Najas 41
Narcissus poëticus 92
   pseudonarcissus 90
Nemophila menziesii 154
Nepenthaceae 84, 104
Nephelium 23
Neptunia 41
Nicandra 31, 137

Nicotiana rustica 67
Nidularia 9
Nigritella 62, 202–**203**
Nymphaea (-ceae) 23, 40, 47, 57, 86, 94, 137, 171
Nymphoides 33
Nypa 21

Oak see *Quercus*
Oats (*Avena sativa*) 137
Ochroma grandiflora, lagopus 132
Ocymoideae 91, 149
Oenothera (-ceae) 45, 52, 92, 137, 156
Oil palm see *Elaeis*
Oleaceae 32
Olyroidae (subfam. Gramineae) 135
Oncidium 40, 75
Onion see *Allium*
Ophrys 6, 45, 66, 78, 80, 84, 109, 139, 152, 205
    apifera 139, 154
    insectifera 74–5
    speculum 47, 74
Opuntia monacantha 65
Orchidaceae 18, 22, 30, 52, 58, 69, 70, 91, 94, 103, 120, 137, 139, 141, 163, **199–204**
Orchis 51, 58, 70, **203–4**
    maculata 106, **200**
    papilionacea 76
    ustulata 201
Orlaya grandiflora 81
Ornithidium 70, 111
Oroxylum 132
Orthocarpus pusillus 110, 112
Orthosiphon 147
Osmanthus fragrans 77
Oxalis (-daceae) 32–3, 139
    acetosella 138

Pachylobus caespitosus 105
Paeonia 20
Palmae 35, 73, 86
Pancratium maritimum 118
Papaver (-aceae) 16, 40, 60, 117, 153
    rhoeas 84, 136
Paphiopedium 86, 200
Papilionaceae 29, 65, 91, 95, 110, 113–14, 136, **142–6**, 151, 165, **180–7**
Pariana 135
Parietaria 27
Parkia 60, 129, 132–3
    auriculata **179**
    clappertoniana **179**
Parmentiera 133
Parnassia 66
Passiflora (*quadragularis*) 58, 90, 163
Pavetta fulgens, javanica 19
Pear-tree (*Pyrus communis*) 124
Pedaliaceae 65
Pedicularis 84, 145, 189–91
    canadensis 149
    capitata **195**
    densiflora 149

    groenlandica 149, 157, **194**
    hirta 149
    lanceolata 53, **196**
    lapponica 42, 149, **193**
    oederi **191**
    palustris 149–50, **191–2**
    racemosa 194
    sceptrum-carolinum 94, 149, **195–6**
    semibarbatum 149
    silvatica 149, **192**
Pedilanthus 22, 91, 127–8
Pelargonium 91–2
Penstemon spp. 154
Peronospora 9
Persea 28
Petalostemon (*pinnatum, violaceus*) 145, **186**
Phallus 9, 77, 104
Phaseolus caracalla 145, **186**
Phaseolus multiflorus **185**
Phaseolus vulgaris 138, 145
Philodendron 74
Phlox 32, 49, 120
Phoenix dactylifera 159
Phormium 128
Phyteuma 90
Picea 39
Pilobolus 9
Pimenta dioica 27
Pinellia 57, 156
Pinguicula alpina 84–94
Pinus (*ponderosa*) 18, 37, 38
Pithecellobium 64
Plantago (-inaceae) 32, 35, 38, 89, 134
    lanceolata 35–6, 103
    major 28
    media 35, 37
Platanthera (*bifolia, chlorantha*) 52, 118, 153
Plumbaginaceae 32–5
Poinsettia 65, 82, 128
Polemoniaceae 45, 53
Polemonium viscosum 157
Polycarpicae 28, 82, 104
Polycarpon succulentum 110
Polygala chamaebuxus **165**
Polygala comosa 52
Polygonaceae 32
Polygonum bistorta 36
    cascadense 110
    viviparum 50
Polytrichum 81
Pontederiaceae 33
Pontederia crassipes 33
Poppy see *Papaver rhoeas*
Populus 29, 37
Potamogeton 35, 40
Potentilla (*glandulosa*) 140
Primroses see *Primula*
Primula (-ceae) 32, 92, 137
    angustifolia 157
    veris 32
    vialii 82
Proboscidea (*arenaria*) 68

Prosopanche burmeisteri   71–2
Protea (barbigera)   76, 127–8
Proteaceae   17, 90, 122, 126, 136
Prunus cerasus   42
Pseudodatura   117
Pseudosolanae (subfam. Solanaceae)   147
Psidium guajava   132
Pteridophyta   10
Pteris   64
Pterostylis   23, 102
Puccinia   61
Puya (alpestris)   119, 123, 127
Pyrrhopappus carolinianus   43–4, 112
Pyrus communis   124

Quassia   127
Quercus (petraea)   37, 39, 60, 91, 134

Rafflesiaceae   103
Rafflesia (arnoldii)   74, 81
Ranales   133
Ranunculus (-aceae)   28, 60, 115
Red Clover see *Trifolium pratense*
Reseda odorata   78, 83, 89, 98
Rhinanthoideae   52, 98, 149
Rhododendron   17, 67, 125
Ricinus   37
Rohdea   119
Rosa (Pimpinellifolia)   60, 90
Rosaceae   27
Roscoea   17, 151
Rubiaceae   17, 32, 90
Rubus (-i)   140
    fruticosi   69, 90
    idaeus   69, 91
Rumex   35
Ruppia   41
Ruta graveolens   165
Rutaceae   22
Rye see *Secale cereale*

Saguaro cactus see *Carnegiea gigantea*
Saintpaulia   83
Salix   37, 90, 113, 135
Salverform   92
Salvia   17, 91, 126, 160, 189
    apiana   147
    aurea   148
    coccinea   148
    gesneriaefolia   148
    glutinosa   53, 147
    heeri   148
    horminum   147
    involucrata   148
    lasiantha   148
    mellifera   147
    nutans   149
    officinalis   147
    patens   147, **196**
    pratensis   17, 91, 147, 150
    splendens   126, 146, 148, **197**
    verbenacea   150

    verticillata   147
Sanguisorba officinalis   36
Sapotaceae   132–3
Saprolegnia   9
Sarracenia (-ceae)   19, 84, 104
Sausage tree see *Kigelia*
Saussurea alpina   78
Saxifraga (aizoides)   27–8, 33, **164**
Scaevola   17
Scrophulariaceae   69, 91, 113–14, 146–50,
    188–90
Scrophularia canina   189
Scrophularia nodosa   28, 109, **189**
Scutellaria   147
Secale cereale   137, 159, 160, 205
Selaginella   7
Serapias   71
Sickmannia   135
Sievekingia   154
Silene noctiflora   138
Silene otites   106
Silenoideae   92
Solanum (-aceae)   17, 136
Solidago canadensis   56
Sonneratiaceae   132
Sorbus aucuparia   101
Spathiphyllum   70
Spathodea (campanulata)   124, 127–8, 132
Sphacelia stage   9
Sphaerobolus   9
Sphagnum   9
Spiraea latifolia   56
Splachnaceae   9, 77
Stachys   147
Stanhopea   26, 94, 154
    graveolens   124
    tricornis   154
Stapelia   57, 104
Sterculiaceae   103
Strelitzia (nicolai)   128, 130
Strobilanthus   58
Subularia   137
Sundal malam (*Cestrum nocturnum*)   117
Sunflower (*Helianthus*)   170
Symphoricarpus (racemosus)   109–10
Symphytum asperum   154
Symphytum officinale   52, 92, 154
Symphytum uplandicum   154

Taccaceae   103
Taraxacum   81, 87, 140, 158, 160–1
Taxus   19
Tecoma   121
Tetracera   27
Teucrium   147
Thalictrum   34–5, 37, 90
    alpinum, aquilegifolium, dipterocarpum   91
Thallophyta   9
Theobroma cacao   15–16, 73, 102, 110, 120
Thunbergia (grandiflora)   17, 110, 113, **168**
Thymelaeaceae   27
Thymus   198

Tilia (*cordata*)   36, 65
Tilletia tritici   9
Tomato (*Solanum lycopersicum*)   163
Trentepohlia iolithus   78
Trifolium   81, 92, 94
    medium   182
    pratense   20, 51, 68, 121, 145, 160
    repens   50
Triglochin   37
Tripogandra grandiflora   62
Triticum sativum   137, 160
Trollius (*europaeus*)   72–3, 94, 105, **174**, 177
Tropaeolum   128
Tsuga   12, 19, 121
Tuber   9
Tuberoses (*Poliantus tuberosa*)   117
Tubiflorae   146–50
Tulipa (*silvestris*)   20, 87, 96
Turneraceae   32
Typha   36

Ulva   7
Umbelliferae   42, 87, 101–2, 117
Uredinales   9
Urtica (-ceae)   34, 37, 89
Ustilago (*violaceae*)   9
Uvularia perfoliata   99

Vaccinium   17
Vallisneria (*spiralis*)   41, **165**

Vanilla   78, 159, 162–3
Verabascum thapsus   61
Veratrum album var. labelianum   79, 83
Verbenaceae   32
Veronica   85, 102
Viburnum   81, 90, 101
Vicia   144–5
    faba   160
    lathyroides   145
    tetrasperma   115
Victoria (*amazonica*)   23, 70, 77, 101, 171
Viola (*mirabilis*)   66, 92, 138
Viscaria vulgaris   109
Vitis   27, 78, 89

Watermelon (*Citrullus vulgaris*)   113
Welwitschia   11
Wheat see *Triticum sativum*
Winteraceae   81, 101
Wistaria   143

Xanthium   35

Yucca   15, 42, 45, 72–3, 98, **175**, 177, 205

Zamia   11, 73
Zanthedeschia   20, 90
Zea mais   137, 159–60, 205
Zizyphus spina-christi   33
Zostera   41, 89

INDEX OF ANIMAL NAMES

For the most frequent pollinators, reference is only made to entries of particular interest. Bold page numbers refer to Case histories or definitions/discussions.

Acanthorhynchus 125
Acherontia 119
Aculeata 108
Aëdes 106
Agaonidae 177–8
Alkali bees see *Nomia*
Andrena (idae) 108, 111, 114, 200
Antechinus apicalis 122
Anthocoridae 72
Anthophoridae 69
Anthostreptes phoenicotis 121
Ants *see* Formicidae
Aphida 63
Apis (idae) 35, 50–1, 60, 62, 95, 97, 108, **111–15**, 153, 158, 160
Arachnothera 125
Autographa gamma 118
Azteca chartifex spiriti 110

Bats 47, 79–80, 97, 121–2, **129–133**
Bees see *Apis*
Beeflies *see* Bombylidae
Beetles *see* Coleoptera
Birds 93, 97, 121–2, **123–9**, 133
Blackbird (*Merula merula*) 123, 127
Blastophaga psenes 178
Bombus (idae) 53, 62, 93, 108, 111–5, 123, 162
    affinis 68
    agrorum 59
    edwardii 56
    hortorum 161
    jonellus 191
    lapidarius 152
    lucorum 59, 190
    mastrucatus 69
    pratorum 191
Bombylius (-idae) 102–3, 119
    fuliginosus, medius 102
Bulbul (*Pycnonotus*) 125
Bumblebees see *Bombus*
Bush baby (*Galago*) 122
Butterflies 97, 115–16

Calliphora 103
Calypta anna, costae 154
Campsomeris 75
Campsoscolia ciliata 47

Capsidae 99
Carpenter bee see *Xylocopa*
Castnia eudesmia 119
Centris 46, 75
Ceratopogonidae 106
Chaffinch (*Fringilla coelebs*) 123
Chalcidoideae 108
Chiastochaeta (*trolli*) 11, 175
Chiroptera *see* Bats
Chloroperla Torrentium 99, 120
Cinnyris 125
Coerebidae 124, 157
Coleoptera 11, 20, 52, 72, 74, 76–7, 79, 94, 97–8, **99–102**, 104, 163, **170**
Colletida 61
Colopterus truncatus 170
Culex (icidae) 106
Curculionidae 73
Cyclocephala castanea 172
    hardyi 101, 171
Cynipoidea 108
Cynopterus 129

Dendrocopus analis 123
Diamesus 101
Dianthoecia 72
Dicaeidae 125
Diptera *see also* Flies 97, 102–3, 198
Drepanididae 124, 126, 140

Earwigs (Forficulidae) 120
Elaeina 127
Empusa 9
Epomophorus gambianus 179
Epomops franqueti 130
Eristalis (*intricarius*) 106, 202
Eucera tuberculata 72, 76
Euglossinae 76, 80, 114

Flies *see* Muscidae
Flower parakeets (Trichoglossidae) 23, 125
Flowerpeckers (Dicaeidae) 125
Forcipomya 73
Formicidae 10, 65, 68, **107–10**
Frankliniella parvula 120
Fringilla coelebs 123
Fungus gnats *see* Mycetophilidae 105

Galago crassicaudata   122
Gall wasps   47
Glaucis hirsuta   124
Glossophaga (-inae)   130−1
Glossopitta porphyrocephala   121
Grasshoppers   120

Hadena (*bicruris*)   11, 72−3
Halictus   114
Hawkmoths *see* Sphingidae
Heliconius   46, 115
Hemihalictus lustrans   43−4, 112
Hemiptera   99
Heterocera *see* Moths
Heteroptera   121
Hive bees *see Apis*
Homoptera   120
Honeybees *see Apis*
Honeyeaters (Meliphagidae)   124
Honeycreepers (Drepanidida)   124
Honey mouse *see Tarsipes spencerae*
Honey parrots (Trichoglossidae)   124
Horseflies (*Tabanus*)   106
Hover-flies *see* Syrphidae
Hummingbirds *see* Trochilidae
Hymenoptera   107

Ichneumonidae   108
Icteridae   127

Lacewing flies (Neuroptera)   120
Leafcutter bees (Megachilidae)   108
Lepidoptera   115−19
Leptoncyteris (*nivalis*)   130−2, 179
Liponeura cinerascens   106
Loriculus   125
Lorikeets (Trichoglossidae)   121, 124
Lucilia   103
Lycaena   72

Macroglossa   82, 85, 95, 117−18, 189
Macroglossinae   129−31, 133
Macroglossus minimus   130
Megachile (*willoughbyella*)   53, 191
Megachilidae   108, 111, 162
Megachiroptera   129−30, 132
Megandrena mentzelii   75
Meliphagidae   124
Melipona   111
Microchiroptera   131−2
Micropterygidae   115
Midges   79
Mosquitoes *see* Culicidae
Moths   93−4, **115−19**, 126
Muscidae   97, **102−6**, 178
Musonycteris harrisonii   130
Mycetophilidae   105

Nectarinidae   124, 127
Nemestrinidae   102, 119
Nemognathus   100, 119

Neuroptera   120
Noctuidae   85, 117
Nomia   162

Opistomya elegans   106
Oriolidae   125
Osmia (*rufa*)   60, 149

Passer domesticus   123
Peponapis   71
Picidae   125
Pieridae   117
Pigeons   121, 123
Plocoglottis   23
Plusia (*gamma*)   117−18
Polistes   107
Pompilidae   108
Prosopi(di)dae   108, 111
Prostemadera novaeseelandiae   146
Pseudomasaris vespoides   154
Pteropus (-inae)   129-30, 132−3
Pycnonotus (-idae)   123, 125

Quits (Coerebidae)   124

Rats, *Rattus fuscipes, hawaiensis*   122
Rhopalocera   115

Sapsuckers (*Sphyrapicus*)   123
Sawflies (Tenthredinidae)   108
Scatophaga   103
Silphidae   101
Solitary bees   47
Sparrows *see Passer domesticus*
Sphaerocera subsultans   173
Sphecoidea   108
Sphegidae   107
Sphingidae   81, 105, 117, 153
Sphodium triste   173
Sphyrapicus   123
Staphylinidae   99
Starling (*Sturnus*)   125
Sturnidae   125
Sugarbirds (Coerebidae)   124
Sunbirds (Nectarinidae)   124
Symphyta   108
Syrphidae   96, 102−3, 144, 183, 206

Tabanus (-idae)   102
Taeniothrips (*ericae*)   68, 72, 174
Tarsipes spencerae   122
Tegiticula (*yuccasella*)   42, 72, 176
Terebrantes   108
Thynnidae   75
Thysanoptera   120
Tordo (*Airaeus aterrimus*)   127
Trichoglossidae   23, 124
Trichosteta fascicularis   76
Trigona   111−12
Trochilidae   53, 119, **121−9**, 157
Trochilus colubris   121
Tui (*Prostemadera novaeseelandiae*)   146

Tyora 120
Tyrannidae 127

Vespa (-idae) 107–9

Wasps 28, 97, 107–8
Woodpeckers 123, 125

Xanthopan morgani f. praedicta 118

Xylocopa 53, 69, 109, 113–14, 148, 157, 163, 184, 186
  augusti 145
  californica 154
  latipes 168
  orpifex 154

Zosterops 124

SUBJECT INDEX

The bold page numbers refer to the Case Histories or definitions/discussions.

Abeilhas   114
Abiotic pollination   34, 89
Absolute odour   78
Accidental utilization   15
Actinomorphic (Leppik)   88
Activity spectrum   43
Adaptation   3, 13, 16
Aggregation instinct (flies)   104
Aggregation (flowers)   117
Alae   142, 180, 184, 186
Alfalfa   80, 111, 120–1, 144, 160–2
Algae   9
Aliens: autogamy   136
Alkali bees   117
Allogamy   24
Allophily   48, 160
Allotropy   48, 97, **164–5**
Amino acids (nectar)   67
Amorphic (Leppik)   88
Androeceum   14
Androgynophore   117
Anemophily   27, **34–40**, 60, 134, **173**
   pollen attractant   60
   secondary   11, 134
   syndrome   40
Anemotaxis   118
Angiospermy   19, 34
Angiovuly   19
Ant guard   64, 109, **169**
Anther   14, 17, 60
Anthesis   13
Anthium   20
Ants   9, 65, 107–10
   pollination   **109–10**
   syndrome   110
Apomixis   140
Approach   44, 77
Arctic   53, 86, 103, 106, 139
Arresting devices   32, 137
Ascendent visit   28
Attractant   55
Attraction unit   21, 44
Autogamy   24, 27, 90, **135**, 139–40, 150, 154,
      165–6, 186, 191
   in submerged blossoms   40, 137

Back-crossing   25

Baker's law   136
Balsa (*Ochroma*)   132
Banana (*Musa sapientium*)   132
Baobab (*Adansonia*)   60, 132
Barley   37, 137, 159
Barriers in reproduction   9, 24–6
Basicaulicarpi   133
Bats   47, 79, 97, 121–2, **129–33**
Bautypus   126
Beaker blossom   97
Beans (*Phaseolus*)   137
Bee blossoms
   pollination   **110–15**
   syndrome   113
Beeflies   102–3, 119
Bees   50–1, 60, 62, 93, 97, 108, **111–15**, 160
Beetles   11, 20, 52, 72, 74, 76–7, 79, 94,
      97–8, **99–102**, 104, 163, 170
Beetle pollination   77, **99–102**, 170
   syndrome   102
Bell blossom   61, 89, 90, 92, 97, **166, 173, 189**
Biocoenose   46, 54, 73, **156–8**
Biotic pollination   **42–54**
   syndrome   53
Birds   47, 93, 97, 121–2, **123–9**, 133
Blackbird (*Merula merula*)   123, 127
Blood-sucking insects   79, 106
Blossom   21
   classes   88–9
   "intelligence"   95, 99, 195
   types   21
Blüte, Blume   21
Bowl blossom   89, 90, 97, **173**
Bracts   IX, 16, 21–2, 147
Broodplace   **71, 173–8**
Brush blossom   36, 89, 90, 93, 97, **186, 198, 203**
Bud pollination   30
Bulbul (*Pycnonotus*)   125
Bumblebees see *Bombus*
Buoyancy of pollen   38
Bush baby (*Galago*)   122
Butterflies   97, 115–19

γ-cadinene   80
Calyx   14, 16
   nectaries   65
Camerarius, R. J.   1

239

Cantharophily　**99–102**, 170
　syndrome　102
Capillitium　9
Caprificus (-ication)　159, 162, 178
Carina　142, 180, 184, 186
Carpenter bees　59, 69, 109, 113–14, 145, 148,
　157, 163, 184, 186
Carrion beetles　77, 101, 104
　blossoms　85, 90
　insects　78, 103–5
　odour　79
Carrots　162
Caudiculae　201–2
Cauliflory　132
Chaffinch (*Fringilla coelebs*)　123
Changing colours　83
Chemical interaction pollen/stigma　29–30
Chestnut　37, 91, 137
　honey　134
Chiropterochory　133
Chiropterophily　60, **129–33, 178–80**
　syndrome　131
Choripetaly　92, 117
Claw marks　112, 132, 194
Cleistogamy　13, 15, 89, 138
Cleistopetaly　89
Clements, F.　3
Clines　152
Closed blossoma　13, 22, 89, 94, 149, **174, 189,**
　195
Closed buds opened　94
Closing movements　13, 17
Clustering　56
Cocoa　15–16, 73, 102, 110, 120
Coevolution　15
Colonizators　136, 154
Colour change　83
Colour preference　84, 97, 103
Colour vision　82–7
Communication system (bees)　112, 161
Community influence *see* Biocoenose
Competition (between blossoms/pollinators)　49,
　53, 69, 139, 154, 157, 161
Concealed pollen　17
Conditioned insect　66, 84
Conducting tissue (style)　14
Conifers　26, 29, 38
Connective　14, 17, 147, 197
Conspicuous blossom, classes　89
Constancy　45, 152
Contour influence　85, 89
Copro- *see* Sapro-
Corbicula　18, 51, 62, 163
Cornflower (*Centaurea cyanus*)　136
Corolla　14, 16
Cotton　61, 161
Courtship territory　80
Cranberry (*Oxycoccus*)　160
Cross-breeding, -fertilization,
　-pollination, crossing　24
Cucumber (*Cucumis*)　162–3
Cycads　73, 77, 81

Dama de noche (*Cestrum nocturnum*)　117
Dandelion　42, 104
Darwin, Ch.　2
Darwin–Knight's law　2
Date palm (*Phoenix dactylifera*)　159
Deceit attraction　57, 79, **199**
Deciduous anthers　27, **164**
Deciduous style　27
Dehiscense, anthers　17, 60
Delpino, F.　3
Deposition, pollen　62
Depth of blossom　82, 95
Descendent visit　28
Dichogamy　25, 27
Dicliny　26
Differentiation in blossom　22, 62
Dimorphic heterostyly　32
Dioecy　26, **169**
　in anemophiles　35
Diptera *see also* Flies　97, 102, 103, 198
Dish blossoms　66, 89, 90, 93, 97, 164
Dobbs, A.　1
Dominating flowering　50
Dry pollen
　anemophiles　38, 52
　entomophiles　52, 147
Dung *see* Carrion
Dystropic　48, 82

Earwigs (Forficulidae)　120
Ecotypes　152
Effective bee population　162
Elaiophors　69
Energy budget　55–7, 123
Entomophily　98, 164–5
Ephemeral blossoms　22, 31
Ephemeral style　27
Ephydrophily　40, **169**
Epigyny　20
Erinosyce　178
Ethodynamic pollination　43, 175–6
Euanthium　21
Eulectic　48
Euphilic　48, **175**
Eutropic　48, 97, **174, 176**
　blossoms　112, **182–3**
　pollen transfer　37
Explosive pollen transfer　37
Extra-floral nectar　64, 168
Extra-nuptial nectar (ies)　64, 135

Feeding anthers　62
Ferns　9
Fertilization　7, 25
Fidelity　45, 47, 49
Filament　14
Filiform appendages　103–4
Filiform blossom　114
Filmmerkörper　104
Flag blossom　85, 91–3, 97, 142, 151, 180–9, 199
Flagelliflory　132
Flies *see* Muscidae

Flower *see also* Blossom 14, 21, 24
Flower beetles 100
Flower parakeets 23
Flowerpeckers 125
Fly pollination **102–6**
  syndrome 103
Food bodies 70, **170**
Food pollen 27, 62
Food tissues 70
Foreign pollen effect 30
Form numerals (Leppik) 89
Frisch, K. v. 3
Fruit tree, bird pollinated 124, 126
Functional structure classes 88
Fungus gnats, Mimesis 105
Funnel blossom 89, 90, 93

Galea 146
Gametes 7
-gamy, -gamous 34
Gastrilegic 111
Geitonogamy 24, 140
Gestalt 13
Gleitfallen 94
Grasses 34, 36–9, 64, 103, 134
Grasshoppers 120
Gravity pollination 137
Gregarious flowering 58
Guava (*Psidium guajava*) 132
Guiding structures 149
Gullet blossom 22, 89, 91, 93, 98, 140
  184, 188–9, 191–6, 200
Gymnospermy 20
Gynoeceum 14
Gynodiocey 26
Gynophore 117
Gynostemium 18, **201**

Hair in blossom 18
Hanging blossom 18
Haplomorphic (Leppik) 88
Haptogamy 140
Harmonic relations 96–7
Hawkmoths 81, 105, 117, 153
Hayfever 163
Head-shaped blossom 89
Hemilectic 48
Hemiphilic 48, **186**
Hemitropic 48, 97, **173**
Herkogamy 25, 27, 175
Hermaphroditism, gymnosperms 12
Heteranthery (-andry), 62, **167**
Heterodichogamy 26
Heteromorphy 25, 32
Heterostyly 31–2
Hildebrand, F. 3
Hive bees *see* Bees 27
Homogany 27
Homostylic dimorphy 32
Homozygosity 153
Honeybees *see* Bees
Honey creepers 124

Honeydew 63
Honeyeaters (Meliphagidae) 124
Honey guide *see* Nectar guide
Honey mouse 122
Honey parrots 124
Honey production 160
Horseflies (*Tabanus*) 106
Hover-flies 96, 102–3, 144, 183, 206
Hummeln 114
Hummingbirds 59, 118, 121–9, 157
Hybridization 24
Hydrogen peroxide reaction 19
Hydrophily 40, **169**
Hyphydrophily 40–1
Hyperocraterimorphous 92
Hypopharyngeal gland 60

Illegitimate pollination 32
Imitative odour 78
Inbreeding 24
  depression 31
Incompatibility 10, 19, 25–9, 31, 33, 150
Inconspicuous blossoms 89
Indirect attraction 42
Inflorescence 28, 56, 81
Insect imitation 74
Insects of prey in blossom 42, 67, 124
Integration in blossom 21
"Intelligence" *see* Blossom i.
Interaction pollen/stigma 19, 29
Interdependence 16, 96
Interspecific symbiosis 157
Intra-ovarian pollination 30
Inversion 145–6, 149, **184, 199**
Invertebrates 98

Kapok 40, 43, 131–3
Kesselfallenblumen 94
Kippfallen 94
Knight, T. A. 2
Knight–Darwin's law 23, 138
Knoll, F. 3
Koelreuter, J. G. 2
Kugler, H. 3
Kullenberg syndrome 79

Labelling empties 81
Lactose 67
Lathyrus type 135
Leading structures 84
Leafcutter bees (Megachilidae) 108
Learning in insects 49
-lecty 45
Legitimate pollination 32
Light windows 86
Linnaeus, C. 1
Lipids in nectar 67
Long, F. L. 3
Lorikeets (Trichoglossidae) 124

Macrospores 7
Maize 137, 159–60, 205

Major pollinators   43
Malacophily   119
Mannose   67
Massula   52
Melittophily   **110–15, 165, 168, 180–5,**
    188–96
    syndrome   113
Memory, pollinators   49
Meranthium   21
Mess and soil pollination   51
Micromelittophily   87, 114
Microspores   7
Midges   79
Minor pollinators   43
Mixed pollen loads   51
Monocliny   26
Monoecy   26, **176**
Monolecty   48
Monophily   48
Monotropy   48–9
Mosquitoes   106
Moth blossom   116
Moths   93–4, 97, **115–19**, 120
Movements
    attraction   87
    blossom parts   18, 22–3
Müller, F.   3
Müller, H.   1
Müller, Hermann   3
Mutual adaptations   3
Myophily   **102–6, 198**
    syndrome   103
Myrmecophily   109–10
    syndrome   110

Natural crossing   24
Nectar
    attraction   63
    guides   5, 22, 85, 128
    hoarding   59, 113
    leaves   175
    presentation   59, 66
    theft   68, 190
Nectaries   65–6, **164–5**, 167–8, 174, 179–80
Nest material   70
Nocturnal bees   119
Nototriby   62, 91

Oak   39, 60, 91, 134
Oats (*Avena sativa*)   137
Obdachblume   71
Ocymoideae   91, 149
Odour
    attraction   77–80, 118
    classification   78
    guides   85
    production   78, 85, 117
    traces   81
Oil attraction   69
Oil on pollen surface *see* Pollenkitt
Oil palm   73, 163
Olfactoric attraction   77

Oligolecty   48
Oligophily   48
Oligotropy   48
Ombrophily   41
Onion   162
Opening of closed buds by bees, birds   94, 123
Optimum blossom size   82
Ornithophily   20, **123–9**, 132, 145, 175, **184,**
    197
    syndrome   127
Orthogenesis   13
Osmophores   78, 170
Outbreeding, -crossing   24
Ovary   14
Oviposition in blossom   74
Ovules   14

Parallel development   95
Parasitic mimesis   51
    pollination   73
Pattern (Stil, Vogel)   96
Pear-tree   127
Pecking order   53
Pedicel
    attraction   21
    orchids   203
Penduliflora   132
Perch   127
Perfume attraction   57, **70**, 80
Perigyny   20
Periodicity   43, 152
    nectar secretion   66
    odour production   78, 117
    pollen presentation   60
Petal   14
Phaleophily   **115–19, 175**
    syndrome   116
Pheromones   80
-phily, -philous   34, 45
Photomultiplicators   4
Pigeons   121, 123
Pili honey   135
Pin flowers   31
Pistil   14
Piston mechanism   144, **181**
Planta-cruel (*Araujoa*)   115
Pleomorphic (Leppik)   89
Podilegy   111
Poisonous *see* Toxic
Polarized light orientation   112
Polemoniaceae   45, 53
Pollen
    attraction   58
    collecting devices   61–2
    dimorphism   52
    dispersal distance   39
    economy   26, 51
    fall pictures   39
    flowers   36, 60, 63, **167, 182–3**
    germination   18, 62
    imitation   71
    loads of bees   35

Pollen (*cont'd*)
longevity   18
odour   60
pockets   177
presentation   17, 28, 59, 164
presentation, secondary   17
protection   17
rain   39
rejection   61
sensitivity, water   18, 41
substitute   61
tube   14, 30
viability   18
Pollenkitt   52
Pollin(ari)um   21, 52, 129,
Pollinating agent   43
Pollination   10, 13
drop   11, 63
spectrum   39
symbiosis   157
unit   20, 44
Polylectic   48
Polyphilic   48
Polyphyly of pollination   10
Polytropy   48
Post-floral nectar   66
Precision in pollen transfer   8, 51
Primary attractant   57
Primary entomophily   11, 34
Probable pollination distance   39
Protandry   27, **164–6, 173, 181, 196–8**
Proteins in pollen coat   18
Protogyny   27, **170–3, 189, 191, 194**
Pseudanthium   21, 81
Pseudoaggression   75
Pseudocompatibility   31
Pseudocopulation   74
Pseudonectaries   66, 75
Pseudoparasitism   75
Pseudosexuality   74
Psychophily   **115–19, 203**
syndrome   116
Pteris nectaries   64
Pump mechanism *see* Piston

Quits (Coerebidae)   124

Rain forest, pollination   139
Rain pollination   41
Rain splash dispersal   9
Rats (*Rattus fuscipes, hawaiensis*)   122
Recognition
pollinators   49, 112
substances   52
Relationship blossom/pollinator   42
Release of grass pollen   37
Rendezvous attraction   75, 80, 101
Repellent odours   80
Resupination *see* Inversion
Revertence   134
Rhythm *see* Periodicity
Robertson, C.   3

Rostellum   202
Rye   137, 160, 205

Saprocantharophily *see* Sapromyophily
Sapromyophily   **103–5, 172–3**
syndrome   105
Saturation, colours   85
Saurochory   133
Sausage tree   129, 131–3
Sawflies   108
Secondary attractants   57
Secondary nectar presentation   66, **189**
Secondary pollen presentation   17, 165–6
Second-order brush
flag blossom   146
dichogamy, herkogamy   28
protogyny   **172–3, 176**
Seed   8
Segmentation, flowers   82, 89
Self-(in)compatibility   10, 24, 29, 33
Self-fertility (-zation)   24
Self-sterility   24
Semaphyll   17, 16
Semi-traps   94, **199**
Sepal   14
Sex in blossoms   25
Sexual attraction   74
Sexuality in plants   1
Shelter blossom   71, 86
Sieve effect of style   19
Size of blossom   81
Slugs, snails   42, 119
Sparrows   123
Speciation   151
Specificity of odours   80, 118
Spermatic odour   100–1
Spiders   124
Spore dispersal   9
Sporophyll   13, 16
Sprengel, C. K.   2
Spurs   66, 92, **189**
Staggered flowering   157
Stamen   14, 16
Standard (*Iris*)   21
Starling (*Sturnus*)   125
Stereomorphic (Leppik)   89
Sternotriby   62, 91
Sticky pollen   52
Stielteller   92
Stigma   14, 19, 21, 29
anemophiles   36
ephemeral   31
Stigmatic surface   19, 30
Stil (Vogel)   96
Streukegel   52
Structural blossom classes   88
Style   14, 96
Sugarbirds (Coerebidae)   124
Sunbirds (Nectarinidae)   121, 124
Sundal malam (*Cestrum nocturnum*)   117
Sunflower (*Helianthus*)   120
Surface texture, blossoms   84

Symbiotic ants   10
Sympetaly   92
Syndrome   16, 23
Synsepaly   92

Teilblüte   21
Television techniques   4, 84
Temperature attraction   11, 86
Temperature relations   56
Territoriality   56, 75, 80
Texture of blossom surface   84
Throat scales   52
Thrum flowers   31
Time memory   43, 152
Tomato (Solanum lycopersicum)   163
Topocentric pollination   43
Tordo (Airaeus aterrimus)   127
Total form effect   95
Toxic nectar/pollen   61, 67
Transport distance, pollen   39, 159
Trap blossom   22, 89, 94, **170–3**
Traplining   56–7, 70, 139
Trelease, W.   3
Trimorphic heterostyly   33
Tripping   162
-tropy   45–6
Trumpet blossom   92–3
Tube flower   89, 91–3, 97, **182, 197**
Tui (Prostemadera novaeseelandiae)   146
Tulip   20, 87, 96
Turbulence   37

Ueberblüte   21
Ultraviolet, visibility   83–5, 117, 119
Unisexuality   26
Unisexuality in anemophiles   35
Utilization   15

Vegetative reproduction   141
Venter-collecting bees   111
Vertebrates   121
Vexillum   142, 180, 184, 186
Viability of pollen   38, 62
Vibrating (wings)   53
Viscidium   52, 202
Viscin   52
Visit   5, 44
Visual attraction   81

Warming up   56
Wasps   28, 97, 107–8
    blossom   107, **189**
Water pollination   40
Wax attractant   70
Weeds   136
Wheat   137, 160
Wind pollination   **34–40**
Window openings   104
Window panes   86, **200**

Xenogamy   24

Zygomorphy   22, 62, 95, 113, 128
    Leppik   89